The Method of Newton's Polyhedron in the Theory of Partial Differential Equations

Mathematics and Its Applications (*Soviet Series*)

Volume 86

The Method of Newton's Polyhedron in the Theory of Partial Differential Equations

by

S. Gindikin
Department of Mathematics,
The State University of New Jersey (Rutgers),
New Brunswick, New Jersey,
U.S.A.

and

L. R. Volevich
Keldysh Institute of Applied Mathematics,
Moscow, Russia

SPRINGER-SCIENCE+BUSINESS MEDIA, B.V.

Library of Congress Cataloging-in-Publication Data

Gindikin, S. G. (Semen Grigor'evich)
The method of Newton's polyhedron in the theory of partial differential equations / by S. Gindikin and L.R. Volevich.
p. cm. -- (Mathematics and its applications. Soviet series ; v. 86)
Includes bibliographical references and index.
ISBN 978-94-010-4794-4 ISBN 978-94-011-1802-6 (eBook)
DOI 10.1007/978-94-011-1802-6
1. Newton diagrams. 2. Differential equations, Partial.
I. Volevich, L. R. (Leonid Romanovich), 1934- . II. Title.
III. Series: Mathematics and its applications (Kluwer Academic Publishers). Soviet series ; 86.
QA341.G56 1992
515'.353--dc20 92-35072

ISBN 978-94-010-4794-4

Printed on acid-free paper

SERIES EDITOR'S PREFACE

'Et moi, ..., si j'avait su comment en revenir, je n'y serais point allé.'

Jules Verne

The series is divergent; therefore we may be able to do something with it.

O. Heaviside

One service mathematics has rendered the human race. It has put common sense back where it belongs, on the topmost shelf next to the dusty canister labelled 'discarded non-sense'.

Eric T. Bell

Mathematics is a tool for thought. A highly necessary tool in a world where both feedback and non-linearities abound. Similarly, all kinds of parts of mathematics serve as tools for other parts and for other sciences.

Applying a simple rewriting rule to the quote on the right above one finds such statements as: 'One service topology has rendered mathematical physics ...'; 'One service logic has rendered computer science ...'; 'One service category theory has rendered mathematics ...'. All arguably true. And all statements obtainable this way form part of the raison d'être of this series.

This series, *Mathematics and Its Applications*, started in 1977. Now that over one hundred volumes have appeared it seems opportune to reexamine its scope. At the time I wrote

> "Growing specialization and diversification have brought a host of monographs and textbooks on increasingly specialized topics. However, the 'tree' of knowledge of mathematics and related fields does not grow only by putting forth new branches. It also happens, quite often in fact, that branches which were thought to be completely disparate are suddenly seen to be related. Further, the kind and level of sophistication of mathematics applied in various sciences has changed drastically in recent years: measure theory is used (non-trivially) in regional and theoretical economics; algebraic geometry interacts with physics; the Minkowsky lemma, coding theory and the structure of water meet one another in packing and covering theory; quantum fields, crystal defects and mathematical programming profit from homotopy theory; Lie algebras are relevant to filtering; and prediction and electrical engineering can use Stein spaces. And in addition to this there are such new emerging subdisciplines as 'experimental mathematics', 'CFD', 'completely integrable systems', 'chaos, synergetics and large-scale order', which are almost impossible to fit into the existing classification schemes. They draw upon widely different sections of mathematics."

By and large, all this still applies today. It is still true that at first sight mathematics seems rather fragmented and that to find, see, and exploit the deeper underlying interrelations more effort is needed and so are books that can help mathematicians and scientists do so. Accordingly MIA will continue to try to make such books available.

If anything, the description I gave in 1977 is now an understatement. To the examples of interaction areas one should add string theory where Riemann surfaces, algebraic geometry, modular functions, knots, quantum field theory, Kac-Moody algebras, monstrous moonshine (and more) all come together. And to the examples of things which can be usefully applied let me add the topic 'finite geometry'; a combination of words which sounds like it might not even exist, let alone be applicable. And yet it is being applied: to statistics via designs, to radar/sonar detection arrays (via finite projective planes), and to bus connections of VLSI chips (via difference sets). There seems to be no part of (so-called pure) mathematics that is not in immediate danger of being applied. And, accordingly, the applied mathematician needs to be aware of much more. Besides analysis and numerics, the traditional workhorses, he may need all kinds of combinatorics, algebra, probability, and so on.

In addition, the applied scientist needs to cope increasingly with the nonlinear world and the extra mathematical sophistication that this requires. For that is where the rewards are. Linear models are honest and a bit sad and depressing: proportional efforts and results. It is in the nonlinear world that infinitesimal inputs may result in macroscopic outputs (or vice versa). To appreciate what I am hinting at: if electronics were linear we would have no fun with transistors and computers; we would have no TV; in fact you would not be reading these lines.

There is also no safety in ignoring such outlandish things as nonstandard analysis, superspace and anticommuting integration, *p*-adic and ultrametric space. All three have applications in both electrical engineering and physics. Once, complex numbers were equally outlandish, but they frequently proved the shortest path between 'real' results. Similarly, the first two topics named have already provided a number of 'wormhole' paths. There is no telling where all this is leading - fortunately.

Thus the original scope of the series, which for various (sound) reasons now comprises five subseries: white (Japan), yellow (China), red (USSR), blue (Eastern Europe), and green (everything else), still applies. It has been enlarged a bit to include books treating of the tools from one subdiscipline which are used in others. Thus the series still aims at books dealing with:

- a central concept which plays an important role in several different mathematical and/or scientific specialization areas;
- new applications of the results and ideas from one area of scientific endeavour into another;
- influences which the results, problems and concepts of one field of enquiry have, and have had, on the development of another.

The shortest path between two truths in the real domain passes through the complex domain.

J. Hadamard

La physique ne nous donne pas seulement l'occasion de résoudre des problèmes ... elle nous fait pressentir la solution.

H. Poincaré

Never lend books, for no one ever returns them; the only books I have in my library are books that other folk have lent me.

Anatole France

The function of an expert is not to be more right than other people, but to be wrong for more sophisticated reasons.

David Butler

Bussum, March 1992

Michiel Hazewinkel

CONTENTS

PREFACE

Newton's polyhedron of a polynomial in several variables is the convex hull of the set of exponents of its monomials completed in some way. The boundary of Newton's polyhedron can be interpreted as one of the possible generalizations of the degree of a polynomial in one variable to the case of several variables. This is a much more informative notion than that of the ordinary degree. At the same time, it is not invariant with respect to linear transformations of coordinates and thus is related to a fixed coordinate system. An intermediate notion between the degree and Newton's polyhedron is the notion of a weighed degree in which the calculation of the "degree" of a monomial is performed by means of summation of the degrees of the various variables with different weights. Newton's polyhedron accumulates information about the various weighted degrees and principal parts which correspond to its faces.

Recently it has been revealed that Newton's polyhedron is a suitable technical means in extremely versatile mathematical problems. In this book we develop the method of Newton's polyhedron for some problems in the theory of partial differential equations. It splits into two parts, Chapters 1 to 4 and Chapters 5 to 7, where Newton's polygon and Newton's polyhedron are considered. The case of polygon not only makes it possible to consider general constructions in the simpler two-dimensional case but also has some natural multidimensional applications.

In Cauchy's problem there is a canonical decomposition into two groups of variables, namely spatial variables and time. Accordingly, the degrees with respect to these variables can be taken into account separately. For the first time this was understood for the heat conduction equation where the weight of the order with respect to time is twice as great as the weight of the order with respect to the spatial variables. In I. G. Petrovskii's general theory of parabolic equations a weight $2b$ is ascribed to time ($2b$-parabolic operators). Considering Newton's polygon with respect to the pairs of (temporal and spatial) degrees we arrive at a natural generalization of parabolicity in Petrovskii's sense and also at a class of dominantly correct operators. In these operators the principal part related to Newton's polygon is dominant, and they include $2b$-parabolic operators (in which the weighted principal part is dominant) and strictly hyperbolic operators. The dominantly correct operators admit of a generalization to variable coefficients.

Attention is focused on three problems in the theory of partial differential equations, namely a special class of hypoelliptic operators defined using Newton's polyhedron (by analogy with the way the ellipticity or quasi-elliptically is defined using the degree or weighted degree of the symbol), energy estimates in Cauchy's problem relating to Newton's polyhedron, and the generalized operators of principal type defined by means of the principal part associated with Newton's polyhedron. Priority is given to the presentation of the algebraic technique which, in our opinion, can be applied in many other problems as well. However, we tried to make the presentation sufficiently autonomous by giving the necessary analytical results.

For a more detailed motivation and description of the content the reader is referred to the introductions at the beginnings of each chapter.

The authors express their deep gratitude to professor V. M. Volosov for translating the book and for valuable and various refinements that he made in the translation.

S. G. Gindikin
L. R. Volevich

CHAPTER I

TWO–SIDED ESTIMATES FOR POLYNOMINALS RELATED TO NEWTON'S POLYGON AND THEIR APPLICATION TO STUDYING LOCAL PROPERTIES OF PARTIAL DIFFERENTIAL OPERATORS IN TWO VARIABLES

Introduction

When studying local smoothness of solutions to partial differential equations, there naturally arises a question that we formulate here for the case of two variables. Consider a polynomial

$$P(\xi,\eta) = \sum_{(\alpha,\beta)\in\nu(P)} a_{\alpha\beta}\xi^\alpha\eta^\beta, \tag{1}$$

where $\nu(P)$ is a finite set of pairs of nonnegative integers. The question is what are the conditions under which there exist constants c, $c_0 > 0$ such that the inequality

$$|\xi^\alpha\eta^\beta| \leqslant c|P(\xi,\eta)| \quad \forall(\alpha,\beta)\in\nu(P), \qquad \xi^2+\eta^2 > c_0, \tag{2}$$

holds.

It is the study of this purely algebraic question and its application to local theory of differential operators in two variables that constitutes the main subject in this chapter. The generalization of the results of this chapter to the case of many variables is discussed in Chapter 5. We note that these problems were considered by Mikhaĭlov [1, 2], Volevich and Gindikin [2], and Gindikin [1] (also see the monograph by Bryuno [1] related to this range of questions).

We note that polynomial inequalities (2) outside a large circle $\xi^2+\eta^2 = c_0^2$ appear just in relation to some local problems for partial differential equations. Other questions in the theory of partial differential equations lead to analogs of estimates (2) in different type of regions. For instance, in Chapter 2, in relation to Cauchy's problem, we shall consider inequality (2) for the case where the variable ξ runs over the real line and η varies throughout the complex half-plane $\operatorname{Im}\eta \leqslant \gamma_0$, where γ_0 is sufficiently large.

If the estimate for (1) holds for all $(\alpha,\beta)\in\nu(P)$, then it is automatically fulfilled for all points belonging to the convex hull $\operatorname{conv}(\nu(P))$ of the set $\nu(P)$. As will be seen later, it is natural to modify slightly the problem in question and require that the inequality hold not only for monomials corresponding to the convex hull of $\nu(P)$ but also for some "subordinate" monomials. The character of the "subordinate" monomials is determined by the problem under consideration for differential equations. In the present chapter we shall require that, along with (2), the inequality

$$|\xi^{\alpha'}\eta^{\beta'}| \leqslant c_1|P(\xi,\eta)| \quad \text{for} \quad \xi^2+\eta^2 \geqslant c_0^2, \tag{2'}$$

be fulfilled, where $0 \leqslant \alpha' < \alpha$, $0 \leqslant \beta' < \beta$, and $(\alpha, \beta) \in \operatorname{conv}(\nu(P))$. The closure of the set of such pairs (α', β') is a convex polygon $N(P)$ which will be called Newton's polygon of polynomial (1).

We briefly discuss the content of this chapter. In §1 the basic notions and constructions are presented for polynomials in relation to Newton's polygons, namely the classification of monomials as senior and minor and the principal quasi-homogeneous parts corresponding to the various orders of quasi-homogeneity. The set of quasi-homogeneous parts characterizes very completely the behavior of the polynomial at infinity. In particular, the main algebraic result of the chapter to which §2 is devoted consists in that the existence of estimates (2) and (2′) is equivalent to the property that all quasi-homogeneous parts of the polynomial have no zeros. Under an additional condition on Newton's polygon the corresponding class of polynomials, the so-called N quasi-elliptic polynomials, is considered in §3. §4 is devoted to studying differential operators whose symbols are N quasi-elliptic polynomials.

§1. Newton's polygon of a polynomial in two variables

This is an introductory section. We begin with the definition of Newton's polygon, senior and minor monomials, and principal quasi-homogeneous parts of a polynomial. We shall prove that any polynomial in two variables can be represented, to within minor monomials (in the sense of Newton's polygon), as a product of quasi-homogeneous polynomials. This results is the main technical means for constructions in the present chapter. The concluding part of the section is devoted to expansions of the roots of a polynomial in two variables into Puiseux's series and the relationship between these expansions and Newton's polygon.

1.1. Notation. Newton's polygon. We shall deal with polynomials (0.1) in two variables. With each polynomial we associate the plane $\mathbb{R}^2_{(\xi,\eta)}$ of the variables ξ, η and the plane $\mathbb{R}^2_{(\alpha,\beta)}$ whose points with nonnegative integral coordinates will be interpreted as monomial power exponents. We shall also consider the plane of the variables $q = (q_1, q_2)$ dual to the variables (α, β):

$$\langle q, (\alpha, \beta)\rangle = q_1\alpha + q_2\beta. \tag{1}$$

Let $\mathbb{R}^2_+$ be the positive quadrant in the plane of exponents:

$$\mathbb{R}^2_+ = \{(\alpha, \beta) \in \mathbb{R}^2, \alpha > 0, \beta > 0\},$$

and let $\overline{\mathbb{R}^2_+}$ be its closure (i.e. the set of vectors with nonnegative components). In what follows we shall regard $\nu(P)$ as a set lying in $\overline{\mathbb{R}^2_+}$.

We shall partially order the set of exponents. Let (α', β') and (α'', β'') be arbitrary pairs of points in $\mathbb{R}^2_{(\alpha,\beta)}$. We shall write $(\alpha', \beta') < (\alpha'', \beta'')$ if $\alpha' < \alpha''$ and $\beta' < \beta''$.

Denote by $N(P)$ the smallest convex polygon in $\overline{\mathbb{R}^2_+}$ possessing the following properties:

(i) $\nu(P) \subset N(P)$;

(ii) if $(\alpha', \beta') \in \overline{\mathbb{R}^2_+}$ and $(\alpha', \beta') < (\alpha, \beta) \in N(P)$, then $(\alpha', \beta') \in N(P)$ (the condition of completeness).

As was already said in the introduction, the polygon $N(P)$ will be called *Newton's polygon* of the polynomial $P(\xi, \eta)$.

According to the above definition, along with each point $(\alpha, \beta) \in N(P)$, the polygon $N(P)$ contains its projections $(\alpha, 0)$ and $(0, \beta)$ on the coordinate axes, and, consequently, $N(P)$ contains the origin $(0,0)$. It follows that $N(P)$ can be defined equivalently as the convex hull of the set consisting of the point $(0,0)$, the points belonging to the finite set $\nu(P)$, and their projections on the coordinate axes.

Hence, the polygon $N(P)$ of an arbitrary polynomial P contains the origin as one of its vertices, and the two sides adjoining this vertex lie on the coordinate axes (see Figure 1). Denote by $\Gamma_j^{(0)} = (\alpha_j, \beta_j)$, $j = 0, 1, \ldots, m+1$, the vertices

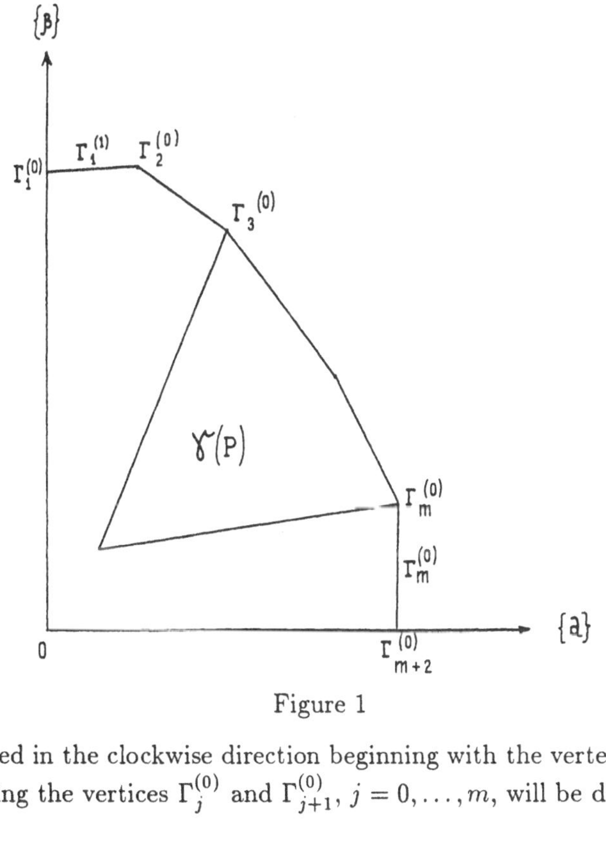

Figure 1

of $N(P)$ indexed in the clockwise direction beginning with the vertex $\Gamma_0^{(0)} = (0,0)$. The sides joining the vertices $\Gamma_j^{(0)}$ and $\Gamma_{j+1}^{(0)}$, $j = 0, \ldots, m$, will be denoted as $\Gamma_j^{(1)}$;

and the side joining $\Gamma_{m+1}^{(0)}$ and $\Gamma_0^{(0)}$ will be denoted $\Gamma_{m+1}^{(1)}$. Thus, the vertex $\Gamma_1^{(0)}$ lies on the axis $\{\beta\}$ and the vertex $\Gamma_{m+1}^{(0)}$ lies on the axis $\{\alpha\}$.

Property (ii) of the polygon $N(P)$ implies that the extensions of its (non-coordinate) sides form obtuse or right angles with the coordinate axes, and the outer normals to the sides form acute (right) angles with the coordinate axes. It follows that if (α_j, β_j) are vertex coordinates, then they must satisfy the inequalities

$$0 = \alpha_0 = \alpha_1 < \cdots < \alpha_m \leqslant \alpha_{m+1}, \tag{2}$$

$$\beta_1 \geqslant \beta_2 > \cdots > \beta_{m+1} = \beta_0 = 0. \tag{3}$$

Definition. A polygon $N(P)$ is said to be regular if it does not contain sides parallel to coordinate axes and not lying on them.

The condition of regularity is equivalent to the property that all inequalities (2) and (3) are strict. If the first relation in (3) involves the sign of equality (i.e. $\beta_1 = \beta_2$), then the side $\Gamma_1^{(1)}$ (see Figure 1) is parallel to the axis $\{\alpha\}$, i.e. is horizontal. If $\alpha_m = \alpha_{m+1}$ in (2), then the side $\Gamma_m^{(1)}$ is vertical.

An integral point $(\alpha, \beta) \in N(P)$ is called a minor point if there is a point $(\alpha', \beta') \in N(P)$[1] such that $(\alpha, \beta) < (\alpha', \beta')$. If otherwise, the integral point (α, β) is said to be a senior. Accordingly, the monomials $\xi^\alpha \eta^\beta$ in the polynomial P are classed as senior and minor.

It is obvious that the senior points of $N(P)$ belong to the boundary of $N(P)$ and lie on the sides $\Gamma_j^{(1)}$, $j = 1, \ldots, m$ not lying on coordinate axes (see Figure 1).

Those vertices of $N(P)$ that are simultaneously senior points of $N(P)$ will be called senior vertices. Obviously, all vertices except for $\Gamma_0^{(0)} = (0,0)$ are senior. The vertices $\Gamma_2^{(0)}, \ldots, \Gamma_m^{(0)}$ necessarily belong to the original set $\nu(P)$. We also note that if the side $\Gamma_1^{(1)}$ is not horizontal (i.e. $\beta_1 > \beta_2$ in (3)), then $\Gamma_1^{(0)} \in \nu(P)$. Similarly, if $\Gamma_m^{(1)}$ is not vertical (i.e. $\alpha_m < \alpha_{m+1}$ in (2)), then $\Gamma_{m+1}^{(0)} \in \nu(P)$.

The convex polygon spanned to the above defined minor points of $N(P)$ will be denoted $\delta(P)$; it satisfies automatically the condition of completeness (ii) in Section 1.1. With the polygons $N(P)$ and $\delta(P)$ we associate two linear spaces of polynomials:

$$L_{N(P)} = \Big\{ Q(\xi, \eta) = \sum_{(\alpha,\beta) \in N(P) \cap \mathbb{Z}^2} b_{\alpha\beta} \xi^\alpha \eta^\beta \Big\}, \tag{4}$$

$$\mathcal{L}_{N(P)} = \Big\{ Q(\xi, \eta) = \sum_{(\alpha,\beta) \in \delta(P) \cap \mathbb{Z}^2} b_{\alpha\beta} \xi^\alpha \eta^\beta \Big\}. \tag{4'}$$

In other words, $L_{N(P)}$ and $\mathcal{L}_{N(P)}$ consist of those and only those polynomials Q for which $N(Q) \subset N(P)$ and $N(Q) \subset \delta(P)$, respectively.

---

[1] We note that, in general, (α', β') is not assumed to be an integral point.

1.2. q-order and q-principal part of a polynomial. We begin with reminding the reader of some well-known geometrical notions. Let N be an arbitrary convex polygon in the plane $\mathbb{R}^2_{(\alpha,\beta)}$ and let $q = (q_1, q_2 \in \mathbb{R}^2_{(q)})$, where $\mathbb{R}^2_{(\alpha,\beta)}$ and $\mathbb{R}^2_{(q)}$ are related according to duality (1). A (closed) half-plane

$$\langle q, (\alpha, \beta)\rangle \leqslant c \tag{5}$$

is called a supporting half-plane if inequality (5) is fulfilled for any point $(\alpha, \beta) \in N$, the sign of equality being attained at at least one point. In other words, (5) can be rewritten as

$$\langle q, (\alpha, \beta)\rangle \leqslant d(N, q), \qquad d(N, q) = \max_{(\alpha,\beta)\in N} \langle q, (\alpha, \beta)\rangle. \tag{5'}$$

The boundary of half-plane (5), i.e. the line

$$\langle q, (\alpha, \beta)\rangle = d(N, q),$$

is called a supporting line of the polygon N, and the vector q is called the direction vector of the supporting half-plane.

If $\Gamma^{(i)}$, $i = 0, 1$, is a vertex or side of N, then $\Gamma^{(i)}$ either belongs to the boundary of half-plane (5) or has no common points with it. The set of direction vectors of half-planes whose boundaries contain $\Gamma^{(i)}$ is called the normal cone for $\Gamma^{(i)}$. If $i = 1$, then $\Gamma^{(1)}$ belongs to the corresponding supporting line, and the normal cone consists of a single ray whose direction is determined by the outer normal to the side $\Gamma^{(1)}$. If $i = 0$, then infinitely many supporting lines pass through the vertex $\Gamma^{(0)}$, and the normal cone is an angle in the plane $\mathbb{R}^2_{(q)}$ between the outer normals to the sides intersecting at the vertex.

The convex polygon N is completely determined by setting the functions $d(N, q)$ in (5′), i.e. by the system of supporting lines. Among inequalities (5′) there are only a finite number of linearly independent relations, and they correspond to the sides of the polygon. Thus, if the outer normals to the sides of the polygon are denoted $q^{(j)}$, $j = 1, \ldots, J$, then it will be determined by the system of inequalities

$$\langle q^{(j)}, (\alpha, \beta)\rangle \leqslant d(N, q^{(j)}), \qquad j = 1, \ldots, J.$$

We come back to polynomial (1). The number

$$d_P(q) = \max_{(\alpha,\beta)\in\nu(P)} \langle q, (\alpha, \beta)\rangle \tag{6}$$

is called the q-order of the polynomial P. If q is a nonnegative vector (and in what follows we shall deal only with such vectors), then the maximum on the right-hand side of (6) does not increase if we add to the points belonging to $\nu(P)$ their projections on the coordinate axes and take the convex hull of all these points. Consequently,

$$d_P(q) = \max_{(\alpha,\beta)\in N(P)} \langle q, (\alpha,\beta)\rangle, \qquad q \in \overline{\mathbb{R}^2_+} \tag{6'}$$

(or $d_P(q) = d(N(P), q)$) in the notation of (5')).

As a result, Newton's polygon is determined by means of the system of inequalities

$$\langle q, (\alpha,\beta)\rangle \leqslant d_P(q), \qquad q \in \overline{\mathbb{R}^2_+},\ (\alpha,\beta) \in \overline{\mathbb{R}^2_+} \tag{7}$$

Let $q^{(j)} = (q^{(j)}, q_2^{(j)})$ be the outer normals to the sides of $N(P)$ not lying on coordinate axes; then the infinite system of inequalities (7) is equivalent to the finite system of inequalities

$$\langle q^{(j)}, (\alpha,\beta)\rangle \leqslant d_P(q^{(j)}), \qquad j = 1,\dots,m,\ (\alpha,\beta) \in \overline{\mathbb{R}^2_+}. \tag{7'}$$

For a given j, the maximum is attained on the side $\Gamma_j^{(1)}$ whose (outer) normal is $q^{(j)}$, i.e.

$$\left\{(\alpha,\beta) \in \Gamma_j^{(1)}\right\} \Leftrightarrow \left\{\langle q^{(j)}, (\alpha,\beta)\rangle = d_P(q^{(j)})\right\}.$$

In terms of functions (6) it is convenient to describe the condition expressing the fact that a polynomial belongs to linear spaces (4) and (4').

Proposition. *Let P be a polynomial with Newton's polygon $N(P)$ and let $q^{(j)}$, $j = 1,\dots,m$, be the outer normals to the sides of $N(P)$ not lying on coordinate axes. Let Q be an arbitrary polynomial. Then*

(i) $Q \in L_{N(P)}$ *if and only if* $d_Q(q^{(j)}) \leqslant d_P(q^{(j)})$, $j = 1,\dots,m$;

(ii) $Q \in \mathfrak{L}_{N(P)}$ *if and only if* $d_Q(q^{(j)}) < d_P(q^{(j)})$, $j = 1,\dots,m$;

A polynomial $P(\xi,\eta)$ is said to be q-homogeneous, $q = (q_1, q_2)$, if there is d such that

$$P(t^{q_1}\xi, t^{q_2}\eta) = t^d P(\xi,\eta). \tag{8}$$

It is clear that d coincides with the q-order $d_P(q)$ of polynomial (8). A polynomial will be called a quasi-homogeneous polynomial if it is q-homogeneous for some $q = (q_1, q_2)$. If $q_1 = q_2$, then polynomial (8) is said to be homogeneous, and the corresponding number $d_P(q)$ is called the order of the polynomial. If $q_1 \neq 0$, $q_2 = 0$, then polynomial (8) has the form $\xi^a Q(\eta)$, where $a = d_P(q)$ and $Q(\eta)$ is an arbitrary polynomial in the variable η. Similarly, $P = \eta^b Q(\xi)$ for $q_1 = 0$, $q_2 \neq 0$.

We note that in the definition of a q-homogeneous polynomial only the ratio of the components q_1 and q_2 plays a significant role.

The q-homogeneous polynomial

$$P_q(\xi,\eta) = \sum_{\substack{(\alpha,\beta)\in\nu(P) \\ \langle q,(\alpha,\beta)\rangle = d_P(q)}} a_{\alpha\beta}\xi^\alpha\eta^\beta, \tag{9}$$

is called the principal q-homogeneous part (or, briefly, the q-principal part) of the polynomial P.

Each vector $q \in \overline{\mathbb{R}^2_+}$ belongs to the normal cone of one of the faces Γ of the polygon $N(P)$. If $\Gamma = \Gamma_j^{(1)}$ is a side of $N(P)$ (i.e. $q = q^{(j)}$, where $q^{(j)}$ is the outer normal to $\Gamma_j^{(1)}$), then the sum in (8) extends over all integral points on the side $\Gamma_j^{(1)}$. If $\Gamma = \Gamma_j^{(0)}$ is a (senior) vertex of $N(P)$, then sum (9) consists of a single monomial $a_{\alpha_j \beta_j} \xi^{\alpha_j} \eta^{\beta_j}$. By what has been said, the right-hand side of (9) will also be denoted as $P_\Gamma(\xi, \eta)$.

Remark. If Γ is a vertical side (not coinciding with the axis of ordinates), then the normal to Γ is the vector $q = (1,0)$, and the q-principal part has the form $\xi^a Q(\eta)$. In case Γ is a horizontal side, the normal is the vector $(0,1)$, and the q-principal part is a polynomial having the form $\eta^b Q(\xi)$.

We present some obvious properties of the q-principal part of a polynomial.

Lemma 1. *For polynomial* (1) *and* $q = (q_1, q_2)$ *we have*

$$P(t^{q_1}\xi, t^{q_2}\eta) = t^{d_P(q)} P_q(\xi, \eta) + o(t^{d_P(q)}), \quad t \to +\infty, \tag{10}$$

i.e.

$$P_q(\xi, \eta) = \lim_{t \to \infty} t^{-d_P(q)} P(t^{q_1}\xi, t^{q_2}\eta). \tag{10'}$$

Lemma 2. *For any polynomials* P *and* Q *and any* $q \in \mathbb{R}^2$ *we have*

$$(PQ)_q = P_q Q_q, \qquad \text{for} \quad d_{PQ}(q) = d_P(q) + d_Q(q), \tag{11}$$

$$(P+Q)_q = P_q + Q_q \qquad \text{for} \quad d_P(q) = d_Q(q), \tag{12}$$

$$(P+Q)_q = P_q \qquad \text{for} \quad d_P(q) > d_Q(q). \tag{12'}$$

1.3. Equivalent polynomials. Factorization.

Definition. Polynomials P and $\widehat{P}$ are said to be equivalent if

$$(i) \qquad N(P) = N(\widehat{P});$$

$$(ii) \qquad P - \widehat{P} \in \mathcal{L}_{N(P)},$$

i.e. the polynomials have the same Newton polygons and their all senior monomials coincide.

Lemma. *A polynomial* $\widehat{P}$ *is equivalent to a polynomial* P *if and only if*

$$P_{q^{(j)}}(\xi, \eta) \equiv \widehat{P}_{q^{(j)}}(\xi, \eta), \qquad j = 1, \ldots, m, \tag{13}$$

where $q^{(j)}$ are the outer normals to non-coordinate sides of the polygon $N(P)$.

Proof. Necessity. If $P \sim \widehat{P}$, then $P = \widehat{P} + Q$ where $Q \in \mathcal{L}_{N(P)}$ and (Proposition 1.2 (ii)) $d_Q(q^{(j)}) < d_P(q^{(j)})$. Applying (12′) we obtain (13).

Sufficiency. By virtue of (13), we have $d_P(q^{(j)}) = d_{\widehat{P}}(q^{(j)})$, $j = 1, \dots, m$, whence (Proposition 1.2 (i)) $\widehat{P} \in L_{N(P)}$. Further, if $Q = P - \widehat{P}$, then $d_Q(q^{(j)}) < d_P(q^{(j)})$, whence (Proposition 1.2 (ii)) $Q \in \mathcal{L}_{N(P)}$, and consequently $N(\widehat{P}) = N(P)$.

We now show that the class of polynomials equivalent to the original polynomial (0.1) contains a polynomial representable as a product of quasi-homogeneous polynomials that are uniquely determined by the principal quasi-homogeneous parts of P.

Theorem. *For each polynomial (0.1) there are vectors $q^{(j)} \in \overline{\mathbb{R}^2_+}$, $j = 1, \dots, m$, and $q^{(j)}$-homogeneous polynomials $P^{[j]}(\xi, \eta)$ such that*

$$P \sim \widehat{P} \quad \text{where} \quad \widehat{P} = P^{[1]} \dots P^{[m]}. \tag{14}$$

If the vectors $q^{(j)}$ are pairwise disproportinate[1]*, then the polynomials $P^{[j]}$ are determined uniquely to within a factor, possibly a monomial (i.e. a factor of the form of* const $\xi^a \eta^b$*). If the polygon $N(P)$ is regular, then the polynomials $P^{[j]}$ are determined to within a constant factor.*

Proof. Take as $q^{(j)}$ the vectors of outer normals to the non-coordinate sides $\Gamma^{(1)}_j$, $j = 1, \dots, m$, of the polygon $N(P)$. The form of $N(P)$ implies that $q^{(j)} \in \overline{\mathbb{R}^2_+}$. Let the polynomial P have the form (0.1) and let (α_j, β_j), $j = 0, \dots, m+1$, be the corresponding vertices. We note that the vertices $(\alpha_2, \beta_2), \dots, (\alpha_m, \beta_m)$ are sure to belong to the original set $\nu(P)$, i.e. the monomials $a_{\alpha_j \beta_j} \xi^{\alpha_j} \eta^{\beta_j}$, $j = 2, \dots, m$, are involved in polynomial (0.1) with nonzero coefficients $a_{\alpha_j \beta_j}$. As the polynomials $P^{[j]}$ in (14) we take

$$P^{[j]}(\xi, \eta) = P_{q^{(j)}}(\xi, \eta)(a'_{\alpha_j \beta_j} \xi^{\alpha_j} \eta^{\beta_{j+1}})^{-1}, \tag{15}$$

where

$$a'_{\alpha_1 \beta_1} = 1, \qquad a'_{\alpha_j \beta_j} = a_{\alpha_j \beta_j}, \qquad j = 2, \dots, m. \tag{15′}$$

By what was said above, $a'_{\alpha_j \beta_j} \neq 0$, $j = 1, \dots, m$. Since the polynomial $P_{q^{(j)}}$ is the sum of the monomials $a_{\alpha\beta} \xi^\alpha \eta^\beta$ corresponding to the integral points of the side $\Gamma^{(1)}_j$ joining the vertices (α_j, β_j) and $(\alpha_{j+1}, \beta_{j+1})$, we have (see (2) and (3))

$$\alpha_j \leqslant \alpha \leqslant \alpha_{j+1}, \quad \beta_{j+1} \leqslant \beta \leqslant \beta_j, \quad (\alpha, \beta) \in \Gamma^{(1)}_j.$$

---

[1]The assumption that the vectors $q^{(j)}$ are pairwise disproportianate does not lead to loss of generality; if some of the vectors were proportional, then we could multiply the corresponding polynomials $P^{[j]}$, i.e. replace them by a single $q^{(j)}$-homogeneous polynomial.

These inequalities imply that the right-hand sides in (15) are polynomials.

In the view of the lemma, the proof of the theorem reduces to the verification of (13). According to (11), we have

$$\widehat{P}_{q^{(j)}} = \prod_{k=1}^{m} P^{[k]}_{q^{(j)}}. \tag{16}$$

We shall find explicit expressions for the factors on the right-hand side of (16). To this end we note that the linear form

$$S_j(\alpha,\beta) = \langle q^{(j)}, (\alpha,\beta)\rangle$$

is constant on straight lines parallel to the side $\Gamma^{(1)}_j$ and increases strictly in the direction of $q^{(j)}$. On the line segment $\Gamma^{(1)}_k$, $k < j$, the form attains a strict maximum at the vertex $\Gamma^{(0)}_{k+1}$, and for $k > j$ it attains maximum at the vertex $\Gamma^{(0)}_k$. Thus with account of (15′), we have

$$P^{[k]}_{q^{(j)}} = P_{q^{(j)}}(a'_{\alpha_j\beta_j}\xi^{\alpha'_j}\eta^{\beta_j})^{-1}, \qquad k = j,$$

$$P^{[k]}_{q^{(j)}} = \frac{a_{\alpha_{k+1}\beta_{k+1}}\xi^{\alpha_{k+1}}\eta^{\beta_{k+1}}}{a'_{\alpha_k\beta_k}\xi^{\alpha_k}\eta^{\beta_{k+1}}} = \frac{a_{\alpha_{k+1}\beta_{k+1}}}{a'_{\alpha_k\beta_k}}\xi^{\alpha_{k+1}-\alpha_k}, \qquad k < j,$$

$$P^{[k]}_{q^{(j)}} = \frac{a_{\alpha_k\beta_k}\xi^{\alpha_k}\eta^{\beta_k}}{a'_{\alpha_k\beta_k}\xi^{\alpha_k}\eta^{\beta_{k+1}}} = \eta^{\beta_k-\beta_{k+1}}, \qquad k > j.$$

Multiplying these relations and using the fact that $a'_{\alpha_1\beta_1} = 1$ we arrive at (13).

The proof of the uniqueness of the polynomials $P^{[k]}$ in (14) reduces in fact to the repetition of the above argument. Let there exist sets of vectors $r^{(k)}$ and $r^{(k)}$-homogeneous polynomials $Q^{[k]}$ such that the polynomial P is equivalent to the polynomial $\widehat{Q} = \prod Q^{[k]}$. If $q^{(j)}$ is the normal vector to the side $\Gamma^{(1)}_j$ of the polygon $N(P)$ not lying on coordinate axes, then, by virtue of (13) and (11), we have

$$P_{q^{(j)}} = \prod_k Q^{[k]}_{q^{(j)}}. \tag{16′}$$

If to the two vertices $\Gamma^{(0)}_j$ and $\Gamma^{(0)}_{j+1}$ of the side $\Gamma^{(1)}_j$ there correspond monomials of the original polynomial, then the left-hand side of (16′) is a polynomial not reducible to a single monomial. On the other hand, each of the polynomials $Q^{[k]}_{q^{(j)}}$ either consists of a single monomial (if the vector $q^{(j)}$ is not parallel to $r^{(k)}$) or coincides with $Q^{[k]}$ if the vector $q^{(j)}$ is parallel to $r^{(k)}$. Thus, for a certain k, say for $k = j$, the vectors $r^{(j)}$ and $q^{(j)}$ are parallel, and (16′) takes the form

$$P_{q^{(j)}} = \text{const}\,\xi^{a_j}\eta^{b_j}Q^{[j]}$$

The above construction of the polynomials $P^{[j]}$ shows that if to the two vertices of the side $\Gamma_j^{(1)}$ there correspond monomials involved in P with nonzero coefficients, then the polynomial $P^{[j]}$ is determined to within a factor, which is what proves the uniqueness of the expansion in the case of a regular Newton polygon.

Remark. If the polygon $N(P)$ has a horizontal side $[\Gamma_1^{(0)}, \Gamma_2^{(0)}]$ and none of the nonzero monomials of the polynomial P corresponds to the vertex $\Gamma_1^{(0)}$, then all senior monomials of P are divisible by a certain power of ξ, i.e. the polynomial P is equivalent to a polynomial of the form of $\xi^a P_1(\xi, \eta)$. Similarly, if $N(P)$ has a vertical side and none of the nonzero monomials of P corresponds to a vertex lying on a coordinate axis, then P is equivalent to a polynomial of the form of $\eta^b P_2(\xi, \eta)$. Thus, finally, in the case of an irregular polygon $N(P)$ the polynomial P is equivalent to a polynomial $\xi^a \eta^b P_3(\xi, \eta)$. The monomial factors in the expression $\xi^a \eta^b$ can be included in the factors $P^{[j]}$ in expansion (15) in an arbitrary manner, which is what causes the non-uniqueness of the expansion.

1.4. Expanding the roots of a polynomial in two variables into Puiseux's series. We write polynomial (0.1) in the form

$$P(\xi, \eta) = P_0(\xi)\eta^M + \sum_{j \geqslant 1} P_j(\xi)\eta^{M-j}, \tag{17}$$

where $P_j(\xi)$ is a polynomial in ξ. Polynomial (17) is said to be solved with respect to the highest power of η if $P_0(\xi) \equiv \text{const}$. In this case we normalize the polynomial by means of the condition $P_0(\xi) \equiv 1$, and it will have the form

$$P(\xi, \eta) = \eta^M + \sum_{j \geqslant 1} P_j(\xi)\eta^{M-j}. \tag{17$'$}$$

We note that the polygon $N(P)$ of polynomial (17′) has no horizontal sides not belonging to the coordinate axis.

All constructions presented below can be realized for polynomials (17) as well, in which $P_0(\xi) \not\equiv \text{const}$ (i.e. $N(P)$ has a horizontal side), but for the sake of simplicity we confine ourselves to the case (17′). We shall need the following classical fact of the theory of algebraic functions (see Hörmander [1, vol. 2, Supplement A]).

Proposition. *Polynomial* (17′) *can be factorized as*

$$P(\xi, \eta) = \prod_{l=1}^{M} (\eta - \lambda_l(\xi)), \tag{18}$$

where each expression $\lambda_j(\xi)$ is a function of $\xi^{1/p}$ for a certain integer p in the neighborhood of $\xi = \infty$, the function having no essential singularity at infinity, i.e.

$$\lambda_l(\xi) = \sum_{-\infty}^{k_l} \lambda_{lk} (\xi^{1/p})^k, \tag{19}$$

where the integer k_l can be positive or negative or zero. The various values of the root $\xi^{1/p}$ give rise to p different expansions (19), *i.e. p different roots of the polynomial P having the form* (19) *in the neighborhood of the point at infinity.*

We introduce the notation that will constantly be used. If a root $\lambda_l(\xi)$ is not identically zero, then k_l in (19) can be chosen so that $\lambda_{lk_l} \neq 0$. We set

$$\widehat{\lambda}_j(\xi) = \begin{cases} \lambda_{jk_j}\xi^{k_j/p} & \text{for } k_j > 0, \\ 0 & \text{for } k_j \leqslant 0. \end{cases} \tag{20}$$

Functions (20) will be called the principal parts of the roots λ_j.

Denote by b_j the different numbers among the exponents k_l/p in (19) and index them so that

$$b_1 > b_2 > \cdots > b_m \geqslant 0, \tag{21}$$

and let μ_j be the number of times the exponent b_j, $j = 1, \dots, m$, is found among k_l/p so that

$$\mu_1 + \cdots + \mu_m = M. \tag{22}$$

It will be convenient to supply the roots of P with two indices, namely to denote the roots of degree b_j as λ_{jl}:

$$\lambda_{jl}(\xi) = c_{jl}\xi^{b_j} + o(|\xi|^{b_j}), \qquad j = 1, \dots, m,\ l = 1, \dots, \mu_j. \tag{19$'$}$$

The principal parts of these roots will be denoted $\widehat{\lambda}_{jl}(\xi)$. There is a close interrelation between the numbers b_j and μ_j characterizing expansions (19) and Newton's polygon $N(P)$ of polynomial (17′).

Theorem. (i) *Newton's polygon $N(P)$ of polynomial* (17′) *coincides with the polygon with vertices (α_j, β_j), $j = 0, \dots, m+1$, where*

$$\alpha_0 = \alpha_1 = 0, \quad \alpha_j = \mu_1 b_1 + \cdots + \mu_{j-1}b_{j-1}, \quad j = 2, \dots, m+1, \tag{23}$$

$$\beta_j = \mu_j + \cdots + \mu_m, \quad j = 1, \dots, m, \quad \beta_{m+1} = \beta_0 = 0. \tag{23$'$}$$

(ii) *If $q^{(j)} = (1, b_j)$, $j = 1, \dots, m$, and $P_{q^{(j)}}$ is the principal $q^{(j)}$-homogeneous part of P, then* (*cf.* (15))

$$P_{q^{(j)}}(\xi,\eta)(a_{\alpha_j\beta_j}\xi^{\alpha_j}\eta^{b_{j+1}})^{-1} = \prod_{l=1}^{\mu_j}(\eta - \widehat{\lambda}_{jl}(\xi)) \overset{\text{def}}{=} P^{[j]}(\xi,\eta).^{1)} \tag{24}$$

Remarks. 1) Proposition (i) of the theorem indicates in fact a constructive method for determining the senior exponents $b_l = k_l/p$ in expansions (19) and

---

1) We remind the reader that for polynomial (17′) solved with respect to the highest power of η we have $a_{\alpha_1\beta_1} = 1$ so that the numbers $a'_{\alpha_j\beta_j}$ in (15) and (15′) coincide with $a_{\alpha_j\beta_j}$.)

their multiplicities μ_l without solving the algebraic equation $P(\xi,\eta)=0$. Indeed, solving system (23), (23′) with respect to the unknowns μ_j and b_j, $j=1,\dots,m$, we find

$$\mu_j = b_j - b_{j-1}, \quad b_j = (\alpha_{j+1}-\alpha_j)/(\beta_j - \beta_{j+1}), \quad j=1,\dots,m. \tag{25}$$

2) Assertion (ii) in the theorem provides a means for determining the coefficients c_{jl} in (19′); for a fixed j the numbers c_{jl}, $l=1,\dots,\mu_j$, are the roots of the algebraic equation

$$\lambda^{-\beta_{j+1}} P_{q^{(j)}}(1,\lambda)=0.$$

3) The method of determining the subsequent exponents in (19) and the corresponding coefficients is quite elementary and reduces to linear systems of equations (e.g. see Fuks and Levin [1] or Chebotarev [1]).

4) In view of the results in Section 1.3, the right-hand side of (24) is a polynomial in the two variables η and ξ. Consequently, the expression

$$\widehat{P} = \prod_{j=1}^{m}\prod_{l=1}^{\mu_j}(\eta - \widehat{\lambda}_{jl}(\xi)), \tag{26}$$

is a polynomial; it differs from P only in minor monomials (Theorem 1.3).

The proof of the theorem. As was indicated in Section 1.3, among the set of q-principal parts P_q, $q\in\overline{\mathbb{R}^2_+}$, there are all monomials corresponding to the senior vertices and all polynomials corresponding to the sides of $N(P)$. Since polynomial (17′) is solved with respect to the highest power of η (i.e. $N(P)$ has no non-coordinate horizontal sides), we can confine ourselves to $q=(q_1,q_2)$, $q_1>0$, $q_2\geqslant 0$. Since it is the ratio of the components q_1 and q_2 that is important for us, we can assume that

$$q=(1,b), \qquad b\geqslant 0.$$

We begin with the case when b is different from the numbers in (21), and let, for definiteness, $b_{j-1}>b>b_j$. When calculating P_q we replace the roots $\lambda_l(\xi)$ in (18) by their expansions (19). We must assign the exponent b to the variable η and the exponent 1 to the variable ξ and retain the terms of maximum degree in each of the factors $\eta-\lambda_l(\xi)$. Therefore in the factor $\eta-\lambda_{kl}(\xi)$ we retain the term η for $k\geqslant j$ and the term $-\widehat{\lambda}_{kl}(\xi)$ for $k<j$. As a result, we obtain

$$P_q(\xi,\eta) = \prod_{k=1}^{j-1}\prod_{l=1}^{\mu_k}(-\widehat{\lambda}_{kl}(\xi))\prod_{k=j}^{m}\prod_{l=1}^{\mu_k}\eta$$
$$= c_j\xi^{\mu_1 b_1+\dots+\mu_{j-1}b_{j-1}}\eta^{\mu_j+\dots+\mu_m} \qquad \Big(c_j=\prod_{k<j}(-c_{kl})\Big). \tag{27}$$

Using (23) and (23′) we find

$$P_q(\xi,\eta) = c_j \xi^{\alpha_j} \eta^{\beta_j} \quad (c_j \neq 0,\ q = (1,b),\ b_{j-1} > b > b_j). \tag{28}$$

If $b = b_j$, i.e. $q = q^{(j)} = (1, b_j)$, then the factors $\eta - \lambda_{jl}(\xi)$ should be replaced by $\eta - \widehat{\lambda}_{jl}(\xi)$, whence

$$P_{q^{(j)}}(\xi,\eta) = c_j \xi^{\alpha_j} \eta^{\beta_j+1} \prod_{l-1}^{\mu_j} (\eta - \widehat{\lambda}_{jl}(\xi)). \tag{29}$$

We also note that if $b_m > 0$, then taking $b_m > b$ we retain the term $-\widehat{\lambda}_{jl}$ in the factors $\eta - \widehat{\lambda}_{jl}$ and obtain relation (28) for $j = m+1$.

We have thus shown that the points (α_j, β_j), $j = 1, \ldots, m$, and $(\alpha_{m+1}, \beta_{m+1})$ (for $b_m > 0$) are senior vertices of $N(P)$, and, by virtue of (28), the constants c_j in (27) coincide with $a_{\alpha_j \beta_j}$. Therefore (29) goes into (24). The theorem is proved.

§2. Polynomials admitting of two-sided estimates

Let $P(\xi, \eta)$ be a polynomial (0.1) and let $N(P)$ be its Newton's polygon. In the introduction to the present chapter we stated the following problem: to find necessary and sufficient conditions on the polynomial P under which estimates (0.2) and (0.2′) hold, i.e. there are $c, c_0 > 0$ such that

$$|\xi^\alpha \eta^\beta| \leqslant c|P(\xi,\eta)| \quad \forall (\alpha,\beta) \in N(P), \quad \xi^2 + \eta^2 > c_0^2. \tag{1}$$

The present section is devoted to the solution of this problem, and we present two types of conditions guaranteeing the fulfilment of (2), namely condition on the principal quasi-homogeneous parts of P (cf. Mikhaĭlov [1, 2]) and conditions on the behavior of the complex zeros of the polynomial P (cf. Volevich and Gindikin [2] and Gindikin [1]).

2.1. Formulation of the main result. With every polygon $N \subset \overline{\mathbb{R}^2_+}$ we associate the function

$$\Xi_N(\xi,\eta) = \sum_{j=0}^{m+1} |\xi|^{\alpha_j} |\eta|^{\beta_j}, \tag{2}$$

where (α_j, β_j), $j = 0, \ldots, m+1$, are the vertices of the polygon N. We now show that function (2) admits of replacement of the set of estimates (1) by a single inequality:

$$\Xi_{N(P)}(\xi,\eta) \leqslant c|P(\xi,\eta)| \quad \text{for} \quad \xi^2 + \eta^2 > c_0^2. \tag{3}$$

Lemma. (i) *If* $(\alpha,\beta)\in N(P)$, *then*

$$|\xi|^{\alpha}|\eta|^{\beta}\leqslant \Xi_{N(P)}(\xi,\eta)\quad \forall(\alpha,\beta)\in N(P); \tag{4}$$

(ii) *there is* $\varkappa>0$ *such that*

$$|\xi|^{\alpha}|\eta|^{\beta}< c'(1+|\xi|+|\eta|)^{-\varkappa}\Xi_{N(P)}(\xi,\eta)\quad \forall(\alpha,\beta)\in \delta(P), \tag{4'}$$

where $\delta(P)$ *is the polygon of minor (integral) points of the polygon* $N(P)$ *(see the definition in Section* (1.1)*).*

Proof. (i) Since (α,β) belongs to the convex hull of the vertices (α_j,β_j), there are nonnegative numbers a_j, $j=0,\dots,m+1$ such that

$$\alpha=\sum_{j=0}^{m+1}a_j\alpha_j,\qquad \beta=\sum_{j=0}^{m+1}a_j\beta_j,\qquad \sum_{j=0}^{m+1}a_j=1.$$

If we set $x_j=|\xi^{\alpha_j}\eta^{\beta_j}|$ in the well-known inequality

$$x_0^{a_0}\dots x_{m+1}^{a_{m+1}}\leqslant\sum_{j=0}^{m+1}a_jx_j,\qquad \sum a_j=1,\quad x_j,\ a_j>0$$

(see Hardy, Littlewood, and Pólia [1, inequality (2.5.2)]), we obtain

$$|\xi|^{\alpha}|\eta|^{\beta}=\prod|\xi^{\alpha_j}\eta^{\beta_j}|^{a_j}\leqslant\sum a_j|\xi^{\alpha_j}\eta^{\beta_j}|\leqslant\Xi_{N(P)}(\xi,\eta).$$

(ii) We shall show that for each minor point $(\alpha,\beta)\in N(P)$ there is $\varkappa(\alpha,\beta)>0$ such that for $\varkappa=\varkappa(\alpha,\beta)$ inequality (4′) is fulfilled. Taking the minimum of $\varkappa(\alpha,\beta)$ over all the vertices of the polygon $\delta(P)$ we obtain (4′).

So, let (α,β) be a minor point of $N(P)$. Then either (α,β) is an interior point of $N(P)$ or it lies on one of the coordinate axes. If (α,β) is an interior point, then for some $\varepsilon>0$ the points $(\alpha+\varepsilon,\beta)$ and $(\alpha,\beta+\varepsilon)$ also belong to $N(P)$. According to (i), we have

$$|\xi|^{\alpha+\varepsilon}|\eta|^{\beta}+|\xi|^{\alpha}|\eta|^{\beta+\varepsilon}\leqslant 2\,\Xi_{N(P)}(\xi,\eta),$$

and, consequently,

$$\frac{|\xi|^{\alpha}|\eta|^{\beta}}{\Xi_{N(P)}(\xi,\eta)}\leqslant\frac{2|\xi|^{\alpha}|\eta|^{\beta}}{|\xi|^{\alpha+\varepsilon}|\eta|^{\beta}+|\xi|^{\alpha}|\eta|^{\beta+\varepsilon}}\leqslant\frac{2}{|\xi|^{\varepsilon}+|\eta|^{\varepsilon}}.$$

Now let the point (α,β) lie on one of the coordinate axes; for definiteness, let $\alpha=0$ and $\beta<\beta_1$. Since the polygon $N(P)$ has the vertices $(0,\beta_1)$ and $(\alpha_{m+1},0)$, $\alpha_{m+1}>0$, we have

$$|\eta|^{\beta}\Xi_{N(P)}^{-1}(\xi,\eta)\leqslant|\eta|^{\beta}(|\xi|^{\beta_1}+|\xi|^{\alpha_{m+1}})^{-1}. \tag{5}$$

We set $k = \alpha_{m+1}/\beta_1$ and

$$\omega = \min_{|\eta|+|\xi|^k=1} \left(|\eta|^{\beta_1} + |\xi|^{\alpha_{m+1}}\right)^{1/\beta_1}.$$

Then the right-hand side of (5) does not exceed

$$\omega^{-\beta_1}|\eta|^{\beta}(|\eta| + |\xi|^k)^{-\beta_1} \leqslant \omega^{-\beta_1}(|\eta| + |\xi|^k)^{\beta-\beta_1}.$$

Since $(0, \beta)$ is a minor point, we have $\beta_1 > \beta$, whence follows the assertion of the lemma.

Hence, by virtue of condition (i) of the lemma, inequalities (1) can be replaced by (2). In view of (ii), it is more natural to state the conditions for the existence of this estimate in terms of senior monomials of the polynomial P. We have the following

Theorem. *For a polynomial $P(\xi, \eta)$ the conditions below are equivalent.*

(I) *There are $c, c_0 > 0$ such that (3) holds.*

(II) *For any side $\Gamma_j^{(1)}$ (Figure 1) not lying on coordinate axes the condition*

$$P_{q^{(j)}}(\xi, \eta) \neq 0 \quad \text{for} \quad \xi \neq 0,\ \eta \neq 0 \tag{6}$$

holds, where $q^{(j)}$ is the outer normal to the side $\Gamma_j^{(1)}$.

(III) *There is a set of vectors $q^{(j)} \in \overline{\mathbb{R}^2_+}$, $j = 1, \dots, m$, and a set of $q^{(j)}$-homogeneous polynomials $P^{[j]}(\xi, \eta)$ such that*

$$P^{[j]}(\xi, \eta) \neq 0 \quad \forall(\xi, \eta) \neq (0, 0) \tag{7}$$

and

$$P \sim \widehat{P}, \qquad \widehat{P} = P^{[1]} \dots P^{[m]}. \tag{8}$$

2.2. Necessary conditions for the existence of estimate (3). In this section we show that conditions (II) and (III) in Theorem 2.1 are consequences of inequality (3).

(I)$\Longrightarrow$(II). Assume that for some $j, \overline{\xi} \neq 0$, and $\overline{\eta} \neq 0$ condition (6) is violated, i.e.

$$P_{q^{(j)}}(\overline{\xi}, \overline{\eta}) = 0, \qquad q^{(j)} = (q_1^{(j)}, q_2^{(j)}). \tag{6'}$$

Consider inequality (3) for the points (ξ, η) lying on the curve

$$\xi(t) = \overline{\xi} t^{q_1^{(j)}}, \qquad \eta(t) = \overline{\eta} t^{q_2^{(j)}}. \tag{9}$$

In view of (1.10) and (6′), for $t \to +\infty$ we have

$$P(\xi(t), \eta(t)) = P_{q^{(j)}}(\overline{\xi}, \overline{\eta}) t^{d_P(q^{(j)})} + o(t^{d_P(q^{(j)})}) = o(t^{d_P(q^{(j)})}). \tag{10}$$

On the other hand, (2) contains a term $|\xi^{\alpha_j}\eta^{\beta_j}|$, where $\alpha_j q_1^{(j)} + \beta_j^{(j)} = d_P(q^{(j)})$, whence

$$\Xi_{N(P)}(\xi(t), \eta(t)) > \text{const}\, t^{d_P(q^{(j)})}. \tag{10'}$$

Comparing (10) and (10′) we set that (6′) contradicts (3), i.e. (6) holds.

(II)$\Longrightarrow$(III). According to Theorem 1.3, condition (8) is fulfilled, where $P^{[j]}$, $j = 1, \dots, m$, are $q^{(j)}$-homogeneous polynomials ($q^{(j)}$ is the outer normal to the side $\Gamma_j^{(1)}$ of the polygon $N(P)$) that are expressed in terms of the polynomials $P_{q^{(j)}}$ by means of explicit formulas (1.15). The formulas imply that if (6) is fulfilled, then

$$P^{[j]}(\xi, \eta) \neq 0 \quad \text{for} \quad \xi \neq 0,\ \eta \neq 0. \tag{7'}$$

It remains to show that the polynomials $P^{[j]}$ do not vanish on the rays $\{|\xi| > 0, \eta = 0\}$ and $\{\xi = 0, |\eta| > 0\}$.

If the side $\Gamma_j^{(1)}$ is not parallel to coordinate axes, then the polynomial $P^{[j]}$ has the form

$$P^{[j]}(\xi, \eta) = \sum b_{\alpha\beta}\xi^{\alpha-\alpha_j}\eta^{\beta-\beta_{j+1}},$$

where

$$\langle q^{(j)}, (\alpha, \beta)\rangle = \langle q^{(j)}, (\alpha_j, \beta_j)\rangle = \langle q^{(j)}, (\alpha_{j+1}, \beta_{j+1})\rangle.$$

The sequence of the exponents α strictly increases from zero to $\alpha_{j+1} - \alpha_j$, and the sequence of the exponents β strictly decreases from $\beta_j - \beta_{j+1}$ to zero. Hence,

$$P^{[j]}(\xi, 0) = \text{const}\, \xi^{\alpha_{j+1}-\alpha_j}, \qquad P^{[j]}(0, \eta) = \text{const}\, \eta^{\beta_j - \beta_{j+1}}.$$

With account of (7′), we arrive at (7).

If $\Gamma_j^{(1)}$ is a horizontal side, then the polynomial $P^{[j]}(\xi, \eta)$ does not in fact depend on η, and therefore condition (7) for the polynomial is implied by (7′). Similarly, if the side $\Gamma_m^{(1)}$ is vertical, then the polynomial $P^{[m]}(\xi, \eta)$ does not depend on ξ, and condition (7) for it also follows from (7′).

The sufficiency of conditions (II) and (III) for inequality (3) to be valid will be proved in the next section. We present a lemma that will be of use later on.

Lemma. *Let $P(\xi, \eta)$ be an arbitrary polynomial. If for some $(\alpha^0, \beta^0) \in \overline{\mathbb{R}_+^2}$ and some $c, c_0 > 0$ the inequality*

$$|\xi^{\alpha^0}\eta^{\beta^0}| < c|P(\xi, \eta)| \quad \textit{for} \quad \xi^2 + \eta^2 > c_0^2 \tag{11}$$

is fulfilled, then $(\alpha^0, \beta^0) \in N(P)$.

Proof. If $(\alpha^0, \beta^0) \notin N(P)$, then there is a vector $q \in \mathbb{R}_+^2$ such that for some $k > 0$ we have

$$q_1\alpha + q_2\beta < k \quad \forall(\alpha, \beta) \in N(P); \qquad q_1\alpha^0 + q_2\beta^0 > k.$$

Replacing ξ and η in (11) by ξt^{q_1} and ηt^{q_2}, respectively, and passing to the limit for $t \to +\infty$ we arrive at a contradiction to inequality (11).

2.3. Sufficient conditions for the existence of estimate (3). The proof of (III)$\Longrightarrow$(I) in Theorem 2.1 is based on the following auxiliary assertions.

Lemma 1. *If P and Q are two arbitrary polynomials, then*

$$N(PQ) = N(P) + N(Q), \tag{12}$$

where the right-hand side is the arithmetical sum of polygons.

Proof. If a monomial $h_{\alpha\beta}\xi^\alpha\eta^\beta$ is contained in the polynomial PQ, then it is a linear combination of products of monomials involved (respectively) in the polynomials P and Q. Consequently, there are pairs of nonnegative integers $(\alpha', \beta') \in N(P)$ and $(\alpha'', \beta'') \in N(Q)$ such that $\alpha = \alpha' + \alpha''$, $\beta = \beta' + \beta''$, i.e.

$$N(PQ) \subset N(P) + N(Q). \tag{13}$$

On the other hand, in view of the second relation (1.11), $\forall q \in \mathbb{R}^2$ we have

$$\max_{(\alpha,\beta)\in N(P)+N(Q)} \langle(\alpha,\beta), q\rangle \leqslant \max_{(\alpha',\beta')\in N(P)} \langle(\alpha',\beta'), q\rangle + \max_{(\alpha'',\beta'')\in N(Q)} \langle(\alpha'',\beta''), q\rangle$$
$$= d_P(q) + d_Q(q) = d_{PQ}(q),$$

whence $N(P) + N(Q) \subset N(PQ)$. Comparing this with (13) we obtain (12).

Lemma 2. *for any polynomials P and Q there is $c = c(PQ) > 0$ such that*

$$c\Xi_{N(P)}(\zeta,\eta)\Xi_{N(Q)}(\zeta,\eta) \leqslant \Xi_{N(PQ)}(\zeta,\eta) \leqslant \Xi_{N(P)}(\zeta,\eta)\Xi_{N(Q)}(\zeta,\eta). \tag{14}$$

Proof. The right-hand inequality is a trivial consequence of (13). The left-hand inequality follows from the opposite inclusion relation by virtue of Lemma 2.1 (i).

Lemma 3. *Let $q \in \overline{\mathbb{R}^2_+}$ and let $Q(\xi,\eta)$ be a q-homogeneous polynomial satisfying the condition*

$$Q(\xi,\eta) \neq 0 \quad \text{for} \quad (\xi,\eta) \neq (0,0). \tag{15}$$

Then the polynomial $Q(\xi,\eta)$ satisfies an inequality of type (3):

$$\Xi_{N(Q)}(\xi,\eta) \leqslant c_\delta |Q(\xi,\eta)| \quad \text{for} \quad \xi^2 + \eta^2 > \delta^2 > 0. \tag{3'}$$

Proof. 1) We begin with the case $q \in \mathbb{R}^2_+$, i.e. $q = (q_1, q_2)$, $q_1, q_2 > 0$.

If $k = \deg d_Q(q)$, then, according to the definition of a q-homogeneous polynomial, we have

$$Q = \sum_{\alpha q_1 + \beta q_2 = k} h_{\alpha\beta}\xi^\alpha\eta^\beta.$$

As was already mentioned above (see the proof of (II)$\Longrightarrow$(III) in the foregoing section), condition (15) implies that the monomials ξ^{k/q_1} and η^{k/q_2} are involved in

Q with nonzero coefficients. Consequently, $N(Q)$ is a triangle with vertices $(0,0)$, $(0,k/q_2)$, and $(k/q_1,0)$, and we have

$$\Xi_{N(Q)}(\xi,\eta) = 1 + |\xi^{k/q_1}| + |\eta^{k/q_2}|. \tag{16}$$

With the vector $q \in \mathbb{R}^2_+$ we associate the q-homogeneous function

$$\rho(\xi,\eta) = |\xi|^{1/q_1} + |\eta|^{1/q_2}. \tag{17}$$

It can easily be seen that (15) is equivalent to

$$c^{-1}\rho^k(\xi,\eta) \leqslant |Q(\xi,\eta)| < c\rho^k(\xi,\eta). \tag{15'}$$

To prove (15′) we introduce the generalized polar coordinates

$$\xi = \bar{\xi}\rho^{q_1}(\xi,\eta), \qquad \eta = \bar{\eta}\rho^{q_2}(\xi,\eta), \qquad \rho(\bar{\xi},\bar{\eta}) = 1.$$

In view of the q-homogeneity, we have

$$Q(\xi,\eta) = \rho^k(\xi,\eta)Q(\bar{\xi},\bar{\eta}). \tag{18}$$

Since the positive continuous function $|Q(\bar{\xi},\bar{\eta})|$ has positive supremum and infimum on the "unit circle" $\rho(\bar{\xi},\bar{\eta}) = 1$, there is $c > 0$ such that

$$c^{-1} < |Q(\bar{\xi},\bar{\eta})| < c, \qquad \rho(\bar{\xi},\bar{\eta}) = 1.$$

Using (18) we obtain inequality (15′) which immediately implies (3′) since (16) can be estimated from above and below by means of $1 + \rho^k(\xi,\eta)$.

2) Now let one of the components of the vector q be zero, say $q = (1,0)$. Then the polynomial Q has the form $Q(\xi,\eta) = \xi^k Q_0(\eta)$ where, generally, $Q_0(\eta)$ is a polynomial in η of an arbitrary degree m. It follows from (15) that $k = 0$ and

$$Q_0(\eta) \neq 0 \quad \text{for} \quad \eta \neq 0, \tag{19}$$

i.e. the polynomial $Q_0(\eta)$ has no nonzero real roots. It follows that $|Q_0(\eta)| > c_\delta(1+|\eta|)^m$ for $|\eta| > \delta$, i.e. $Q = Q_0$ satisfies (3′). The lemma is proved.

We now consider the proof of Theorem 2.1, i.e. show that (III)$\Longrightarrow$(I). According to Lemma 2.1 (ii), when deriving (3) we can assume, without loss of generality, that $P = \widehat{P} = P^{[1]} \dots P^{[m]}$. By Lemma 3, we have the inequalities

$$\Xi_{N(P^{[j]})}(\xi,\eta) < c|P^{[j]}(\xi,\eta)| \quad \text{for} \quad \xi^2+\eta^2 > c_0^2. \tag{3''}$$

Multiplying these inequalities and using Lemma 2 we find:

$$\Xi_{N(P)}(\xi,\eta) \leqslant \prod \Xi_{N(P^{[j]})}(\xi,\eta) \leqslant c^m|\widehat{P}(\xi,\eta)|, \quad \xi^2+\eta^2 > c_0.$$

Remarks. 1) If a polynomial $P(\xi, \eta)$ satisfies the equivalent conditions of Theorem 2.1, then

$$N(P) = N(P^{[1]}) + \cdots + N(P^{[m]}), \tag{20}$$

where $P^{[j]}$ are the quasi-homogeneous polynomials in condition (III). Indeed, according to (III), we have $N(P) = N(\widehat{P})$. Using Lemma 2 we obtain (20).

We note that, according to Lemma 3, the polygons $N(P^{[j]})$ are either right triangles or horizontal or vertical line segments.

2) The class of polynomials satisfying the conditions of Theorem 2.1 is closed relative to multiplication. Indeed, let polynomials P_1 and P_2 satisfy these conditions and let $\widehat{P}_1$ and $\widehat{P}_2$ be the corresponding products of quasi-homogeneous polynomials in condition (III). The above argument implies that the polynomial $\widehat{P} = \widehat{P}_1\widehat{P}_2$ satisfies (3), and we have

$$N(\widehat{P}) = N(\widehat{P}_1) + N(\widehat{P}_2) = N(P_1) + N(P_2). \tag{21}$$

Let us show that $\widehat{P} \sim P$. Setting $P_i = \widehat{P}_i + Q_i$, $Q_i \in \mathcal{L}_{N(P_i)}$, $i = 1, 2$, we obtain $P - \widehat{P} = Q_1\widehat{P}_1 + Q_2\widehat{P}_2 + Q_1Q_2$. It follows from (21) that $P - \widehat{P}$ is linear combination of minor monomials of the polynomial $\widehat{P}$, i.e. $P \sim \widehat{P}$.

2.4. Stability of polynomials admitting of estimate (3). Let a polynomial

$$P(\xi, \eta) = \sum a_{\alpha\beta}\xi^\alpha\eta^\beta \tag{22}$$

satisfy the conditions of Theorem 2.1. Then the polynomial

$$P_\delta(\xi, \eta) = \sum (a_{\alpha\beta} + \delta_{\alpha\beta})\xi^\alpha\eta^\beta, \qquad |\delta_{\alpha\beta}| < \varepsilon, \tag{22'}$$

whose coefficients are close to those in (22) for a sufficiently small ε, possesses the same property. Indeed, if (3) holds, then for $\xi^2 + \eta^2 > c_0^2$ we have

$$\begin{aligned} |P_\delta(\xi, \eta)| &> |P(\xi, \eta)| - \sum |\delta_{\alpha\beta}||\xi^\alpha\eta^\beta| \\ &> (c^{-1} - \varepsilon)\Xi_{N(P)}(\xi, \eta) > \frac{1}{2}c^{-1}\Xi_{N(P)}(\xi, \eta) \end{aligned}$$

provided that $K\varepsilon < c^{-1}/2$, where K is the number of integral points of $N(P)$.

Thus, the polynomials belonging to the finite-dimensional space $L_{N(P)}$ of polynomials with given Newton's polygon $N(P)$ and satisfying (3) form an open set in $L_{N(P)}$. We now show that this is a characteristic property, i.e. it can be included in the conditions of Theorem 2.1.

Denote by M_P the subspace of $L_{N(P)}$ consisting of those $Q \in L_{N(P)}$ which tend uniformly to infinity as $\xi^2 + \eta^2 \to \infty$, i.e.

$$\{Q \in M_P\} \Rightarrow \{Q \in L_{N(P)}, |Q(\xi, \eta)| \to \infty, \xi^2 + \eta^2 \to \infty\}. \tag{23}$$

We note that, by definition, the polygon $N(P)$ contains the vertices $(0, \beta_1)$ and $(\alpha_{m+1}, 0)$ lying on the coordinate axes so that

$$\Xi_{N(P)}(\xi, \eta) > |\xi^{\alpha_{m+1}}| + |\eta^{\beta_1}| \to \infty \quad \text{for} \quad \xi^2 + \eta^2 \to \infty.$$

It follows that the polynomials P admitting estimate (3) belong to the subspace M_P.

Theorem. *The following conditions are equivalent for polynomial* (22).

(I) *Inequality* (3) *holds* (*or the equivalent conditions* (II) *and* (III) *in Theorem* 2.1 *are fulfilled*).

(IV) *There exists a sufficiently small* $\varepsilon > 0$ *such that all the polynomials* (22′) *belong to the class* M_P *and tend uniformly* (*relative to the coefficients* $\delta_{\alpha\beta}, |\delta_{\alpha\beta}| < \varepsilon$) *to infinity as* $\xi^2 + \eta \to \infty$.

(V) *There are* $c, c_0 > 0$ *such that for any* $(\xi, \eta) \in \mathbb{R}^2$, $\xi^2 + \eta^2 > c_0^2$, *the polynomial* $P(z, w)$ *is nonzero in the bicylinder*

$$|z - \xi| < c|\xi|, \qquad |w - \eta| < c|\eta|, \qquad (z, w) \in \mathbb{C}^2. \tag{24}$$

Proof. (I)$\Longrightarrow$(IV). We have already proved that inequality (3) for P implies an analogous inequality for polynomials (22′) and, the more so, the fact that they belong to M_P.

(IV)$\Longrightarrow$(V). Every point (z, w) of bicylinder (24) can be represented as

$$z = (1 + d_1)\xi, \qquad w = (1 + d_2)\eta, \quad |d_1|,\ |d_2| < c.$$

Therefore

$$P(z, w) = \sum a_{\alpha\beta}(1 + d_1)^\alpha (1 + d_2)^\beta \xi^\alpha \eta^\beta = \sum (a_{\alpha\beta} + \delta_{\alpha\beta})\xi^\alpha \eta^\beta, \tag{25}$$

where

$$|\delta_{\alpha\beta}| = |a_{\alpha\beta}||(1 + d_1)^\alpha (1 + d_2)^\beta - 1| \leqslant |a_{\alpha\beta}|((1 + c)^{\alpha+\beta} - 1).$$

It is obvious that $\forall \varepsilon > 0$ there is c such that $|\delta_{\alpha\beta}| < \varepsilon$. If ε is sufficiently small, then all polynomials on the right-hand side of (25) tend to infinity (uniformly relative to $\delta_{\alpha\beta}$) for $\xi^2 + \eta^2 \to \infty$. Consequently, we can choose c_0 such that for $\xi^2 + \eta^2 > c_0^2$ all the polynomials are nonzero, i.e. the polynomial $P(z, w)$ is nonzero in bicylinder (24).

(V)$\Longrightarrow$(I). More precisely, we shall prove that condition (V) implies condition (II) in Theorem 2.1. This fact will be proved by contradiction. Assume that the condition is violated, i.e. there is a vector $q = (q_1, q_2) \in \overline{\mathbb{R}^2_+}$ and a point (ξ^0, η^0), $\xi^0 \neq 0$, $\eta^0 \neq 0$, such that

$$P_q(\xi^0, \eta^0) = 0. \tag{26}$$

The vector q is the outer normal to a side not lying on coordinate axes.

We select a vector $r = (r_1, r_2)$ (r_1, r_2 are natural numbers), so that $R_r(\xi^0, \eta^0) \neq 0$, $R = P_q$. Such a vector exists. Indeed, one of the vertices (α, β) lying on the side $\Gamma^{(1)}$ does not lie on coordinate axes, and therefore its normal cone is an angle belonging to $\mathbb{R}^2_+$. Among the vectors in this angle there is a vector with positive rational components, and hence a vector with natural components as well. In this case the r-principal part $R_r(\xi, \eta)$ consists of a single monomial $a_{\alpha\beta}\xi^\alpha\eta^\beta$, whence $R_r(\xi^0, \eta^0) \neq 0$ for $\xi^0 \neq 0$, $\eta^0 \neq 0$.

Consider an auxiliary polynomial in $\zeta \in \mathbb{C}$:

$$Q_\rho(\zeta) = \rho^{-d_P(q)} P(\rho^{q_1}\zeta^{r_1}\xi^0, \rho^{q_2}\zeta^{r_2}\eta^0). \tag{27}$$

Since

$$Q_\rho(\zeta) = P_q(\zeta^{r_1}\xi^0, \zeta^{r_2}\eta^0) + o(1), \qquad \rho \to \infty,$$

the coefficients of Q_ρ are continuous functions of ρ for $\rho \to +\infty$. The coefficient in the highest power of ζ in polynomial (27) is equal to $a_{\alpha\beta}(\xi^0)^\alpha(\eta^0)^\beta$ and does not depend on ρ. Hence, the roots of polynomial (27) are continuous functions of ρ in the neighborhood of $\rho = +\infty$ and are close to the roots of the limiting polynomial $Q_\infty(\zeta) = P_q(\zeta^{q_1}\xi^0, \zeta^{q_2}\eta^0)$. By virtue of (26), this polynomial has a root $\zeta = 1$. Therefore polynomial (27) possesses a root $\zeta(\rho)$ such that $\forall\varepsilon > 0$ there is $\rho(\varepsilon)$ such that $|\zeta(\rho) - 1| < \varepsilon$ for $\rho > \rho(\varepsilon)$. Thus,

$$P(\rho^{q_1}\zeta(\rho)^{r_1}\xi^0, \rho^{q_2}\zeta(\rho)^{r_2}\eta^0) = 0 \quad \text{for} \quad \rho > \rho(\varepsilon). \tag{28}$$

On the other hand, according to (24), we have

$$P(\rho^{q_1}z, \rho^{q_2}w) \neq 0, \qquad |z - \xi^0| < |\xi^0|, \qquad |w - \eta^0| < c|\eta^0| \tag{29}$$

provided that ρ is so large that

$$\rho^{2q_1}(\xi^0)^2 + \rho^{2q_2}(\eta^0)^2 > c_0^2.$$

We take ε such that $|\zeta(\rho)^{r_i} - 1| < c$, $i = 1, 2$, for $\rho > \rho(\varepsilon)$. Then we see that (29) contradicts (28), i.e. assumption (26) leads to a contradiction.

Remark. The space (23) is not exhausted by polynomials admitting of an estimate from below via the sum of the absolute values of its monomials. The asymptotic behavior of these polynomials is determined by senior monomials and does not depend on the behavior of minor monomials. However, it is easy to give examples of polynomials whose uniform growth at infinity is determined by minor monomials. For example, consider the polynomial

$$P(\xi, \eta) = (\xi^3 - \xi^2\eta)^2 + \eta^4 + \xi^2 + 1. \tag{30}$$

If $q = (1, 1)$, then the q-principal part

$$P_q(\xi, \eta) = (\xi^3 - \xi^2\eta)^2$$

vanishes for $\xi = \eta \neq 0$, i.e. polynomial (30) does not satisfy condition (II) of Theorem 2.1. Nevertheless, it is obvious that $P \in M_P$.

§3. *N* Quasi-elliptic polynomials in two variables

In the theory of partial differential operators an important part is played by elliptic operators whose basic properties are determined by the principal homogeneous part. Quasi-elliptic (or, more specifically, q quasi-elliptic) operators are a generalization of elliptic operators. Their basic properties are determined by the principal quasi-homogeneous part, and some of the results in the theory of elliptic operators are retained for them. It turns out that if compositions of q quasi-elliptic operators with disproportionate values of q are considered, then it is still possible to obtain for the resulting operators (with both constant and variable coefficients) some results of a local character analogous to those in the theory of quasi-elliptic equations. As is suggested by Theorem 2.1, to include this case as well use should be made of the set of all quasi-homogeneous parts of the operator, each of which (as was already said above) is related to one of the sides of Newton's polygon. We shall call the corresponding operators N quasi-elliptic[1)] operators (the letter N indicates that their definition involves Newton's polygon). The theory of these operators will be presented in §4. In this section we shall consider N quasi-elliptic polynomials in two variables which can be regarded as symbols of the corresponding differential operators with constant coefficients.

3.1. Quasi-elliptic polynomials. Let $q = (q_1, q_2) \in \mathbb{R}^2_+$. Polynomial (0.1) is said to be q quasi-elliptic if the principal q-homogeneous part P_q of the polynomial satisfies the condition

$$P_q(\xi, \eta) \neq 0 \quad \text{for} \quad (\xi, \eta) \neq 0. \tag{1}$$

Theorem. *A polynomial $P(\xi, \eta)$ is q quasi-elliptic if and only if*

(a) *the polygon $N(P)$ is a triangle with vertices $(0,0)$, $(0, m_2)$, and $(m_1, 0)$, where $m_i = d_P(q)/q_i$, $i = 1, 2$;*

(b) *there are constants $c, c_0 > 0$ such that the inequalities*

$$\begin{aligned} &|\xi^\alpha \eta^\beta| < c|P(\xi, \eta)|, && \xi^2 + \eta^2 > c_0^2, \\ &(\alpha, \beta) \in \overline{\mathbb{R}^2_+}, && \alpha q_1 + \beta q_2 \leqslant d_P(q), \end{aligned} \tag{2}$$

hold.

Remark. 1) In view of (a), estimate (2) is equivalent to the inequality (cf. (2.3′))

$$\Xi_{N(P)}(\xi, \eta) < c|P(\xi, \eta)| \quad \text{for} \quad \xi^2 + \eta^2 > c_0^2. \tag{3}$$

Hence, q quasi-elliptic polynomials are separated out from the entire class of polynomials satisfying the conditions of Theorem 2.1 by means of the additional condition (a) on the shape of Newton's polygon.

---

1)In literature such operators are also called multiquasi-elliptic operators (see Friberg [1])

2) If P is a q quasi-elliptic polynomial for some $q \in \mathbb{R}^2_+$, then, as was in fact already noted in the proof of Theorem 2.1, the polynomials $P_q(\xi, 0)$ and $P_q(0, \eta)$ are not identically zero, and therefore their degrees coincide, respectively, with the degrees m_1 and m_2 of the polynomials $P(\xi, 0)$ and $P(0, \eta)$. Since

$$m_1 q_1 = m_2 q_2 = d_P(q),$$

the numbers q_1 and q_2 are determined by the numbers m_1 and m_2 up to within an inessential common factor. Thus, a given polynomial can be q quasi-elliptic only for a single (to within a common factor) set of weights (q_1, q_2). In view of this, when the weight is not specified, we shall simply speak of quasi-elliptic polynomials.

3) Since $q_1/q_2 = m_2/m_1$ is a rational number, the weights q_1 and q_2 can be normalized so that they are rational numbers (and even integers).

The proof of the theorem. The necessity of condition (a) was in fact established in the proof of Lemma 3 in Section 2.3 (also see Remark 2). We shall prove the necessity of condition (b). Put $k = d_p(q)$. As was already shown in Lemma 3 in Section 2.3 (see inequality (2.15′)), condition (1) implies the inequality

$$\rho^k(\xi, \eta) \leqslant c|P_q(\xi, \eta)|, \qquad \xi^2 + \eta^2 > 0,$$

where $\rho(\xi, \eta)$ is the function in (2.17). This inequality implies that

$$\rho^k(\xi, \eta) \leqslant c|P(\xi, \eta)| + c|(P - P_q)(\xi, \eta)|. \tag{4}$$

From the definition of the q-principal part it follows that $P - P_q$ is a linear combination of monomials $\xi^\alpha \eta^\beta$, where $\alpha q_1 + \beta q_2 < k$. Since there are only a finite number of nonnegative integers satisfying these inequalities, there is $\varepsilon > 0$ such that $\alpha q_1 + \beta q_2 < k - \varepsilon$. Passing to the generalized polar coordinates (see the proof of Lemma 3 in Section 2.3) we obtain

$$c|(P - P_q)(\xi, \eta)| < c \sum |a_{\alpha\beta} \bar{\xi}^\alpha \bar{\eta}^\beta| \rho(\xi, \eta)^{\alpha q_1 + \beta q_2} \leqslant c_1 \rho^{k-\varepsilon} + c_2.$$

If ρ is sufficiently large, then the right-hand side can be estimated by means of $\rho^k(\xi, \eta)/2$. In view of (4), we arrive at the inequality

$$1 + \rho^k(\xi, \eta) < c'|P(\xi, \eta)|, \qquad \rho(\xi, \eta) > c_0'. \tag{4′}$$

As was already noted in the proof of Lemma 3 in Section 2.3, inequality (4′) is equivalent to (3).

We now show that conditions (a) and (b) imply (1). If $N(P)$ contains the points $(k/q_1, 0)$ and $(0, k/q_2)$, then (2) implies the inequality

$$|\xi|^{k/q_1} + |\eta|^{k/q_2} \leqslant 2c|P(\xi, \eta)|, \qquad \xi^2 + \eta^2 > c_0^2.$$

Replace ξ and η by $t^{q_1}\xi$ and $t^{q_2}\eta$, respectively, where $t > 0$, divide both sides of the inequality by t^k, and pass to the limit for $t \to \infty$. With account of (1.10′), this results in

$$|\xi|^{k/q_1} + |\eta|^{k/q_2} \leqslant 2c|P_q(\xi, \eta)|, \quad \text{for} \quad \xi^2 + \eta^2 > 0,$$

whence follows (1).

3.2. Polynomials with regular Newton's polygons. Hypoelliptic polynomials. Let P be a polynomial in two variables, let $N(P)$ be Newton's polygon, and let $\Gamma_j^{(1)}$, $j=1,\dots,m$, be its sides not lying on coordinate axes (Figure 1). As was already mentioned in Section 1.1, $N(P)$ is called a regular Newton polygon if its side $\Gamma_1^{(1)}$ is not horizontal and the side $\Gamma_m^{(1)}$ is not vertical.

This definition can be reformulated thus: the vectors $q^{(j)}$ of outer normals to the sides $\Gamma_j^{(1)}$, $j=1,\dots,m$, not lying on coordinate axes have positive components. As to the latter definition, it is equivalent to the property that the extensions of the sides $\Gamma_j^{(1)}$ intersect the coordinate axes at obtuse angles.

In terms of the polynomial $P(\xi,\eta)$, the regularity condition for $N(P)$ means that P can be solved in both the highest power of ξ and the highest power of η (cf. (1.17')):

$$P(\xi,\eta)=a_0\xi^{m_1}+\sum_{j>0}A_j(\eta)\xi^{m_1-j}=b_0\eta^{m_2}+\sum_{j>0}B_j(\xi)\eta^{m_2-j}. \tag{5}$$

For the polynomials in two variables the regularity property of $N(P)$ closely relates to hypoellipticity.

According to Hörmander (see [1, Chapter 11]), a polynomial $P(\xi_1,\dots,\xi_n)$ is said to be hypoelliptic if

$$\frac{P^{(\gamma)}(\xi_1,\dots,\xi_n)}{P(\xi_1,\dots,\xi_n)}\to 0 \quad \text{for} \quad |\xi_1|+\dots+|\xi_n|\to\infty, \quad \gamma>0. \tag{6}$$

Here we use the following standard notation:

$$P^{(\gamma)}(\xi_1,\dots,\xi_n)=\partial^{\gamma_1+\dots+\gamma_n}P(\xi_1,\dots,\xi_n)/\partial\xi_1^{\gamma_1}\dots\partial\xi_n^{\gamma_n}.$$

Lemma 1. *If $P(\xi_1,\dots,\xi_n)$ is a hypoelliptic polynomial, then it is solved relative to the highest power of any of the variables $\xi_1,\dots,\xi_n$.*

Proof. Let m_1 be the maximum degree of ξ_1 in the polynomial P and let the polynomial contain a monomial const $\xi_1^{m_1}\xi_2^{\beta_2}\dots\xi_n^{\beta_n}$. If we set $\gamma=(0,\beta_2,\dots,\beta_n)$ and $\xi_2=\dots=\xi_n=0$ in the left-hand side of (6), then we arrive at a contradiction since the ratio $(P^{(\gamma)}P^{-1})(\xi_1,0,\dots,0)$ cannot tend to zero as $|\xi_1|\to 0$. Hence, it is proved that $\beta_1=\dots=\beta_n=0$.

As a direct consequence of Lemma 1, we obtain

Lemma 2. *An arbitrary hypoelliptic polynomial in two variables has a regular Newton polygon $N(P)$.*

Thus, condition (6) (for $n=2$) implies that $N(P)$ is regular. On the other hand, for an arbitrary polynomial P the regularity of $N(P)$ implies a somewhat weaker condition (as compared to (6)). Namely, we have

Lemma 3. *If a polynomial $P(\xi,\eta)$ has a regular Newton polygon $N(P)$, then*

$$\frac{P^{(\gamma)}(\xi,\eta)}{\Xi_{N(P)}(\xi,\eta)} \to 0 \quad \textit{for} \quad |\xi|+|\eta| \to \infty,\ \gamma > 0. \tag{6'}$$

Comparing Lemmas 2 and 3 we obtain the following

Proposition. *A polynomial $P(\xi,\eta)$ admitting of an estimate from below (2.3) is hypoelliptic if and only if Newton's polygon $N(P)$ is regular.*

The proof of Lemma 3. By virtue of Lemma 2.1 (ii), it suffices to verify the inclusion

$$P^{(\gamma)}(\xi,\eta) \in \mathcal{L}_{N(P)} \quad \forall\gamma > 0. \tag{7}$$

Let $\xi^\alpha\eta^\beta$ be a monomial contained in $P^{(\gamma)}$. Then in the original polynomial P there is a monomial const $\xi^{\alpha'}\eta^{\beta'}$ the differentiation of which results in the original monomial. In other words, $\alpha < \alpha'$, $\beta < \beta'$, where $(\alpha,\beta) \in N(P)$ and one of the inequalities is strict. To prove (7) we have to show that both the inequalities are strict. Therefore (7) and, accordingly, Lemma 3 are consequences of the geometrical fact below.

Denote by $\delta_0(P)$ the convex hull of the integral points $(\alpha,\beta) \in N(P)$ possessing the following property: there is a point $(\alpha',\beta') \in N(P)$ such that $\alpha \leqslant \alpha'$ and $\beta \leqslant \beta'$, one of the inequalities being strict.

The polygon $\delta_0(P)$ is obviously an extension of the polygon $\delta(P)$ of minor points, i.e. $\delta(P) \subset \delta_0(P)$. As was already noted in §1, the points of the polygon $\delta(P)$ either belong to the interior of $N(P)$ or lie on coordinate axes. The points of the polygon $\delta_0(P)$ may also belong to the sides of $N(P)$ that are parallel to coordinate axes but do not belong to the axes. We have thus proved

Lemma 4. *Newton's polygon $N(P)$ of a polynomial P is regular if and only if $\delta_0(P) = \delta(P)$.*

3.3. *N* Quasi-elliptic polynomials. Polynomial (0.1) is said to be N quasi-elliptic if it is hypoelliptic and satisfies the equivalent conditions of Theorem 2.1

As was already noted in Proposition 3.2, for polynomials estimated from below by means of the sum of the moduli of the monomials contained in them the condition of hypoellipticity is equivalent to the regularity condition for the polygon $N(P)$. Consequently, adding the regularity condition to the formulation of Theorem 2.1 we obtain a set of equivalent definitions of N quasi-elliptic polynomials. We have the following

Theorem. *For polynomial* (0.1) *the conditions below are equivalent.*

(I) *There are $c, c', c_0 > 0$ and $\varkappa > 0$ such that*

$$|\xi^\alpha\eta^\beta| \leqslant c|P(\xi,\eta)|, \qquad \xi^2+\eta^2 > c_0^2, \quad \forall(\alpha,\beta) \in N(P), \tag{8}$$

$$|\xi^\alpha\eta^\beta|(1+|\xi|+|\eta|)^\varkappa \leqslant c'|P(\xi,\eta)|, \qquad \xi^2+\eta^2 > c_0^2, \quad \forall(\alpha,\beta) \in \delta_0(P). \tag{9}$$

(I′) *P is an N quasi-elliptic polynomial (i.e. inequalities (8) are fulfilled and P is a hypoelliptic polynomial).*

(II) *The polygon $N(P)$ is regular, and for any side $\Gamma_j^{(1)}$ not lying on coordinate axes we have (cf. (2.6))*

$$P_{\Gamma_j^{(1)}}(\xi, \eta) \neq 0 \quad \textit{for} \quad \xi \neq 0,\ \eta \neq 0. \tag{10}$$

(III) *Let $\widehat{\lambda}_j(\xi) = c_j \xi^{b_j}$ be the principal parts of Puiseux's expansions for $|\xi| \to \infty$ of the roots of the polynomial P. Then*

$$b_j > 0, \qquad \operatorname{Im} c_j \neq 0. \tag{11}$$

(IV) *There is a set of pairwise disproportionate positive vectors $q^{(j)}$, $j = 1, \dots,$ m, and a set of $q^{(j)}$ quasi-elliptic polynomials $P^{[j]}$ such that $P \sim \widehat{P}$, where $\widehat{P} = P^{[1]} \dots P^{[m]}$, the polynomials $P^{[j]}$ being determined to within a constant factor.*

Proof. (I)$\Longrightarrow$(I′). As has been mentioned, $P^{(\gamma)}$, $\gamma > 0$, is a linear combination of monomials with exponents belonging to $\delta_0(P)$. Hence, condition (9) implies the hypoellipticity of P.

(I′)$\Longrightarrow$(I). According to Proposition 3.2, the polygon $N(P)$ is regular. Therefore (Lemma 4 in Section 3.2) $\delta_0(P) = \delta(P)$, and conditions (9) follow from Lemma 2.1 (ii).

(I)$\Longrightarrow$(II) is true by virtue of Theorem 2.1 and Proposition 3.2.

(II)$\Longrightarrow$(III). By Theorem 1.4, $\widehat{\lambda}_j(\xi)$ are the roots of the $q^{(j)}$ quasi-homogeneous polynomial $P^{[j]}(\xi, \eta)$ related to $P_{\Gamma_j^{(j)}}$ by (1.15) (here relation (1.15) contains $P_{q^{(j)}} = P_{\Gamma_j^{(1)}}$, where $q^{(i)}$ is the outer normal vector to the side $\Gamma_j^{(1)}$). By virtue of (10), the polynomial $P^{[j]}$ has no real roots for $\xi \neq 0$, whence $\operatorname{Im} c_j \neq 0$. Moreover, by the regularity of $N(P)$, the outer normals $q^{(j)} = (q_1^{(j)}, q_2^{(j)})$ have positive components, whence $b_j = q_2^{(j)}/q_1^{(j)} > 0$.

(III)$\Longrightarrow$(IV). According to Section 1.4, the polygon $N(P)$ is completely determined by the numbers b_j and their "multiplicities" μ_j, the vectors $(1, b_j)$ being outer normals to the non-coordinate sides of $N(P)$. Thus, by virtue of (10), the normals $q^{(j)}$ have positive components, and it remains to apply Theorem 2.1.

(IV)$\Longrightarrow$(I′). Inequalities (8) follow from Theorem 2.1. The polynomials $P^{[j]}$ are $q^{(j)}$ quasi-elliptic and, consequently, hypoelliptic. It follows from definition (6) that the class of hypoelliptic polynomials is closed relative to multiplication, whence $\widehat{P} = P^{[1]} \dots P^{[m]}$ is a hypoelliptic polynomial. Therefore Newton's polygon $N(P) = N(\widehat{P})$ is regular and the polynomial P (Proposition 3.2) is hypoelliptic.

Remarks. 1) The class of polynomials satisfying the equivalent conditions of Theorem 2.1 is closed relative to multiplication (Remark 2 in Section 2.3). As has been noted in the proof of the theorem, the class of hypoelliptic polynomials is closed relative to multiplication. It follows that the class of N quasi-elliptic polynomials is also closed relative to multiplication.

2) It follows from inequalities (9) that the polynomial P remains N quasi-elliptic when minor monomials with arbitrary coefficients are added to it. According to Theorem 2.4, senior monomials can also be added but with sufficiently small coefficients. Moreover, an N quasi-elliptic polynomial can be defined as a hypoelliptic polynomial satisfying the equivalent conditions of Theorem 2.4.

In conclusion we note that N quasi-elliptic polynomials may remain hypoelliptic under addition of monomials not corresponding to the points of $N(P)$[1]. The construction of examples of this kind is based on the following obvious

Proposition. *Let $P(\xi,\eta)$ be an N quasi-elliptic polynomial and let a polynomial $Q(\xi,\eta)$ possess the property that*

$$N(Q^{(\gamma)}) \subset \delta(P) \quad \forall\gamma > 0. \tag{12}$$

Then $P + Q$ is a hypoelliptic polynomial.

Example. Take $P(\xi,\eta) = \xi^6 + \eta^4$ and $Q(\xi,\eta) = a\xi^5\eta$ with an arbitrary complex a. Here P is a $(1,3/2)$ quasi-elliptic polynomial and $N(P)$ is a triangle with vertices $(0,0)$, $(0,4)$, and $(6,0)$. The point $(5,1)$ lies outside the triangle since $5\cdot 1 + 1\cdot 1.5 = 6.5 > 6$. However, the derivatives $Q^{(\gamma)}(\xi,\eta)$ are linear combinations of the monomials $\xi^k\eta$ $(k \leqslant 4)$ and ξ^l $(l \leqslant 5)$, where $(k,1)$ and $(l,0)$ are interior points of $N(P)$. Hence, condition (12) is fulfilled, and, consequently, the polynomial $\xi^6 + \eta^4 + a\xi^5\eta$ is hypoelliptic for any a.

§4. *N* Quasi-elliptic differential operators

In this section we consider the properties of differential operators in two variables, whose symbols are N quasi-elliptic polynomials defined in §3. As was already mentioned in the Introduction, in this book emphasis will primarily be laid upon algebraic and geometrical aspects relating to Newton's polygons. As to analytical questions, they will be presented more schematically, and use will be made of well-known results on differential operators without presenting their proofs. The reader is supposed to be familiar with basic notions of distribution theory and fundamentals of the theory of pseudodifferential operators. All the necessary facts can be found in the initial chapters of the books by Hörmander [1] and Volevich and Gindikin [1] or, for instance, in the paper by Kohn and Nirenberg [1]. To fix the notation we remind the reader of the main elementary facts relating to differential operators, and, for the sake of convenience of the further references, all definitions are given for the general case of n variables (and not only for two variables).

---

[1] Under the addition, the polynomials no longer remain N quasi-elliptic.

4.1. Some definitions and notation. In this section $x = (x_1, \ldots, x_2)$ are the points in $\mathbb{R}^n$ and $\xi = (\xi_1, \ldots, \xi_n)$ are the points of the dual space, the duality being induced by means of the form

$$\langle x, \xi \rangle = x_1\xi_1 + \cdots + x_n\xi_n.$$

We set

$$D = (D_1, \ldots, D_n), \qquad D_j = -i\partial/\partial x_j.$$

If $\alpha = (\alpha_1, \ldots, \alpha_n)$ is a set of natural numbers (a multiindex), then $\xi^\alpha = \xi_1^{\alpha_1} \cdots \xi_n^{\alpha_n}$, $D^\alpha = D_1^{\alpha_1} \cdots D_n^{\alpha_n}$, $|\alpha| = \alpha_1 + \cdots + \alpha_n$, and $\alpha! = \alpha_1! \ldots \alpha_n!$.

By a differential operator with variable coefficients will be meant an expression

$$P(x; D) = \sum_\alpha a_\alpha(x)\xi^\alpha. \tag{1}$$

The coefficients $a_\alpha(x)$ are assumed to be infinitely differentiable functions, and in what follows we shall not stipulate this condition. The function

$$P(x; \xi) = \sum_\alpha a_\alpha(x)\xi^\alpha$$

is called the symbol of operator (1). If the coefficients $a_\alpha(x)$ do not depend on x, we shall write (accordingly) $P(D)$ and $P(\xi)$.

We shall use the following standard notation: if α and β are two multiindices, then

$$a_{(\beta)}^{(\alpha)}(x; \xi) = \partial_\xi^\alpha D_x^\beta a(x; \xi).$$

If $\beta = 0$ or $\alpha = 0$, we shall write $a^{(\alpha)}$ or $a_{(\beta)}$, respectively. Leibniz' formula (see Hörmander [1, vol. 1])

$$P(x; D)(u(x)v(x)) = \sum D^\alpha v(x) P^{(\alpha)}(x; D)u(x)/\alpha! \tag{3}$$

holds.

Consider the bilinear form

$$(u, v) = \int u(x)v(x)\, dx. \tag{4}$$

Integrating by parts we obtain

$$(D^\alpha u, v) = (u, (-D)^\alpha v).$$

By the transpose of an operator $P(x; D)$ is meant an operator ${}^tP(x; D)$ such that

$$(P(x; D)u, v) = (u, {}^tP(x; D)v). \tag{5}$$

In the case of operators with constant coefficients we have ${}^tP(D) = P(-D)$, and in the case of variable coefficients the symbol of the operator ${}^tP$ can be calculated by means of integration by parts.

If instead of bilinear form (4) the Hermitian form

$$[u,v] = \int u(x)\overline{v(x)}\,dx, \tag{4'}$$

is considered, then the adjoint operator $P^*(x;D)$ is determined by the relation

$$[P(x;D)u,v] = [u,P^*(x;D)v]. \tag{5'}$$

The operators ${}^tP$ and P^* are related by the formula

$${}^tP(x;D)\overline{u(x)} = \overline{P^*(x;D)u(x)}. \tag{6}$$

We note that the symbol of P^* has the form

$$P^*(x,\xi) = \overline{P}(x;\xi) + \sum_{\alpha>0} \overline{P}^{(\alpha)}_{(\alpha)}(x;\xi)/\alpha!. \tag{7}$$

We now remind the reader of the notation for some well-known spaces of functions and distributions.

By $\mathcal{D}(\Omega)$ ($\mathcal{D} = \mathcal{D}(\mathbb{R}^n)$) we shall denote the space of infinitely differentiable functions of compact support in $\Omega(\mathbb{R}^n)$ endowed with the natural topology. The symbol $\mathcal{S}$ will denote Schwarz' space of infinitely differentiable functions decreasing stronger than any power of $|x|$; this space is also equipped with the natural topology. By $\mathcal{D}'(D)$, $\mathcal{D}'$, and $\mathcal{S}'$ the respective conjugate spaces of $\mathcal{D}(\Omega)$, $\mathcal{D}$, and $\mathcal{S}$ are dencted. The elements of $\mathcal{D}'$ are usually called distributions and the elements of $\mathcal{S}'$ are called tempered distributions.

The expression (u,v) will be retained to denote the values of distributions $u \in \mathcal{D}'(\Omega)$, $\mathcal{D}'$, $\mathcal{S}'$ on the test functions $v \in \mathcal{D}(\Omega)$, $\mathcal{D}$, $\mathcal{S}$; formulas (5) and (6) serve as definitions of derivatives and differential operators for distributions.

If $u(x) \in \mathcal{D}$, $\mathcal{S}$, then by the Fourier transform of $u(x)$ is meant the function

$$\widehat{u}(\xi) = 2\pi^{-n/2} \int \exp(-i\langle x,\xi\rangle)u(x)\,dx, \tag{8}$$

and we have

$$\widehat{D^\alpha u}(\xi) = \xi^\alpha \widehat{u}(\xi) \tag{9}$$

so that if $P(D)$ is an operator with constant coefficients, then

$$(\widehat{P(D)}u)(\xi) = P(\xi)\widehat{u}(\xi). \tag{9'}$$

For the Fourier transform (8) the inversion formula

$$u(x) = (2\pi)^{-n/2} \int \exp(i\langle x, \xi\rangle)\widehat{u}(\xi)\, d\xi \tag{8'}$$

holds. Applying operator (1) to both sides of (8′) and inserting differentiation under the integral sign we obtain

$$P(x; D)u(x) = (2\pi)^{-n/2} \int \exp(i\langle x, \xi\rangle)P(x; \xi)\widehat{u}(\xi)\, d\xi. \tag{10}$$

Relation (10) elucidates definition (2) for the symbol of operator (1).

For Fourier transforms the forms (4) and (4′) are expressed in the following way:

$$(u, v) = (\widehat{u}, I\widehat{v}), \qquad I\widehat{v}(\xi) = \widehat{v}(-\xi), \tag{11}$$

$$[u, v] = [\widehat{u}, \widehat{v}]. \tag{11'}$$

Setting $u = v$ in (11′) we derive the classical Parseval relation:

$$\int |u(x)|^2\, dx = \int |\widehat{u}(\xi)|^2\, d\xi. \tag{12}$$

The Fourier operator defined by (8) transforms into itself the space $\mathcal{S}$ and, by the conjugacy, the space $\mathcal{S}'$ as well, i.e. for the elements $f \in \mathcal{S}'$ the generalized Fourier transform is defined. We denote by $H^{(s)}$ the set of $f \in \mathcal{S}'$ such that $\widehat{f}$ is a locally square summable function and the norm

$$\|f\|_{(s)} = \Big(\int (1 + |\xi|^2)|\widehat{f}(\xi)|^2\, d\xi\Big)^{1/2} \tag{13}$$

is finite. For $s = 0$, in view of Parseval's relation (12), formula (13) defines the ordinary L_2 norm. For natural values of s the norm (13) is equivalent to Sobolev's norm

$$\Big(\sum_{|\alpha| \leqslant s} \|D^\alpha f\|^2\Big)^{1/2} \qquad (s \in \mathbb{Z}_+). \tag{13'}$$

4.2. *N* Quasi-elliptic differential operators with constant coefficients. In this section and below we shall deal only with differential operators acting on functions of two variables. The variables will be denoted (accordingly) as x and y, and the dual variables as ξ and η:

$$\langle (x, y), (\xi, \eta)\rangle = x\xi + y\eta.$$

We write the differential operator (1) in the form

$$P(x, y; D_x, D_y) = \sum a_{\alpha\beta}(x, y) D_x^\alpha D_y^\beta, \tag{14}$$

where $D_x = -i\partial/\partial x$, $D_y = -i\partial/\partial y$ and α and β are natural numbers. We first of all consider operators with constant coefficients

$$P(D_x, D_y) = \sum a_{\alpha\beta} D_x^\alpha D_y^\beta \tag{14'}$$

whose symbols $P(\xi,\eta)$ are polynomials in Theorem 3.3.

Let N be a polygon lying in $\overline{\mathbb{R}^2_+}$ and having integral vertices (say $N = N(P)$, $\delta(P)$, or $\delta_0(P)$, where P is a polynomial). We associate the norm

$$\|u\|_{N,(s)} = \Big(\sum_{(\alpha,\beta)\in N\cap\mathbb{Z}^2} \|D_x^\alpha D_y^\beta u\|^2_{(s)} \Big)^{1/2} \tag{15}$$

with this polygon. By virtue of (9) and Lemma 2.1, this norm is equivalent to the norm

$$\Big(\iint \Xi_N(\xi,\eta)(1+\xi^2+\eta^2)^s |\hat{u}(\xi,\eta)|\, d\xi\, d\eta \Big)^{1/2} \tag{15'}$$

Theorem. *For differential operator* (14′) *the following conditions are equivalent,*

(I) *The symbol $P(\xi,\eta)$ is an N quasi-elliptic polynomial (i.e. the equivalent conditions of Theorem 3.3 are fulfilled).*

(II) *The inequalities below hold:*

(i) *$\forall s \in \mathbb{R}$ there is $c_s > 0$ such that*

$$\|u\|_{N(P),(s)} \leqslant c_s(\|P(D_x,D_y)u\|_{(s)} + \|u\|_{(s)}) \quad \forall u \in H^\infty(\mathbb{R}^2). \tag{16}$$

(ii) *there is $\varkappa > 0$ such that $\forall s \in \mathbb{R}$ $\exists c'_s > 0$ such that*

$$\|u\|_{\delta_0(P),(s+\varkappa)} \leqslant c'_s(\|P(D_x,D_y)u\|_{(s)} + \|u\|_{(s)}), \qquad u \in H^\infty(\mathbb{R}^2). \tag{17}$$

(III) *For any region $\Omega \subset \mathbb{R}^2$ with compact closure $\overline{\Omega}$ the following inequalities hold.*

(i) *$\forall s$ there is $c_s(\Omega) > 0$ such that*

$$\|u\|_{N(P),(s)} \leqslant c_s(\Omega)\|P(D_x,D_y)u\|_{(s)} \quad \forall u \in D(\Omega). \tag{16'}$$

(ii) *There is $\varkappa > 0$ such that $\forall s \in \mathbb{R}$ $\exists c'_s(\Omega)$ such that*

$$\|u\|_{\delta_0(P),(s+\varkappa)} \leqslant c'_s(\Omega)\|P(D_x,D_y)u\|_{(s)} \quad \forall u \in D(\Omega). \tag{17'}$$

Remark. We once again remind the reader that the polygon $\delta_0(P)$ (see Section 3.2) involved in (17) and (17′) does not coincide with the polygon $\delta(P)$ when the polygon $N(P)$ is not regular. Since we shall subsequently deal with differential operators whose symbols are N quasi-elliptic polynomials, in sufficient conditions for the validity of the inequalities we may confine ourselves to using only polygons $\delta(P)$. However, in the statements of necessary and sufficient conditions it is more suitable to define the norms with the aid of $\delta_0(P)$.

The proof of the theorem. (I)$\Longrightarrow$(II). Inequalities (3.9) in Theorem 3.3 can be rewritten as

$$\sum_{(\alpha,\beta)\in N(P)} |\xi^\alpha\eta^\beta|^2 \leqslant c_1(|P(\xi,\eta)|^2+1) \quad \forall(\xi,\eta)\in\mathbb{R}^2. \tag{18}$$

If both sides of (18) are multiplied by $(1+\xi^2+\eta^2)^s|\widehat{u}(\xi,\eta)|^2$ and integrated over $\mathbb{R}^2_{(\xi,\eta)}$, then we obtain inequality (16) (with a constant not depending on s).

Similarly, inequalities (3.9) are equivalent to

$$(1+\xi^2+\eta^2)^\varkappa\Big(\sum_{(\alpha,\beta)\in\delta_0(P)} |\xi^\alpha\eta^\beta|^2+1\Big) \leqslant c_1'(|P(\xi,\eta)|^2+1), \tag{19}$$

whence follows (17).

(II)$\Longrightarrow$(III). This implication is based on the fact that for any differential operator with constant coefficients $Q(D_1,\dots,D_n)$, any region $G\subset\mathbb{R}^n$ with compact closure, and any $s\in\mathbb{R}$ the inequality

$$\sum_{\alpha>0} \|Q^{(\alpha)}(D_1,\dots,D_n)u\|_{(s)} \leqslant K_s(G)\|Q(D_1,\dots,D_n)u\|_{(s)} \quad \forall u\in\mathcal{D}(G) \tag{20}$$

holds. For $s=0$ this inequality was proved by Hörmander [2, Chapter 2]. The proof of (20) for arbitrary real s can be found in the paper by Paneyakh [1]. In particular, (20) implies the inequality

$$\|u\|_{(s)} \leqslant K_s(\Omega)\|P(D_x,D_y)u\|_{(s)} \quad \forall u\in D(\Omega).$$

Substituting this inequality into the right-hand sides of (16) and (17) we arrive at (16′) and (17′).

(III)$\Longrightarrow$(I). The above inequality (20) is part of a more general theorem (see Hörmander [2] and Paneyakh [1]): for some $s\in\mathbb{R}$ and operators $P(D_1,\dots,D_n)$ and $Q(D_1,\dots,D_n)$ the inequality

$$\|Q(D_1,\dots,D_n)u\|^2_{(s)} \leqslant K\|P(D_1,\dots,D_n)u\|^2_{(s)}, \qquad u\in D(G), \tag{21}$$

holds, where G is a region with compact closure, if and only if

$$\sum_{\alpha\geqslant 0} |Q^{(\alpha)}(\xi_1,\dots,\xi_n)| \leqslant K\sum_{\alpha\geqslant 0} |P^{(\alpha)}(\xi_1,\dots,\xi_n)| \quad \forall(\xi_1,\dots,\xi_n)\in\mathbb{R}^n. \tag{21′}$$

Applying this result we conclude that the inequalities

$$\sum_{(\alpha,\beta)\in N(P)} |\xi^\alpha \eta^\beta| < c\Big(|P(\xi,\eta)| + \sum_{\alpha>0} |P^{(\alpha)}(\xi,\eta)|\Big), \tag{22}$$

$$\sum_{(\alpha,\beta)\in \delta_0(P)} |\xi^\alpha \eta^\beta| < c'(1+|\xi|+|\eta|)^{-\varkappa}\Big(|P(\xi,\eta)| + \sum_{\alpha>0} |P^{(\alpha)}(\xi,\eta)|\Big), \tag{23}$$

are consequences of (16′) and (17′). Since the polynomials $P^{(\alpha)}(\xi,\eta)$, $\alpha > 0$, consist of monomials corresponding to the points in $\delta_0(P)$, inequality (23) implies that

$$\sum_{(\alpha,\beta)\in \delta_0(P)} |\xi^\alpha \eta^\beta| < c'(|\xi|+|\eta|)^{-\varkappa}|P(\xi,\eta)| + c''(1+|\xi|+|\eta|)^{-\varkappa} \sum_{(\alpha,\beta)\in \delta_0(P)} |\xi^\alpha \eta^\beta|.$$

Taking $R > 0$ satisfying the condition

$$c''(1+R)^{-\varkappa} < 1/2$$

we arrive at the inequality (cf. (3.9))

$$\sum_{(\alpha,\beta)\in \delta_0(P)} |\xi^\alpha \eta^\beta| < 2c'(|\xi|+|\eta|)^{-\varkappa}|P(\xi,\eta)|, \qquad |\xi|+|\eta| \geqslant R.$$

By virtue of this inequality, for $|\xi|+|\eta| \geqslant R$ we have

$$c\sum_{\alpha>0} |P^{(\alpha)}(\xi,\eta)| < c'''|P(\xi,\eta)|.$$

Substituting this into (22) we obtain (3.9), i.e. the polynomial P is Nquasi-elliptic.

Definition. Differential operator (14) is said to be N quasi-elliptic if it satisfies the equivalent conditions of the theorem.

4.3. *N* Quasi-elliptic differential operators with variable coefficients. Consider differential operator (14). The symbol

$$P(x,y;\xi,\eta) = \sum a_{\alpha\beta}(x,y)\xi^\alpha \eta^\beta \tag{24}$$

of the operator is a polynomial in the variables ξ, η for fixed (x,y). We denote by $\nu(x,y)$ the set of integral points $(\alpha,\beta) \in \mathbb{R}^2$ for which $a_{\alpha\beta}(x,y) \neq 0$. Using the procedure presented in §1 we construct Newton's polygons that will be denoted $N(P(x,y))$ and the corresponding polygons $\delta(P(x,y))$ and $\delta_0(P(x,y))$. By $N(P)$, $\delta(P)$, and $\delta_0(P)$ we shall denoted the convex hulls of the set-theoretical unions of the polygons $N(P(x,y))$ etc., where the point (x,y) runs over the plane $\mathbb{R}^2$ or over region $\Omega \subset \mathbb{R}^2$.

Definition. Differential operator (14) is said to be N quasi-elliptic in a region Ω if its symbol (24) satisfies the following conditions.

(i) $N(P(x,y)) = N(P)\ \forall (x,y) \in \overline{\Omega}$ (i.e. Newton's polygons corresponding to the various polynomials $p_{x,y}(\xi,\eta) = P(x,y;\xi,\eta)$ coincide).

(ii) For any fixed $(x^0,y^0) \in \overline{\Omega}$ the polynomial $P(x^0,y^0;\xi,\eta)$ is N quasi-elliptic.

Proposition. *Let symbol (24) satisfy the conditions of the definition in a bounded region $\Omega \subset \mathbb{R}^2$ and let the coefficients $a_{\alpha\beta}(x,y)$ belong to C^∞. Then*

(i) *the coefficients of the symbols $P^{[j]}(x,y,\xi,\eta)$ defined for any (x,y) by formulas (1.15) belong to C^∞;*

(ii) *there are constants $c, c_0 > 0$ such that for all $(x,y) \in \overline{\Omega}$ an inequality of the type (3) holds:*

$$\Xi_{N(P)}(\xi,\eta) < c|P(x,y;\xi,\eta)| \quad \text{for } \xi^2+\eta^2 > c_0^2. \tag{25}$$

Proof. (i) Since Newton's polygons of the polynomials $P(x,y;\xi,\eta)$ do not depend on (x,y), we have $a_{\alpha_j\beta_j}(x,y) \neq 0$ for any vertex $(\alpha_j,\beta_j) \in N(P)$ distinct from $(0,0)$.[1)] If Ω is a bounded region, then its closure $\overline{\Omega}$ is a compact set on which all the functions $|a_{\alpha_j\beta_j}(x,y)|$ have positive infima, and, consequently, the functions $a^{-1}_{\alpha_j\beta_j}(x,y)$ are bounded in $\overline{\Omega}$ together with all their derivatives. In particular, a trivial consequence of this property is the fact that the coefficients of the polynomials $P^{[j]}(x,y;\xi,\eta)$ defined by means of formulas (1.15) are smooth. Further, according to Theorem 3.3, we have

$$P^{[j]}(x,y;\xi,\eta) \neq 0 \quad \forall (x,y) \in \overline{\Omega}, \qquad \xi^2+\eta^2 > 0.$$

Therefore the positive continuous function $|P^{[j]}(x,y;\xi,\eta)|$ on the compact set $\overline{\Omega} \times \{(\xi,\eta) \in \mathbb{R}^2, |\xi|^{1/q_1^{(j)}} + |\xi|^{1/q_2^{(j)}} = 1\}$ attains its positive supremum and infimum. In view of the $q^{(j)}$-homogeneity, it follows that there exists $\varepsilon > 0$ such that the two-sided estimate

$$\varepsilon \leqslant |P^{[j]}(x,y;\xi,\eta)|\,(|\xi|^{1/q_1^{(j)}} + |\eta|^{1/q_2^{[j]}})^{-m_j} \leqslant \varepsilon^{-1}, \quad \forall (\xi,\eta) \in \mathbb{R}^2, \tag{26}$$

holds, where m_j is the $q^{(j)}$-order of $P^{[j]}$ and $q^{(j)} = (q_1^{(j)}, q_2^{(j)})$.

(ii) A careful examination of the proof of the implication (III)$\Longrightarrow$(I) in Section 2.3 shows that the constants c and c_0 in (2.3) depend in fact on the constant in inequality (2.3″) (i.e. in our situation on the constant ε in (26)) and on the maximum of the moduli of the coefficients of the original polynomial P. Since all coefficients $a_{\alpha\beta}(x,y)$ are uniformly bounded in $\overline{\Omega}$, the proposition is proved.

---

1) Recall that, by virtue of the N quasi-ellipticity of the polynomials $P(x,y;\xi,\eta)$, the polygon $N(P)$ is regular, and all its vertices (except for $(0,0)$) belong to the original set $\nu(x,y)$.

Let symbol (24) with smooth coefficients be given and let $N(P)$ be the convex hull of the union of all polygons $N(P(x,y))$. By analogy with spaces of polynomials (1.4) and (1.4'), we define spaces of symbols $SL_{N(P)}$ and $S\mathcal{L}_{N(P)}$ consisting of

$$Q(x,y;\xi,\eta) = \sum b_{\alpha\beta}(x,y)\xi^\alpha\eta^\beta$$

such that the coefficients $b_{\alpha\beta}(x,y)$ belong to C^∞ and (respectively) $N(Q) \subset N(P)$ and $N(Q) \subset \delta(P)$.

Theorem. *If operator* (14) *is* N *quasi-elliptic in a region* Ω, *then there are vectors* $q^{(j)} \in \mathbb{R}^2_+$, $j = 1,\dots,m$, $q^{(j)}$ *quasi-elliptic symbols* $P^{[j]}(x,y;\xi,\eta)$, *and a symbol* $Q(x,y;\xi,\eta) \in S\mathcal{L}$ *such that the operator relation*

$$P(x,y;D_x,D_y) - P^{[1]}(x,y;D_x,D_y)\dots P^{[m]}(x,y;D_x,D_y) \\ = Q(x,y;D_x,D_y) \qquad (27)$$

holds.

Proof. According to Theorem 3.3, at each point (x,y) the representation

$$P(x,y;\xi,\eta) - \prod_{j=1} P^{[j]}(x,y;\xi,\eta) = Q_0(x,y;\xi,\eta) \qquad (28)$$

takes place, where $P^{[j]}$ are $q^{(j)}$-elliptic polynomials and Q belongs to $\mathcal{L}_{N(P)}$. By the proposition, the symbols $P^{[j]}$ have smooth coefficients. Therefore the symbol Q_0 also has smooth coefficients and belongs to $S\mathcal{L}_{N(P)}$. If we set $\widehat{P} = P^{[1]}\dots P^{[m]}$, then relation (28) for the symbols can be rewritten as an operator relation:

$$P(x,y;D_x,D_y) - \widehat{P}(x,y;D_x,D_y) = Q_0(x,y;D_x,D_y). \qquad (27')$$

We now show that there is a symbol $Q_1 \in S\mathcal{L}_{N(P)}$ such that

$$\widehat{P}(x,y;D_x,D_y) - P^{[1]}(x,y;D_x,D_y)\dots P^{[m]}(x,y;D_x,D_y) \\ = Q_1(x,y;D_x,D_y). \qquad (27'')$$

Putting $Q = Q_0 + Q_1$ we arrive at (27).

Relation (27'') is based on Leibniz' formula in Hörmander's form: if A and B are two differential operators, then AB is a differential operator with symbol (see the notation in Section 4.1)

$$\sum_{\alpha \geqslant 0} A^{(\alpha)}B_{(\alpha)}/\alpha!$$

By this formula, the right-hand side of $(27'')$ is a differential operator whose symbol is a sum of expressions of the form of

$$A_{a_1 b_1 \dots a_m b_m}(x,y)(\xi^{a_1}\eta^{b_1})\dots(\xi^{a_m}\eta^{b_m}),$$

where $(a_j, b_j) \in N(P^{[j]})$, $j = 1,\dots,m$, and there exists at least one pair (a_{j_0}, b_{j_0}) belonging to $\delta_0(P^{[j]})$. Since $N(P^{[j]})$ is a triangle and, consequently, a regular polygon, we have $\delta_0(P^{[j]}) = \delta(P^{[j]})$.

By virtue of Remark 1) in Section 2.3, the polygon $N(P) = N(\widehat{P})$ is the arithmetical sum of the triangles $N(P^{[j]})$ (see (2.20)). Whence it follows that the expression $\xi^{a_1+\dots+a_m}\eta^{b_1+\dots+b_m}$ is a minor monomial of the polygon $N(P)$, which proves $(27'')$. The theorem is proved.

4.4. A priori estimates for N quasi-elliptic differential operators. In Section 4.2 we showed that in the case of differential operators with constant coefficients the condition for N quasi-ellipticity of a symbol is equivalent to estimates $(16')$, $(17')$ in any bounded region. We now prove an analog of this assertion for variable coefficients.

Theorem. *For a differential operator (14) with smooth coefficients the following conditions are equivalent.*

(I) *Operator (14) is N quasi-elliptic (i.e. its symbol (24) satisfies conditions (i) and (ii) in definition 4.3).*

(II) *In any sufficiently small region Ω, $\operatorname{diam}\Omega < \varepsilon$, the inequality*

$$\|u\|_{N(P)} \leqslant c\|P(x,y;\mathcal{D}_x,\mathcal{D}_y)u\| \quad \forall u \in D(\Omega) \tag{29}$$

is fulfilled. Moreover, there is $\varkappa > 0$ such that

$$\|u\|_{\delta(P),(\varkappa)} \leqslant c'\|P(x,y;D_x,D_y)u\| \quad \forall u \in D(\Omega). \tag{30}$$

Proof. (II)$\Longrightarrow$(I). Let (29) hold and let, for definiteness, the region Ω contain the origin. Setting $P_0(D_x,D_y) = P(0,0;D_x,D_y)$ we shall show that if the diameter of the region Ω is sufficiently small, then an analogous inequality holds for the operator with constant coefficients $P_0(D_x,D_y)$ as well. Take a truncating function $\psi \in \mathcal{D}$ equal to 1 in Ω and nonzero only in a sufficiently small neighborhood of Ω. Then if $u \in \mathcal{D}(\Omega)$, we have

$$\begin{aligned} P(x,y;D_x,D_y)u(x,y) = {} & P_0(D_x,D_y)u(x,y) \\ & + \sum a'_{\alpha\beta}(x,y)D_x^\alpha D_y^\beta u(x,y), \end{aligned} \tag{31}$$

where

$$a'_{\alpha\beta}(x,y) = \psi(x,y)(a_{\alpha\beta}(x,y) - a_{\alpha\beta}(0,0)). \tag{32}$$

Inequality (29) trivially implies the estimate

$$\|u\|_{N(P)} \leqslant c\|P_0(D_x, D_y)u\| + c \max_{\alpha,\beta,x,y} |a'_{\alpha\beta}| \|u\|_{N(P)}.$$

If the region Ω is sufficiently small, the coefficient in the second term on the right-hand side is less than $\theta < 1$, whence follows the inequality

$$\|u\|_{N(P)} \leqslant c_1 \|P_0(D_x, D_y)u\|, \qquad u \in D(\Omega), \tag{33}$$

where $c_1 = (1-\theta)^{-1}$. According to Hörmander's theorem (cf. Theorem 4.2), inequality (33) implies that

$$\begin{aligned}\sum_{(\alpha,\beta)\in N(P)} |\xi^\alpha \eta^\beta| &\leqslant c_2|P_0(\xi,\eta)| + c_2 \sum_{\gamma>0} |P_0^{(\gamma)}(\xi,\eta)| \\ &\leqslant c_2|P_0(\xi,\eta)| + c_3 \sum_{(\alpha,\beta)\in\delta_0(P_0)} |\xi^\alpha\eta^\beta|.\end{aligned} \tag{34}$$

We shall show that (30) implies the regularity of the polygon $N(P)$. If this property is already proved, then $\delta_0(P_0) = \delta(P)$ (Lemma 4 in Section 3.2), and then, according to Lemma 2.1 (ii), we have

$$\sum_{(\alpha,\beta)\in\delta_0(P_0)} |\xi^\alpha\eta^\beta| \leqslant c_4(|\xi| + |\eta|)^{-\varkappa} \sum_{(\alpha,\beta)\in N(P_0)} |\xi^\alpha\eta^\beta|. \tag{35}$$

In view of (35), there exists c_0 such that

$$\sum_{(\alpha,\beta)\in N(P)} |\xi^\alpha\eta^\beta| \leqslant c_5|P_0(\xi,\eta)|, \qquad \xi^2 + \eta^2 > c_0^2. \tag{34$'$}$$

Since $N(P) \supset N(P_0)$, it follows that the polynomial $P_0(\xi,\eta)$ is N quasi-elliptic. Further, by virtue of Lemma 2.2 and inequality (34′), we have $N(P) \subset N(P_0)$, and, consequently $N(P) = N(P_0)$. Since $(0,0)$ is an arbitrary point in the region Ω, the symbol $P(x,y;\xi,\eta)$ satisfies all conditions of definition 4.3.

To prove that the polygon $N(P)$ is regular it suffices to show that the symbol (24) is solved relative to the highest powers of both ξ and η. Assume the contrary, say

$$P(x,y;\xi,\eta) = a(x,y)\xi^\mu\eta^\nu + \sum_{j\geqslant 1} a_j(x,y,\eta)\xi^{\mu-j}, \quad \nu > 0, \tag{36}$$

where $a_j(x,y,\eta)$ are polynomials in η.

Take a function $u_R(x,y) = u(xR,y)$, where $u(x,y) \in \mathcal{D}$ and $u(x,y) = 0$ for $|y| > \delta$ (δ is sufficiently small). Then for a sufficiently large R the support of u_R

belongs to the region Ω so that (30) must hold for $u = u_R$. We now prove that if symbol (24) has the form of (36), then

$$\|u_R\|_{\delta_0(P),(\varkappa)} \geqslant \text{const}\, R^{\mu+\varkappa-1/2}, \tag{37}$$

$$\|P(x,y;D_x,D_y)u_R\| \leqslant \text{const}\, R^{\mu-1/2}. \tag{38}$$

Inequalities (37) and (38) contradict (30) for $R \to +\infty$, i.e. representation (36) with $\nu > 0$ cannot hold.

The proof of inequality (37). If the symbol has the form (36) with $a(x,y) \neq 0$, i.e. $N(P)$ contains a point (μ,ν), $\nu > 0$, then $(\mu, 0) \in \delta_0(P)$. Therefore, to prove (37) it suffices to show that

$$\|D_x^\mu u_R\| > \text{const}\, R^{\mu+\varkappa-1/2}. \tag{37'}$$

Setting $v(x,y) = D_x^\mu(x,y)$ we calculate the Fourier transform of $D_x^\mu u_R(x,y)$. We have

$$(2\pi)^{-1} \iint \exp(-ix\xi - iy\eta) R^\mu v(Rx,y)\, dx\, dy = R^{\mu-1}\widehat{v}(\xi/R,\eta).$$

Therefore

$$\begin{aligned}
\|D_x^\mu u_R\|^2_{(\varkappa)} &= R^{2\mu-2} \iint (1+\xi^2+\eta^2)^\varkappa |\widehat{v}(\xi/R,\eta)|\, d\xi\, d\eta \\
&= R^{2\mu-1} \iint (1+R^2\xi^2+\eta^2)^\varkappa |\widehat{v}(\xi,\eta)|^2\, d\xi\, d\eta \\
&\geqslant R^{2\mu+2\varkappa-1} \iint |\xi|^{2\varkappa} |\widehat{v}(\xi,\eta)|^2\, d\xi\, d\eta = \text{const}\, R^{2\mu+2\varkappa-1}
\end{aligned}$$

To prove (38) we note that, with account of (36), we have

$$\begin{aligned}
P(x,y;D_x,D_y)u_R(x,y) &= R^\mu b(x,y)(D_x^\mu D_y^\nu u)(Rx,y) \\
&\quad + \sum_{j\geqslant 1} R^{\mu-j}(b_j(x,y,D_y)D^{\mu-j}u)(Rx,y),
\end{aligned}$$

whence

$$\begin{aligned}
|Pu_R|^2 \leqslant{}& K_1 R^{2\mu} |(D_x^\mu D_y^\nu u)(Rx,y)| \\
&+ K_2 \sum_{j\geqslant 1} R^{2\mu-2j} |(b_j(x,y,D_y)D^{\mu-j}u)(Rx,y)|^2.
\end{aligned}$$

Integrating both sides of this inequality with respect to x and y and making change of variable $Rx \mapsto R$ we arrive at (38).

(I)$\Longrightarrow$(II). Write operator (14) in the form (33). By the condition (ii) in Definition 4.3, the operator $P_0(D_x, D_y)$ is N quasi-elliptic, and therefore, by Theorem 4.2, inequalities (16) and (17) are fulfilled. Repeating the argument in the proof of the implication (II)$\Longrightarrow$(I) we derive from the inequality for P_0 the inequality (29). Further, the regularity of $N(P)$ and Lemma 2.1 (ii) readily imply the existence of $\varkappa > 0$ such that

$$\|u\|_{\delta(P),(\varkappa)} \leqslant \text{const}\, \|u\|_{N(P)}.$$

Estimating the right-hand side by means of (29) we derive inequality (30).

Corollary. *Under the conditions of the theorem the following estimate for the transposed operator ${}^tP$ holds:*

$$\|u\|_{N(P)} \leqslant c'\|{}^tP(x,y;D_x,D_y)u\|, \quad \forall u \in D(\Omega). \tag{29'}$$

Proof. Recall that the symbol $P^*(x,y;\xi,\eta)$ is determined by formula (7). Clearly, the symbols $P(x,y;\xi,\eta)$ and $\overline{P}(x,y;\xi,\eta)$ simultaneously satisfy or do not satisfy the conditions of definition 4.3. Moreover, the symbol $\sum_{\alpha>0} P^{(\alpha)}_{(\alpha)}/\alpha!$ is a linear combination of the monomials corresponding to the minor points of $N(P)$. It is obvious that the addition of this symbol to the symbol $\overline{P}(x,y;\xi,\eta)$ does not violate the condition of N quasi-ellipticity. Therefore inequality (29) is fulfilled for the operator ${}^tP = \overline{P^*}$.

4.5. A priori estimates for N quasi-elliptic differential operators in the spaces H^μ. It is convenient to derive a priori estimates for N quasi-elliptic operators not in the scales of $H^{(s)}$ with norm (13) but in scales of a more general type. The corresponding definitions will be presented for the case of functions of n variables.

Denote by $\mathcal{B}$ the class of continuous functions $\mu(\xi_1,\dots,\xi_n)$ possessing the following properties: for each function $\mu \in \mathcal{B}$ there are numbers c and l such that

$$|\mu(\xi_1',\dots,\xi_n')\mu^{-1}(\xi_1'',\dots,\xi_n'')| < c(1 + |\xi_1' - \xi_1''| + \dots + |\xi_n' - \xi_n''|)^l. \tag{39}$$

With the symbol μ we associate the pseudodifferential operator (PDO)

$$\mu(D_1,\dots,D_n)f(x_1,\dots,x_n) = (2\pi)^{-n/2}\int \exp(-i(x_1\xi_1 + \dots, x_n\xi_n)) \times \mu(\xi_1,\dots,\xi_n)\widehat{f}(\xi_1,\dots,\xi_n)\,d\xi_1\dots d\xi_n. \tag{40}$$

Let H^μ denote the set of distributions $f \in H^{(-\infty)}$ for which $\mu(D)f \in L_2$. In the space H^μ the Banach norm

$$\|f\|_\mu = \|\mu(D)f\| = \left(\int |(\mu\widehat{f})(\xi_1,\dots,\xi_n)|^2\,d\xi_1\dots d\xi_n\right)^{1/2} \tag{41}$$

is defined in a natural way.

Owing to condition (39), the space H^μ is a module over $\mathcal{D}$ (see Volevich and Paneyakh [1]). We shall need a more exact estimate for the norm of the multiplication operator, which we derive under an additional condition on $\mu \in \mathcal{B}$. We assume that

$$\begin{aligned}\left|\mu(\xi_1',\dots,\xi_n') - \mu(\xi_1'',\dots,\xi_n'')\right| < \varepsilon\left(1 + |\xi_1' - \xi_1''| + \dots + |\xi_n' - \xi_n''|\right)^l \left|\mu(\xi_1'',\dots,\xi_n'')\right| \\ \forall(\xi_1',\dots,\xi_n'),(\xi_1'',\dots,\xi_n'') \in \mathbb{R}^n. \qquad (39')\end{aligned}$$

Note that since the points $(\xi_1',\dots,\xi_n')$ and $(\xi_1'',\dots,\xi_n'')$ are equivalent, the number l in (39) should be assumed to be nonnegative. Then (39′) implies (39) with $c = 1 + \varepsilon$.

Lemma. *Let a function μ satisfy condition (39′) and let $a(x_1,\dots,x_n) \in \mathcal{D}$. Then the inequality*

$$\|au\|_\mu \leqslant \left(\max|a(x_1,\dots,x_n) + \varepsilon K(a,l)|\,\|u\|_\mu\right) \qquad (42)$$

holds, where the constant K is the same for all functions μ satisfying (39′) with a fixed l.

Proof. Setting $v = \mu(D)u$, i.e. $u = \mu^{-1}(D)v$, we rewrite (42) in the form

$$\|\mu(D)a\mu^{-1}(D)v\| \leqslant (\max|a| + \varepsilon K)\|v\|. \qquad (42')$$

The expression under the sign of the norm on the left-hand side of (42′) is rewritten in the form of the sum

$$av + (\mu(D)a - a\mu(D))\mu^{-1}(D)v = T_1v + T_2v.$$

Clearly, we have

$$\|T_1v\| \leqslant \max|a|\|v\|. \qquad (43)$$

When estimating the operator T_2 it is more suitable to apply the Fourier transformation. Since under the Fourier transformation the operation of multiplication goes into convolution, the composition of the Fourier operator and T_2 is an integral operator

$$\widehat{v}(\xi_1,\dots,\xi_n) \mapsto \int T(\xi_1,\dots,\xi_n,\eta_1,\dots,\eta_n)\widehat{v}(\eta_1,\dots,\eta_n)\,d\eta_1\dots d\eta_n,$$

where

$$\begin{aligned}T(\xi_1,\dots,\eta_n) = \widehat{a}(\xi_1 - \eta_1,\dots,\xi_n - \eta_n)\big(\mu(\xi_1,\dots,\xi_n) \\ - \mu(\eta_1,\dots,\eta_n)\big)\mu^{-1}(\eta_1,\dots,\eta_n) \qquad (44)\end{aligned}$$

By virtue of (39′), we have

$$
\begin{aligned}
|T(\xi_1,\dots,\xi_n)| &\leqslant \varepsilon(1+|\xi_1-\eta_1|+\dots+|\xi_n-\eta_n|)^l|\widehat{a}(\xi_1-\eta_1,\dots,\xi_n-\eta_n)| \\
&\leqslant \varepsilon K(1+|\xi_1-\eta_1|+\dots+|\xi_n-\eta_n|)^{-n-1} \\
&\stackrel{\text{def}}{=} \varepsilon K h(\xi_1-\eta_1,\dots,\xi_n-\eta_n).
\end{aligned}
$$

Here we have used the fact if $a \in \mathcal{D}$, then the function $\widehat{a}(\xi_1,\dots,\xi_n)$ decreases stronger than any power of $|\xi_1|+\dots+|\xi_n|$. Hence, the kernel T is majorized by the function $\varepsilon K h$, where $h \in L_1$ and $\|h, L_1\| \leqslant \varkappa(n)$. Whence follows the inequality

$$\|T_2 v\| \leqslant cK\varkappa'(n)\|v\|, \tag{43′}$$

where K depends on l and on the function a. Comparing (43) and (43′) we derive inequality (42).

The above-mentioned results on the equivalence of (21′) and inequalities (21) in the norms (13) are in fact established for estimates in the norms (41), i.e. if condition (21′) is fulfilled for polynomials $P(\xi_1,\dots,\xi_n)$ and $Q(\xi_1,\dots,\xi_n)$, then for any region Ω with compact closure the inequality

$$\|Q(D)u\|_\mu \leqslant K\|P(D)u\|_\mu \quad \forall u \in D(\Omega) \tag{45}$$

holds. If one carefully examines the proof of this theorem in the paper by Paneyakh [1] (see formulas (3.9) and (3.10) on pages 91–92), then one can see that the constant K in (45) depends on the symbols Q and P, the region Ω, and the constant l in inequality (39). In other words, for different functions μ admitting of estimate (39) with one and the same number l the inequality (45) holds with the same constant K.

Proposition. *Let an inequality of the form of* (39′), *where the number* ε *is sufficiently small, be fulfilled for a function* $\mu(\xi,\eta)$, $(\xi,\eta) \in \mathbb{R}^2$. *Let operator* (24) *be* N *quasi-elliptic. If the diameter of the region* Ω *is sufficiently small,* $\operatorname{diam}\Omega \leqslant d$, *then the inequality*

$$\|u\|_{N(P),\mu} \leqslant c(\Omega)\|P(x,y;D_x,D_y)u\|_\mu \quad \forall u \in D(\Omega) \tag{46}$$

holds, where

$$\|u\|_{N(P),\mu} = \Big(\sum_{(\alpha,\beta)\in N(P)} \|D_x^\alpha D_y^\beta u\|_\mu^2\Big)^{1/2} \tag{47}$$

Proof. Write the operator P in the form (31). Since the expressions $P_0(\xi,\eta)$ and $\xi^\alpha\eta^\beta$, $(\alpha,\beta) \in N(l)$, are related by an inequality of the type (21′), the inequality

$$\|u\|_{N(P),\mu} \leqslant c\|P_0(Dx,D_y)u\|_\mu$$

holds, whence

$$\|u\|_{N(P),\mu} \leqslant c\|P(x,y;D_x,D_y)u\|_\mu + c \sum_{(\alpha,\beta)\in N(P)} \|a'_{\alpha\beta}(x,y)D_x^\alpha D_y^\beta u\|_\mu.$$

Applying the lemma we conclude that the second term on the right-hand side does not exceed

$$c \max_{(\alpha,\beta)\in N(P)} (\max_{x,y} |a'_{\alpha\beta}(x,y)| + \varepsilon K(a'_{\alpha\beta}, l))\|u\|_{N(P)}.$$

Since the functions $a'_{\alpha\beta}(x,y)$ (see (33′)) tend to zero as $\operatorname{diam}\Omega \to 0$, there is d such that

$$c \max_{(x,y)} |a'_{\alpha\beta}(x,y)| < 1/4 \quad \text{if} \quad \operatorname{diam}\Omega < d, \quad (\alpha,\beta) \in N(P).$$

We now fix the region Ω (and, consequently, the truncating function ψ in (33′)) and select ε such that

$$\varepsilon c K(a'_{\alpha\beta}, l) < 1/4.$$

In view of these inequalities, we arrive at (46) with constant $c(\Omega) = 2c$.

We set

$$\nu_{N(P)}(\xi,\eta) = \Big(\sum_{(\alpha,\beta)\in N(P)} \xi^{2\alpha}\eta^{2\beta} \Big)^{1/2} \tag{48}$$

and introduce the norm

$$|[u]|_{N(P),s} = \|\nu^s_{N(P)}(D_x,D_y)u\|. \tag{49}$$

Note that the function (48) can be estimated from above and below by means of $\Xi_{N(P)}(\xi,\eta)$ with constants not depending on ξ and η. It follows that the norm $\| \ \|_{P,1}$ is estimated from above and below in terms of the norm (47) for $\mu(\xi,\eta) \equiv$ const.

Theorem. *Let operator (24) be N quasi-elliptic and let the diameter of the region Ω be sufficiently small. Then $\forall s \in \mathbb{R}$ the inequality*

$$|[u]|_{N(P),s+1} \leqslant K_1(\Omega)|[P(x,y;D_x,D_y)u]|_{N(P),s} \quad \forall u \in \mathcal{D}(\Omega) \tag{50}$$

holds.

Proof. We construct a family of positive functions $\nu_{R,s}(\xi,\eta)$ depending on a large parameter $R \to +\infty$ and satisfying the following conditions.

(i) There is a constant $c = c(R,s)$ such that

$$c^{-1} \leqslant \nu^s_{N(P)}(\xi,\eta)\nu^{-1}_{R,s}(\xi,\eta) \leqslant c.$$

(ii) The function $\nu_{R,s}$ satisfies condition (39′), i.e.

$$|\nu_{R,s}(\xi',\eta') - \nu_{R,s}(\xi'',\eta'')| < \varepsilon_s(R)(1 + |\xi' - \xi''| + |\xi' - \xi''|)^{l_s}\nu_{R,s}(\xi'',\eta''),$$

where

$$\varepsilon_s \to 0 \quad \text{as} \quad R \to \infty.$$

If the function $\nu_{R,s}$ satisfies (ii), then the proposition can be applied to it, i.e. for a sufficiently large R the function $\mu = \nu_{R,s}$ satisfies (46). By virtue of (i), for $\mu = \nu_{s,R}$ the left-hand side of (46) can be estimated from below by means of $\|u\|_{N(P),s+1}$, and the right-hand side of (46) can be estimated from above using the right hand side of (50).

Thus, we have reduced the proof of the theorem to the problem of constructing a function $\nu_{R,s}$ satisfying conditions (i) and (ii).

As was shown in Section 3.3, to within minor terms, an N quasi-elliptic polynomial $P(\xi,\eta)$ is a product of quasi-elliptic polynomials $P^{[j]}(\xi,\eta)$. Let m_j and n_j be the highest powers of ξ and η in these polynomials so that (cf. (2.15′))

$$c^{-1} \leqslant |P^{[j]}(\xi,\eta)|(\xi^{2m_j} + \eta^{2n_j})^{-1/2} \leqslant c. \tag{51}$$

We set

$$\nu_{R,s}(\xi,\eta) = \prod_{j=1}^{m}(\xi^{2m_j} + \eta^{2n_j} + R^2)^{s/2}. \tag{52}$$

Condition (i) for the function (52) can easily be verified. We may confine ourselves to the case $s = 1$. For $s = 1$ the function (52) is estimated from above and below by means of the function

$$\prod_{j=1}^{m}(|\xi|^{m_j} + |\eta|^{n_j} + R) = \Xi_{N(P)}(\xi,\eta) + \cdots,$$

where the dots symbolize monomials $R^\gamma|\xi|^\alpha|\eta|^\beta$ with $(\alpha,\beta) \in \delta(P)$. Consequently, $\nu_{R,s}$ is estimated from above and below by means of $\Xi_{N(P)}(\xi,\eta)$, and hence condition (i) is fulfilled. The verification of condition (ii) for the function (52) is carried out in the Appendix to this section.

4.6. Local solvability of *N* quasi-elliptic differential operators. The first question arising in the study of a differential equation

$$P(x,y;D_x,D_y)u(x,y) = f(x,y) \tag{53}$$

is whether there exist any solutions u for a given right-hand side f.

Equation (53) will be said to be locally solvable at a point, say at $(0,0)$, if there is a neighborhood $\Omega \supset (0,0)$ such that for any right-hand side $f \in \mathcal{D}(\Omega)$ there exists a distribution $u \in \mathcal{D}'(\Omega)$ which is a solution to Equation (53) in the sense of the theory of distributions:

$$(u, {}^tP\varphi) = (f, \varphi) \quad \forall \varphi \in D(\Omega). \tag{54}$$

The local solvability of N quasi-elliptic equations follows from inequalities (50) or, more precisely, from analogous inequalities for the transposed operator:

$$\|[u]\|_{N(P),s+1} \leqslant K'_s \|[{}^tP(x, y; D_x, D_y)u]\|_{N(P),s} \tag{55}$$

in accordance with the traditional methods in functional analysis. Preliminarily, we remind the reader of the definition of the spaces H^μ in a bounded region (see Hörmander [1, Vol. 2, Chapter 10], or Volevich and Paneyakh [1]).

We begin with the definition of support for a distribution $f \in \mathcal{D}'$. We shall say that f is equal to zero in a region $G \subset \mathbb{R}^n$ if $(f, \varphi) = 0 \ \forall \varphi \in \mathcal{D}(G)$. By $\operatorname{supp} f$ is meant the smallest closed set on whose complement the functional f is equal to zero. For ordinary functions (say, belonging to L_1^{loc}) the support in the sense of distributions is the smallest closed set on whose complement the function f is equal to zero (almost everywhere).

Denote by H_G^μ the subspace of H^μ consisting of distributions with support belonging to the closure $\overline{G}$. If the boundary ∂G is assumed to possess some reasonable regularity properties (see Volevich and Paneyakh [1, Lemma 3.3]), then H_G^μ coincides with the completion of $\mathcal{D}(G)$ in the norm (41). Subsequently we shall consider, without special stipulation, only those regions which possess the indicated property.

Let $H^\mu(G)$ denote the quotient space of H^μ relative to the subspace $H^\mu_{\mathbb{R}^n \setminus G}$ of distributions with support lying outside G. We endow this space with the quotient norm

$$\|f, H^\mu(G)\| = \inf_{f_- \in H^\mu_{\mathbb{R}^n \setminus G}} \|f_0 + f_-\|, \tag{56}$$

where f_0 is an arbitrary representative of the residue class f. Definition (41) implies that under the condition $\mu(\xi) = \mu(-\xi)$ we have

$$|(f, \varphi)| = |(\widehat{f}, I\widehat{\varphi})| = |(\mu\widehat{f}, \mu^{-1}I\widehat{\varphi})| \leqslant \|\mu\widehat{f}\| \|I(\mu^{-1}\widehat{\varphi})\| = \|f\|_\mu \|\varphi\|_{1/\mu}. \tag{57}$$

If $\varphi \in \mathcal{D}(\Omega)$, then the left-hand side of (57) does not change when functions with support outside G are added to f. With account of (56), relation (57) implies the inequality

$$|(f, \varphi)| \leqslant \|f, H^\mu(G)\| \|\varphi, H_G^{1/\mu}\|.$$

It follows that the form (f, φ) is continued by continuity to $H^\mu(G) \times H_G^{1/\mu}$ and induces the duality of these spaces:

$$H^\mu(G) = (H_G^{1/\mu})', \qquad H_G^{1/\mu} = (H^\mu(G))',$$

the norms of the left-hand spaces coinciding with the norms in the Banach conjugate spaces of the right-hand spaces.

We come back to the proof of local solvability. Denote by $H^{N(P),s}$ the space of those distributions $u \in H^{(-\infty)}$ for which the norm (49) is finite. We have the following

Theorem. *Let operator (24) be N quasi-elliptic. If the region Ω is sufficiently small, then $\forall s \in \mathbb{R}$ for any right-hand side $f \in H^{N(P),s}(\Omega)$ there is a distribution $u \in H^{N(P),s+1}(\Omega)$ which is a solution to Equation (53) in the sense of distributions (i.e. (54) is fulfilled).*

Proof. Consider the operator

$$P\colon H^{N(P),s+1} \to H^{N(P),s}(u \mapsto P(x,y;D_x,D_y)u).$$

This operator is continuous. Indeed, in view of the definition of the norm (49) and Lemma 4.6, we have

$$\begin{aligned}\|[Pu]\|_{N(P),s} &\leqslant \sum_{(\alpha,\beta)\in N(P)} \|[a_{\alpha\beta}D_x^\alpha D_y^\beta u]\|_{N(P),s} \\ &\leqslant \text{const} \sum_{(\alpha,\beta)\in N(P)} \|[D_x^\alpha D_y^\beta u]\|_{N(P),s} \leqslant \text{const}\,\|[u]\|_{N(P),s+1}.\end{aligned}$$

Since the operator transforms the subspace $(H^{N(P),s+1})_{\mathbb{R}^2\setminus\Omega}$ into $(H^{N(P),s})_{\mathbb{R}^2\setminus\Omega}$, it induces a continuous operator in the quotient spaces:

$$P\colon H^{N(P),s+1}(\Omega) \to H^{N(P),s}(\Omega). \tag{58}$$

In view of duality (57), the operator

$${}^tP\colon (H^{N(P),-s})_\Omega \to (H^{N(P),-s-1})_\Omega. \tag{59}$$

is the adjoint operator of operator (58) while operator (58) is the adjoint of (59).

According to Theorem 4.5, if the region Ω has a sufficiently small diameter, then the inequality

$$\|[v]\|_{N(P),-s} \leqslant \text{const}\,\|[{}^tP(x,y;D_x,D_y)v]\|_{N(P),-s-1}, \quad v \in D(\Omega) \tag{60}$$

holds. Since $\mathcal{D}$ is dense in $(H^{N(P),-s})_\Omega$ in the norm $\|[\ \ ]\|_{N(P),-s}$, inequality (60) is extended to the entire space $(H^{N(P),-s})_\Omega$, i.e. operator (59) has no kernel. By virtue of (60), this operator has a closed image. Indeed, if $f_j = {}^tPv_j$, $v_j \in (H^{N(P),-s})_\Omega$, is a Cauchy sequence in $(H^{N(P),-s-1})_\Omega$, then, by virtue of (60), v_j is a Cauchy sequence in $(H^{N(P),-s})_\Omega$. According to the well-known theorem by Banach (e.g. see Yosida [1 ,Chapter 7, §5]), the image of the operator (58) coincides with $H^{N(P),s}(\Omega)$.

The above theorem can also be deduced from the general results by Hörmander [1, Chapter 3] on local solvability of operators of constant strength. According to Hörmander, operator (24) is called an operator of constant strength in Ω if for any pairs (x',y'), $(x'',y'') \in \Omega$ there is a constant $c = c(x',x'',y'',\Omega)$ such that

$$c^{-1} \leqslant \widetilde{P}(x',y';\xi,\eta)/\widetilde{P}(x'',y'';\xi,\eta) \quad \forall(\xi,\eta) \in \mathbb{R}^2$$

where

$$\widetilde{P}(x,y;\xi,\eta)=\Big(\sum_{\alpha\geqslant 0}|P^{(\alpha)}(x,y;\xi,\eta)|^2\Big)^{1/2}.$$

As a trivial consequence, it follows from Proposition 4.3 (ii) that an N quasi-elliptic operator is an operator of constant strength.

4.7. Hypoellipticity of N quasi-elliptic operators. We shall say that a distribution $u\in\mathcal{D}'(\Omega)$ belongs to $H^{\mu}_{\text{loc}}(\Omega)$ if $\varphi u \in H^{\mu}$ for all $\varphi\in\mathcal{D}(\Omega)$.

Theorem. *Let operator* (24) *be N quasi-elliptic and let the region Ω be sufficiently small. Let $u\in D'(\Omega)$. Then*

$$\{P(x,y;D_x,D_y)u\in H^{\mu}_{\text{loc}}(\Omega)\}\Rightarrow\{D_x^{\alpha}D_y^{\beta}u\in H^{\mu}_{\text{loc}}(\Omega),\quad \forall(\alpha,\beta)\in N(P)\}.$$

Corollary. *An N quasi-elliptic operator P is hypoelliptic, that is*

$$\{u\in\mathcal{D}'(\Omega), Pu\in C^{\infty}(\Omega)\}\Rightarrow\{u\in C^{\infty}(\Omega)\}.$$

The above-stated theorem follows from Hörmander's results on hypoelliptic operators of constant strength (see Hörmander [1, Chapter 13, Theorem 13.4.1]).

Appendix

We now show that function (52) satisfies condition (39′) with constant $\varepsilon=\varepsilon(R)\to 0$, $R\to+\infty$. It is convenient to obtain this estimate as a consequence of more general assertions.

Denote by $\mathcal{B}_0$ the class of smooth functions $\mu(\xi,R)$ depending on $\xi\in\mathbb{R}^n$ and $R>1$ and satisfying the following conditions.

(a) There is $\rho>0$ such that for any $\alpha>0$ we have

$$|\mu^{(\alpha)}(\xi,R)\mu^{-1}(\xi,R)|<c_{\alpha}(R+|\xi|)^{-\rho|\alpha|}.$$

(b) There are R_1, R_2, and $c_1,c_2>0$ such that

$$c_1(R+|\xi|)^{R_1}<|\mu(\xi,R)|<c_2(R+|\xi|)^{R_2}.$$

Lemma 1. *If $\mu(\xi,R)\in\mathcal{B}_0$, then the function $\mu(\xi,R)$ satisfies condition* (39′) *with a constant $\varepsilon=\varepsilon(R)\to 0$, $R\to+\infty$.*

Proof. Expanding $\mu(\xi',R)$ by Taylor's formula at the point ξ' and taking the remainder in Lagrange's form we obtain

$$\mu^{-1}(\xi'',R)(\mu(\xi',R)-\mu(\xi'',R))=\sum_{|\alpha|=1}^{N-1}\frac{(\xi'-\xi'')^{\alpha}}{\alpha!}\frac{\mu^{(\alpha)}(\xi'',R)}{\mu(\xi'',R)}$$

$$+\sum_{|\alpha|=N}\frac{(\xi'-\xi'')^{\alpha}}{\alpha!}\frac{\mu^{(\alpha)}(\zeta,R)}{\mu(\xi'',R)},\quad \zeta=\xi''+t(\xi'-\xi''),\quad 0\leqslant t\leqslant 1.$$

By virtue of condition (a), the first sum on the right-hand side can be estimated by means of

$$\sum_{|\alpha|=1}^{N-1} c_\alpha |\xi' - \xi''|^{|\alpha|}(R + |\xi''|)^{-\rho|\alpha|}/\alpha! \leqslant cR^{-\rho}(1 + |\xi' - \xi''|)^{N-1}.$$

In view of (a) and (b), the absolute value of the second sum does not exceed

$$\mathrm{const}(1 + |\xi' - \xi''|)^N (R + |\zeta|)^{-\rho N + R_2}(1 + |\xi''|)^{-R_1}. \tag{1}$$

Using the elementary inequality

$$\frac{R + |\zeta'|}{R + |\zeta''|} < 1 + |\zeta' - \zeta''|, \quad R > 1,$$

and setting $\zeta'' = \xi''$ and $\zeta' = \zeta$ we find

$$(R + |\zeta|)^{-\rho N + R_2} \leqslant (1 + |\xi' - \xi''|)^{|\rho N - R_2|}(1 + |\xi''|)^{-\rho N + R_2}.$$

Substituting this inequality into (1) we estimate the expression under consideration by means of

$$\mathrm{const}(1 + |\xi' - \xi''|)^{N + |\rho N - R_2|}(1 + |\xi''|)^{-\rho N + R_2 - R_1}.$$

Choosing N from the condition $\rho N > |R_1 - R_2|$ we prove the lemma.

Thus, it suffices to prove that function (52) belongs to the class $\mathcal{B}_0$. The proof is facilitated by the following lemma.

Lemma 2. (i) *If* $\mu_1, \mu_2 \in \mathcal{B}_0$, *then* $\mu_1\mu_2 \in \mathcal{B}_0$. (ii) *If* $\mu(\xi, R) \in \mathcal{B}_0$ *and* $\mu(\xi, R) > 0$, *then* $\mu^s(\xi, R) \in \mathcal{B}_0$ *for any* $s \in \mathbb{R}$.

Proof. The validity of condition (b) for the functions $\mu_1\mu_2$ and μ^s is obvious. We check the fulfilment of condition (a) for them. By Leibniz' formula, we have

$$(\mu_1, \mu_2)^{(\alpha)} = \sum_{\beta \geqslant 0} \frac{\alpha!}{\beta!(\alpha - \beta)!}\mu_1^{(\alpha-\beta)}\mu_2^{(\beta)},$$

and it remains to note that for μ_1 and μ_2 the condition (a) holds. By the chain rule, the derivative $(\mu^s)^{(\alpha)}$ is a linear combination of expressions

$$\frac{\mu^{(\alpha_1)}}{\mu} \cdots \frac{\mu^{(\alpha_k)}}{\mu}\mu^s.$$

These expressions are readily estimated by means of (a).

We now prove that $\nu_{R,s} \in \mathcal{B}_0$. In view of Lemma 2 (ii), it suffices to verify this inclusion relation for $s = 2$, i.e. to prove that

$$(\xi^{2m_1} + \eta^{2n_1} + R^2)\dots(\xi^{2m_m} + \eta^{2n_m} + R^2) \in \mathcal{B}_0.$$

By Lemma 2 (i), it suffices to prove the inclusion for each of the factors, i.e. to show that

$$\mu(\xi, \eta, R) = \xi^{2m} + \eta^{2n} + R^2 \in \mathcal{B}_0. \tag{2}$$

For $R > 1$ the condition (b) for the function $\mu(\xi, \eta, R)$ is obvious. Further, we have

$$\begin{aligned} |\partial^k \mu/\partial \xi^k| &= 2m\dots(2m-k+1)|\xi|^{2m-k} \\ &\leqslant 2m\dots(2m-k+1)(|\xi| + |\eta|^{n/m} + R^{1/m})^{2m-k} \\ &= \mathrm{const}(|\xi| + |\eta|^{n/m} + R^{1/m})^{-k}\mu. \end{aligned}$$

Similarly,

$$|\partial^k \mu/\partial \eta^k| \leqslant \mathrm{const}(|\xi|^{m/n} + |\eta| + R^{1/n})^{-k}\mu.$$

Choosing ρ from the condition $\rho < m/n, 1/n, n/m, 1/m$ we obtain condition (a) for function (2).

CHAPTER II

PARABOLIC OPERATORS ASSOCIATED WITH NEWTON'S POLYGON

Introduction

This chapter is devoted to studying Cauchy's problem for differential operators in several variables. When studying Cauchy's problem, there naturally arises separation of variables into a temporal variable and spatial variables. Taking into account separately the order of differentiation with respect to time and the total order of differentiation with respect to the spatial variables we associate with a differential operator a set of integral points in the plane that are used to construct Newton's polygon and the related set of principal quasi-homogeneous parts of the operator. In terms of these quasi-homogeneous parts we describe a new class of parabolic operators which relates to the classical operators parabolic in Petrovskiĭ's sense in the same way as N quasi-elliptic operators relate to quasi-elliptic operators.

In Chapter 1, in relation to the problem of local solvability and local smoothness for differential operators on functions of two variables, we stated and solved the algebraic problem of describing polynomials in two variables $P(\xi, \eta)$ that are estimated from below outside a large circle $\xi^2 + \eta^2 = c_0^2$ by means of the sum of the moduli of the monomials contained in them. A similar problem arises in the present chapter as well. Consider a polynomial in the variables $\xi = (\xi_1, \dots, \xi_n) \in \mathbb{R}^n$ and $\tau \in \mathbb{C}$:

$$P(\xi_1, \dots, \xi_n, \tau) = \sum a_{\alpha_1 \dots \alpha_n \beta} \xi_1^{\alpha_1} \dots \xi_n^{\alpha_n} \tau^\beta. \tag{1}$$

Denote by ν the set of integral points in the plane $(|\alpha|, \beta)$ such that $a_{\alpha_1 \dots \alpha_n \beta} \neq 0$, where $|\alpha| = \alpha_1 + \dots + \alpha_n$. Adding to the points belonging to ν the origin and the projections of points on the coordinate axes and taking the convex hull we obtain a polygon which we denote as $\Delta(P)$. In the case $n = 1$ it coincides with Newton's polygon $N(P)$. We are interested in finding necessary and sufficient conditions on polynomial (1) under which the inequality

$$\sum_{(\alpha,\beta) \in \Delta(P)} |\xi|^\alpha |\tau|^\beta < c|P(\xi, \tau)|, \qquad \operatorname{Im} \tau \leqslant \gamma_0, \quad (\xi_1, \dots, \xi_n, \operatorname{Re} \tau) \in \mathbb{R}^{n+1}, \tag{2}$$

holds. Here $|\xi| = (\xi_1^2 + \dots + \xi_n^2)^{1/2}$. For $n = 1$ there arises an analog of the problem considered in §1.2, in which the range of variables is changed, namely instead of the exterior of a large circle in $\mathbb{R}^2$ we consider the product of a straight line by the lower complex half-plane.

The presentation of the material in this chapter follows a plan which is in many respects analogous to that of Chapter 1.

§1 is of auxiliary character; it presents well-known Petrovskiĭ's definitions of correct and parabolic polynomials and describes correct polynomials in two variables in terms of Puiseux' series for their roots.

In §2 we consider necessary and sufficient conditions for the validity of inequalities (2) in the case of two variables and introduce the classes of N-parabolic and N-stable correct polynomials. §3 deals with N-stable correct and N-parabolic symbols with variable coefficients and presents the proofs of theorems on correctness of Cauchy's problem for differential operators with such symbols. In §§4 and 5 the corresponding results of §§2 and 3 are generalized to the case of several variables.

§1. Polynomials correct in Petrovskiĭ's sense

1.1. Definitions and some Classes of Polynomials Correct in Petrovskiĭ's sense. We consider Cauchy's problem for a differential operator on functions in $\mathbb{R}^{n+1}$. As was mentioned above, one of the variables is separated as time t, the other variables being denoted $x = (x_1, \dots, x_n)$; the dual variables will be denoted as $\tau \in \mathbb{C}$ and $\xi = (\xi_1, \dots, \xi_n) \in \mathbb{R}^n$.

In the well-known paper by Petrovskiĭ [1], for the differential operator with constant coefficients

$$P\Big(\frac{1}{i}\frac{\partial}{\partial x_1}, \dots, \frac{1}{i}\frac{\partial}{\partial x_n}, \frac{1}{i}\frac{\partial}{\partial t}\Big)$$

corresponding to a polynomial

$$P(\xi_1, \dots, \xi_n, \tau) = \sum_{j=0}^{m} P_j(\xi_1, \dots, \xi_n)\tau^{m-j}, \tag{1}$$

Cauchy's problem

$$P\Big(\frac{1}{i}\frac{\partial}{\partial x_1}, \dots, \frac{1}{i}\frac{\partial}{\partial x_n}, \frac{1}{i}\frac{\partial}{\partial t}\Big)u(x_1, \dots, x_n, t) = f(x_1, \dots, x_n, t), \quad 0 \leqslant t \leqslant T, \tag{2}$$

$$\frac{\partial^k u}{\partial t^k}(x_1, \dots, x_n, 0) = u_k(x_1, \dots, x_n), \qquad k = 0, \dots, m-1, \tag{3}$$

was considered. Following Hadamard, Petrovskiĭ says that problem (2), (3) is correctly posed (well-posed) if for any right-hand sides $f, u_0, \dots, u_{m-1}$ bounded together with a finite number of derivatives there exists a unique solution $u(x, t)$ bounded on the interval $0 \leqslant t \leqslant T$ together with the given finite number of derivatives, the operator $\{f, u_0, \dots, u_{m-1}\} \mapsto u$ being continuous in the corresponding norms.

Under the additional assumption that polynomial (1) is solved with respect to the highest power of τ, i.e.

$$P_0(\xi_1, \dots, \xi_n) \equiv \text{const}, \tag{4}$$

Petrovskiĭ proved that problem (2), (3) is well-posed (in the above sense) if and only if the following algebraic condition is fulfilled: there is a real number γ_0 such that

$$P(\xi_1, \dots, \xi_n, \tau) \neq 0 \quad \text{for} \quad (\xi_1, \dots, \xi_n) \in \mathbb{R}^n, \qquad \operatorname{Im}\tau < \gamma_0. \tag{5}$$

Denoting by $\tau_j(\xi_1,\dots,\xi_n)$, $j=1,\dots,n$, the roots of polynomial (1) we can restate condition (5) as follows: there is a real number γ_0 such that

$$\operatorname{Im}\tau_j(\xi_1,\dots,\xi_n) \geqslant \gamma_0 \quad \text{for} \quad (\xi_1,\dots,\xi_n)\in\mathbb{R}^n, \qquad j=1,\dots,m. \tag{6}$$

The polynomials satisfying (5), (4) are said to be correct in Petrovskiĭ sense.

Remarks. 1) Petrovskiĭ considered in fact general systems of differential operators with coefficients depending on time. However, it is the above-mentioned special case when Petrovskiĭ's result admits of an effective formulation.

2) A detailed analysis of problem (2), (3) was carried out by Volevich and Gindikin in [1]. In particular, it was shown that Petrovskiĭ's result remains valid if (4) is replaced by a weaker condition:

$$P_0(\xi_1,\dots,\xi_n) \neq 0 \quad \text{for} \quad (\xi_1,\dots,\xi_n)\in\mathbb{R}^n. \tag{4'}$$

If Cauchy's problem with zero initial data is considered, then one can discard condition (4′) as well confining oneself to condition (5).

According to Volevich and Gindikin [1], polynomial (1) is said to be *exponentially correct* if for any point $\omega=(\omega_1,\dots,\omega_n)\in\mathbb{R}^n$ the "translated" polynomial

$$P_\omega(\xi_1,\dots,\xi_n,\tau) = P(\xi_1+i\omega_1,\dots,\xi_n+i\omega_n,\tau)$$

is correct in Petrovskiĭ's sense.

As was shown by Volevich and Gindikin [1], the exponential correctness is a necessary and sufficient condition for the correctness of Cauchy's problem (2), (3) in the class of functions of exponential growth (decrease). As will be seen in the end of this chapter, Cauchy's problem can be studied for exponentially correct equations with variable coefficients, which, however, requires an additional condition on the symbols (the condition of constant strength). In the proposition below some equivalent definitions of exponential correctness are collected.

Proposition 1. *For a polynomial $P(\xi_1,\dots,\xi_n,\tau)$ the following conditions are equivalent.*

(i) *The polynomial P is exponentially correct, i.e. $\forall\omega\in\mathbb{R}^n\ \exists\gamma(\omega)$ such that*

$$P_\omega(\xi_1,\dots,\xi_n,\tau)\neq 0, \qquad \operatorname{Im}\tau\leqslant\gamma(\omega), \quad (\xi_1,\dots,\xi_n)\in\mathbb{R}^n.$$

(i′) *$\forall M>0\ \exists\gamma(M)$ such that*

$$P_\omega(\xi_1,\dots,\xi_n,\tau)\neq 0 \quad \text{for} \quad |\omega|\leqslant M, \qquad \operatorname{Im}\tau\leqslant\gamma(M), \quad (\xi_1,\dots,\xi_n)\in\mathbb{R}^n.$$

(ii) *The distance $d_P(\xi_1,\dots,\xi_n,\tau)$ from the point $(\xi_1,\dots,\xi_n,\tau)$ to the manifold of zeros of the polynomial P tends to infinity uniformly with respect to $(\xi_1,\dots,\xi_n,\operatorname{Re}\tau)\in\mathbb{R}^{n+1}$ as $\operatorname{Im}\tau\to\infty$.*

(iii) *$\forall\alpha>0$ we have*

$$P^{(\alpha)}(\xi_1,\dots,\xi_n,\tau)P^{-1}(\xi_1,\dots,\xi_n,\tau)\to 0 \quad \textit{for } \operatorname{Im}\tau\to-\infty \tag{7}$$

uniformly with respect to $(\xi_1,\dots,\xi_n,\operatorname{Re}\tau)\in\mathbb{R}^{n+1}$.

Proof. Conditions (i′) and (ii) are tautologically equivalent. The equivalence of (ii) and (iii) follows from the well-known property of polynomials (see Hörmander [3, formula (4.1.5))]:

$$\sum_{\alpha\neq 0}\left|\frac{P^{(\alpha)}(\xi_1,\dots,\xi_n,\tau)}{P(\xi_1,\dots,\xi_n,\tau)}\right|^{1/|\alpha|} \leqslant \text{const}[d_P(\xi_1,\dots,\xi_n,\tau)]^{-1}.$$

The equivalence of (i) and (i') is a consequence of some facts in multidimentionial complex analysis, namely the convexity of the maximal function $\gamma(\omega)$ (for more detail see Volevich and Gindikin [1, Lemma 3.4.1 and Proposition 3.4.1]).

An important sufficient condition for exponential correctness is contained in the following proposition.

Proposition 2. *If a hypoelliptic polynomial P satisfies Petrovskiĭ's correctness condition* (5), *then it is exponentially correct.*

For the proof see Volevich and Gindikin [1 (Proposition 3.4.2)].

Lemma. *An exponentially correct polynomial is solved with respect to the highest power of* τ *(i.e.* (4) *is fulfilled).*

This assertion is deduced from Proposition 1 (iii) in the same way as Lemma 1 in Section 1.3.2 is deduced from the hypoellipticity condition.

1.2. Quasi-homogeneous polynomials correct in Petrovskiĭ's sense. Let P_0 be a $(1,\dots,1,q)$-homogeneous polynomial, i.e.

$$P_0(\lambda\xi_1,\dots,\lambda\xi_n,\lambda^q\tau) = \lambda^{m(q)}P_0(\xi_1,\dots,\xi_n,\tau) \quad \forall\lambda > 0. \tag{8}$$

In view of the quasi-homogeneity condition (8), the conditions for correctness in Petrovskiĭ's sense (5), (6) for the polynomial P_0 take the form

$$P_0(\xi_1,\dots,\xi_n,\tau)\neq 0 \quad \text{for} \quad \operatorname{Im}\tau<0, \qquad (\xi_1,\dots,\xi_n,\operatorname{Re}\tau)\in\mathbb{R}^{n+1}, \tag{9}$$
$$\operatorname{Im}\tau_{0j}(\xi_1,\dots,\xi_n)\geqslant 0 \quad \text{for} \quad (\xi_1,\dots,\xi_n)\in\mathbb{R}^n, \qquad j=1,\dots,m, \tag{10}$$

where $\tau_{0j}(\xi_1,\dots,\xi_n)$ are the roots of the polynomial P_0.

Proposition. (i) *For polynomial* (8) *solved with respect to the highest power of* τ *condition* (9) *(or* (10)*) is fulfilled if and only if*

(a) *q is natural number;*
(b) *if q is odd (i.e. $q = 2b+1$), then $\operatorname{Im}\tau_{0j}(\xi_1,\dots,\xi_n) = 0$; if q is even (i.e. $q = 2b$), then*

$$\operatorname{Im}\tau_{0j}(\xi_1,\dots,\xi_n)\geqslant 0 \quad \textit{for} \quad (\xi_1,\dots,\xi_n)\in\mathbb{R}^n;$$

(ii) *For polynomial* (8) *solved with respect to the highest power of τ the strict conditions* (9), (10), *i.e.*

$$P_0(\xi_1,\dots,\xi_n,\tau) \neq 0, \quad \operatorname{Im}\tau \leqslant 0, \quad |\tau|+|\xi| \neq 0, \tag{9'}$$
$$\operatorname{Im}\tau_{0j}(\xi_1,\dots,\xi_n) > 0, \qquad (\xi_1,\dots,\xi_n) \in \mathbb{R}^n \setminus \{0\}, \tag{10'}$$

are fulfilled if and only if q is even.

The proof is based on the following

Lemma. *Let k, p be coprime natural numbers and let $z \in \mathbb{C}$. Then*

(i) *the inequality*

$$\operatorname{Im}(z(\xi^{1/p})^k) \geqslant 0, \qquad -\infty < \xi < \infty, \tag{11}$$

holds for all branches of $\xi^{1/p}$ if and only if $p = 1$ and

$$\operatorname{Im} z \geqslant 0 \quad \text{for even} \quad k = 2b, \tag{12}$$
$$\operatorname{Im} z = 0 \quad \text{for odd} \quad k = 2b+1; \tag{13}$$

(ii) *the strict inequality*

$$\operatorname{Im}(z(\xi^{1/p})^k) > 0, \qquad -\infty < \xi < \infty, \tag{11'}$$

holds for all branches of $\xi^{1/p}$ if and only if

$$k = 2b, \qquad \operatorname{Im} z > 0. \tag{12'}$$

Proof. Let, for definiteness, $\xi > 0$. Then

$$\arg(z(\xi^{1/p})^k) = \arg z + 2\pi qk/p, \qquad q = 0,1,\dots,p-1.$$

If $p \geqslant 2$, then there is a natural number q_0 lying between 0 and $p-1$ such that the fractional part of the ratio q_0k/p is greater than $1/2$. Therefore inequality (11) cannot hold simultaneously for the two branches corresponding to $q = 0$ and $q = q_0$ (since the arguments in the left hand side of (11) will differ by more than π). Now let $p = 1$ and let k be even. Then the sign of the left-hand side of (11) coincides with that of $\operatorname{Im} z$, whence follows conditions (12) and (12′). If k is odd, then the left-hand side of (11) has opposite signs for ξ and $-\xi$, which implies (13).

The proof of the Proposition. 1) We first consider the case of a polynomial $P_0(\xi,\eta)$ in two variables $\xi \in \mathbb{R}$ and $\tau \in \mathbb{C}$. In view of the quasi-homogeneity, the roots $\tau_{0j}(\xi)$ have the form $\tau_{0j} = a_j\xi^{1/q}$, the number q being obviously rational. Applying the lemma we prove the desired assertion.

In case $n > 1$ we associate with the polynomial $P_0(\xi_1, \ldots, \xi_n, \tau)$ the polynomial $Q(\lambda, \tau) = P_0(\lambda\xi_1, \ldots, \lambda\xi_n, \tau)$ depending on $\xi_1, \ldots, \xi_n$ as parameters. Applying the already proved proposition to this polynomial and setting $\lambda = 1$ we prove the proposition.

If condition (i) of the proposition is fulfilled and $q = 1$, then P_0 is called a homogeneous *hyperbolic* polynomial. We can also consider polynomials corresponding to odd $q > 1$ (the so-called *q-hyperbolic* polynomials).

If q is a positive even number, then a $(1, \ldots, 1, q)$-homogeneous polynomial satisfying condition (ii) is said to be q-parabolic in Petrovskiĭ's sense. In this case we shall often put $q = 2b$, $b \in \mathbb{Z}_+$, and, accordingly, say that the polynomial is $2b$-parabolic in Petrovskiĭ's sense.

1.3. Inhomogeneous 2b-parabolic polynomials. Polynomial (1) solved with respect to the highest power of τ (i.e. condition (4) holds: $P_0 \equiv 1$) is said to be q-parabolic (q is even) or q-parabolic in Petrovskiĭ's sense if its principal $(1, \ldots, 1, q)$-homogeneous part $P_q(\xi, \tau)$ possesses this property.

As was already said in the Introduction, with a polynomial $P(\xi_1, \ldots, \xi_n, \tau)$, a polygon $\Delta(P)$ is associated in a natural manner, which for $n = 1$ goes into Newton's polygon $N(P)$. We denote by (α_j, β_j), $j = 0, 1, \ldots, m + 1$, the vertices of this polygon, which are indexed counterclockwise beginning with the vertex $(\alpha_0, \beta_0) = (0, 0)$. As in Chapter 1, put

$$\Xi_{\Delta(P)}(\xi_1, \ldots, \xi_n, \tau) = \sum_{j=0}^{m+1} |\xi|^{\alpha_j} |\tau|^{\beta_j}, \qquad \xi = (\xi_1, \ldots, \xi_n). \tag{14}$$

Theorem. *Polynomial* (1) *solved with respect to the highest power of* τ (*i.e.* $P(\xi, \tau) = \tau^m + O(|\tau|^{m-1})$) *is q-parabolic if and only if*

(a) *the polygon* $\Delta(P)$ *is a triangle with vertices* $(0, 0)$, $(0, m)$, *and* $(mq, 0)$;

(b) *there are* $c > 0$ *and* $\gamma_0 < 0$ *such that*

$$\sum_{(\alpha,\beta)\in\Delta(P)} |\xi|^{\alpha} |\tau|^{\beta} < c|P(\xi_1, \ldots, \xi_n, \tau)| \quad \textit{for} \quad \operatorname{Im}\tau \leqslant \gamma_0. \tag{15}$$

Proof. Necessity. Let conditions (a) and (b) be fulfilled. Replace $(\xi_1, \ldots, \xi_n, \tau)$ in (15) by $(\lambda\xi_1, \ldots, \lambda\xi_n, \lambda^q\tau)$, divide both sides by λ^{mq}, and pass to the limits as $\lambda \to +\infty$. This results in the inequality

$$|\tau|^m + |\xi|^{qm} \leqslant c|P_q(\xi_1, \ldots, \xi_n, \tau)| \quad \text{for} \quad \operatorname{Im}\tau \leqslant 0. \tag{16}$$

(Under the passage to the limit inequality (15) goes into inequality (16) in the open half-plane $\operatorname{Im}\tau < 0$. However, since both sides of the inequality are continuous for $\operatorname{Im}\tau \to -0$, the inequality remains valid for $\operatorname{Im}\tau = 0$.) Therefore conditions (9′) and (10′) are sure to hold, i.e. the q-homogeneous polynomial P_q is q-parabolic.

Sufficiency. We show that for a q-parabolic polynomial conditions (a) and (b) are fulfilled. It suffices to verify condition (a) for the q-principal part P_q. Since for $P_0 = P_q$ condition (9′) is fulfilled, we have $P_q(0,\dots,0,\tau) = c\tau^m$, $c \neq 0$, and

$$P_q(\xi_1,\dots,\xi_n,0) = \sum_{|\alpha|=mq} a_{\alpha_1\dots\alpha_n}\xi_1^{\alpha_1}\dots\xi_n^{\alpha_n} \neq 0, \qquad |\xi| \neq 0.$$

Thus, the points $(0,0)$, $(0,m)$, and $(mq,0)$ belong to $\Delta(P_q)$, whence follows that $\Delta(P_q)$ is a triangle with the indicated vertices. We now note that, by virtue of condition (a), we have

$$\sum_{(\alpha,\beta)\in\Delta(P)} |\xi|^\alpha|\tau|^\beta \leqslant \varkappa(1 + |\tau|^m + |\xi|^{mq}).$$

If the principal part $P_q(\xi_1,\dots,\xi_n,\tau)$ satisfies (10′), then, in view of the quasi-homogeneity, (16) holds, whence

$$\begin{aligned}\sum |\xi|^\alpha|\tau|^\beta &\leqslant c\varkappa|P_q(\xi_1,\dots,\xi_n,\tau)| + \varkappa\\ &\leqslant c\varkappa|P(\xi_1,\dots,\xi_n,\tau)| + \{\varkappa + c\varkappa|P(\xi_1,\dots,\xi_n,\tau) - P_q(\xi_1,\dots,\xi_n,\tau)|\}.\end{aligned}$$

There is $\varepsilon > 0$ and a constant c_1 such that the expression in curly brackets does not exceed $c_1(1 + |\tau| + |\xi|^q)^{m-\varepsilon} \leqslant (1/2)\Sigma|\xi|^\alpha|\tau|^\beta$ provided that $\operatorname{Im}\tau < \gamma_0$ and $-\gamma > \gamma_0(\varepsilon)$. This implies inequality (15).

1.4. Expansion into Puiseux's series of the roots of polynomials in two variables correct in Petrovskiĭ's sense.

Proposition 1 (Borok [1]). *A polynomial*

$$P(\xi,\tau) = \tau^m + \sum_{j=1} P_j(\xi)\tau^{m-j}, \qquad \xi \in \mathbb{R}, \tag{17}$$

satisfies Petrovskiĭ's condition (5) (or (6)) if and only if the expansions of the roots $\tau_j(\xi)$, $j = 1,\dots,m$, into Puiseux's series in the neighborhood of $|\xi| = \infty$ have one of the following three forms:

(a) $$\tau_j(\xi) = c_{j1}\xi^{\varkappa_{j1}} + \dots + c_{jl}\xi^{\varkappa_{jl}} + o(\xi^{\varkappa_{jl}}), \qquad |\xi| \to \infty,$$

where $\varkappa_{j1} > \dots > \varkappa_{jl} > 0$ are integers, $\varkappa_{jl}$ is even, and

$$\operatorname{Im} c_{j1} = \dots = \operatorname{Im} c_{j,l-1} = 0, \qquad \operatorname{Im} c_{jl} > 0;$$

(b) $$\tau_j(\xi) = c_{j1}\xi^{\varkappa_{j1}} + \dots + c_{jl}\xi^{\varkappa_{jl}} + O(1), \qquad |\xi| \to \infty,$$

where $\varkappa_{j1} > \cdots > \varkappa_{jl}$ are natural numbers and the coefficients $c_{j1}, \ldots, c_{jl}$ are real numbers;

(c) $$\tau_j(\xi) = O(1) \quad \textit{for} \quad |\xi| \to \infty.$$

Proof. Since conditions (6) obviously hold for the above types of roots, it only remains to prove the necessity of the indicated expansions. As was noted in Section 1.1.4, Puiseux's series for the roots $\tau_j(\xi)$ have, in general, the form

$$\tau_j(\xi) = \sum c_{jl}(\xi^{1/p})^l. \tag{18}$$

Let $\varkappa_{j1} = k/p$ be the greatest power exponent in this expansion. If $\varkappa_{j1} \leqslant 0$, then $\tau_j(\xi)$ is a root of type (c), and the proposition is proved. In case $\varkappa_{j1} > 0$, we have

$$\tau_j(\xi) = c_{j1}\xi^{\varkappa_{j1}} + o(\xi^{\varkappa_{j1}}) \quad \text{for} \quad |\xi| \to \infty.$$

If condition (6) is fulfilled for some γ_0, then there is γ_1 such that

$$\operatorname{Im}(c_{j1}\xi^{\varkappa_{j1}}) \geqslant \gamma_1 \quad \text{for} \quad -\infty < \xi < \infty, \qquad \varkappa_{j1} = k/p.$$

In view of the homogeneity, it follows that $\gamma_1 = 0$. Applying Lemma 1.2 we conclude that $\varkappa_{j1}$ is an integer and one of the two conditions

$$\operatorname{Im} c_{j1} = 0, \tag{19}$$

$$\operatorname{Im} c_{j1} > 0, \qquad \varkappa_{j1} \quad \text{is even}, \tag{20}$$

is fulfilled. Under condition (20) the root $\tau_j(\xi)$ belongs to type (a), i.e. the proposition is proved. Let (19) be fulfilled and let $c_{j2}\xi^{\varkappa_{j2}}$ be the subsequent expansion term. Then, if $\varkappa_{i2} > 0$, we have

$$\operatorname{Im} \tau_j(\xi) = \operatorname{Im}(c_{j2}\xi^{\varkappa_{j2}}) + o(\xi^{\varkappa_{j2}}),$$

and, by virtue of inequality (6), there is γ_2 such that

$$\operatorname{Im}(c_{j2}\xi^{\varkappa_{j2}}) \geqslant \gamma_2.$$

Applying the lemma once again we conclude that $\varkappa_{j2}$ is a natural number and conditions (19) and (20) (where c_{j1} and $\varkappa_{j1}$ should be replaced by c_{j2} and $\varkappa_{j2}$, respectively) hold. Continuing this process we complete the proof of the proposition.

Proposition 2. *Polynomial (17) is exponentially correct if and only if the expansions of the roots $\tau_j(\xi)$, $j = 1, \dots, m$, into Puiseux's series in the neighborhood of $\xi = \infty$ have one of the following three forms:*

$$\text{(a}'\text{)} \qquad \tau_j(\xi) = c_{j1}\xi^{\varkappa_{j1}} + o(\xi^{\varkappa_{j1}}) \quad \textit{for} \quad |\xi| \to \infty,$$

where $\operatorname{Im} c_{j1} > 0$ *and* $\varkappa_{j1}$ *is a positive even integer;*

$$\text{(b}'\text{)} \qquad \tau_j(\xi) = c_{1j}\xi + o(1), \quad c_{j1} \neq 0, \quad \operatorname{Im} c_{j1} = 0, \quad |\xi| \to \infty,$$

$$\text{(c)} \qquad \tau_j(\xi) = O(1), \qquad |\xi| \to \infty.$$

Proof. According to Proposition 1, the polynomial $P(\xi, \tau)$ is exponentially correct if and only if for any real η Puiseux' expansions of the roots $\tau_j(\xi + i\eta)$ of the polynomial $P(\xi + i\eta, \tau)$ have the form (a), (b), or (c).

Necessity. Since the roots $\tau_j(\xi + i\eta)$ are holomorphic functions of $\xi + i\eta$ in the neighborhood of ∞, for large ξ and small η the expansions of the roots $\tau_j(\xi + i\eta)$ of the polynomial $P(\xi + i\eta, \tau)$ are obtained from the expansions of the roots $\tau_j(\xi)$ by replacing ξ by $\xi + i\eta$. Applying the binomial formula we find

$$\begin{aligned}\tau_j(\xi + i\eta) = c_{j1}\xi^{\varkappa_{j1}} &+ i\varkappa_{j1}c_{j1}\eta\xi^{\varkappa_{j1}-1} + c_{j2}\xi^{\varkappa_{j2}} \\ &+ o(\xi^{\varkappa_{j1}-1}) + o(\xi^{\varkappa_{j2}}) \quad \text{for} \quad |\xi| \to \infty. \end{aligned} \tag{21}$$

If $\varkappa_{j1} > 1$ and c_{j1} is a real number, then expansion (21) does not belong to the types (a), (b), and (c). Consequently, either $\varkappa_{j1}$ is even and $\operatorname{Im} c_{j1} > 0$ or $\varkappa_{j1} = 1$ and c_{j1} is real. Thus, the root $\tau_j(\xi)$ can have Puiseux' expansion only of the form of (a'), (b'), or (c').

Sufficiency. As has been in fact proved, expansions (a'), (b'), and (c') go into expansions of the same type when ξ is replaced by $\xi + i\eta$. According to Proposition 1, the polynomial $P(\xi + i\eta, \tau)$ is correct in Petrovskiĭ's sense for any η, i.e. is an exponentially correct polynomial.

§2. Two-sided estimates for polynomials in two variables satisfying Petrovskiĭ's condition. *N*-parabolic polynomials

In this section we shall construct a theory "parallel" to the one presented in §1.2. We shall state necessary and sufficient conditions on a polynomial $P(\xi, \tau)$, $\xi \in \mathbb{R}$, $\tau \in \mathbb{C}$, under which the inequalities

$$|\xi^\alpha \tau^\beta| \leqslant c|P(\xi, \tau)|, \qquad (\alpha, \beta) \in N(P), \qquad \operatorname{Im} \tau \leqslant \gamma_0, \tag{1}$$

hold. These estimates differ from the analogous estimates in Chapter 1 in that the exterior of a large circle in $\mathbb{R}^2$ is replaced by a region $\mathbb{R} \times \{\tau \in \mathbb{C}, \operatorname{Im} \tau < \gamma_0\}$, where γ_0 is a sufficiently large number.

2.1. The main estimate. The change in the region where adequate estimates are considered makes us somewhat modify the notion of minor monomials. The matter is that in the estimates under consideration there is in fact a large parameter, and the classification as senior and minor monomials is performed so that the ratios of the absolute values of minor monomials to the sum of the absolute values of all monomials tends to zero as the corresponding large parameters increases. In Chapter 1 the role of the large parameter was played by the expression $\xi^2+\eta^2$, which is what implied the definition of minor monomials given there (see Section 1.1.1). In the present chapter the role of the large parameter in estimates (1) is played by $-\operatorname{Im}\tau$. Therefore the conditions on minor monomials can be weakened as compared to Chapter 1. An (integral) point $(\alpha,\beta)\in N(P)$ is said to be minor if there is a point $(\alpha',\beta')\in N(P)$ such that $\alpha'\geqslant\alpha$, $\beta'>\beta$. According to this more general definition, the minor points of $N(P)$ are not only the points on the coordinate axes and the interior points of the polygon but also the points belonging to the vertical side not coinciding with the coordinate axis (of course, provided that the polygon $N(P)$ possesses such a side).

If (α_j,β_j), $j=0,\dots,m+1$, are the vertices of $N(P)$, then, as in Section 1.2.1, we define the function

$$\Xi_{N(P)}(\xi,\tau)=\sum_{j=0}^{m+1}|\xi^{\alpha_j}\tau^{\beta_j}|. \tag{2}$$

For function (2) an exact analog of Lemma 1.2.1 takes place.

Lemma. (i) *If $(\alpha,\beta)\in N(P)$, then*

$$|\xi^{\alpha}\tau^{\beta}|<\Xi_{N(P)}(\xi,\tau)\quad\forall(\alpha,\beta)\in N(P). \tag{3}$$

(ii) *There is $\varkappa>0$ such that*

$$|\xi^{\alpha}\tau^{\beta}|<c'|\tau|^{-\varkappa}\Xi_{N(P)}(\xi,\tau)\quad\forall(\alpha,\beta)\in\delta(P), \tag{3'}$$

where $\delta(P)$ is the polygon spanned to the minor integral points of the polygon $N(P)$.

Assertion (i) is proved in the same way as the analogous assertion in Lemma 1.2.1, and assertion (ii) is a trivial consequence of the definition of minor points.

Theorem. *For a polynomial $P(\xi,\eta)$ the following conditions are equivalent.*

(I) *There are constants $c>0$ and γ_0 such that the inequality*

$$\Xi_{N(P)}(\xi,\tau)<c|P(\xi,\tau)|,\qquad \operatorname{Im}\tau<\gamma_0, \tag{4}$$

holds.

(II) *For any side $\Gamma_j^{(1)}$ that does not lie on the coordinate axes and is not vertical we have*

$$P_{q^{(j)}}(\xi,\tau)\neq 0\quad\text{for}\quad\xi\neq 0,\quad|\tau|\neq 0,\quad\operatorname{Im}\tau\leqslant 0. \tag{5}$$

(III) *There exists a set of vectors* $q^{(j)} = (q_1^{(j)}, q_2^{(j)})$, $q_1^{(j)} \geqslant 0$, $q_2^{(j)} > 0$, *and a set of* $q^{(j)}$*-homogeneous polynomials* $P^{[j]}(\xi, \eta)$, $j = 1, \dots, \mu$, *such that*

$$P^{[j]}(\xi, \tau) \neq 0 \quad \text{for} \quad |\xi| + |\tau| > 0, \qquad \operatorname{Im} \tau \leqslant 0, \tag{6}$$

and an integer $b \geqslant 0$ *such that*

$$P \sim \widehat{P}, \qquad \widehat{P} = \tau^b P^{[1]} \dots P^{[\mu]}. \tag{7}$$

Remark. In the statement of condition (II) of the theorem (in contrast to the analogous condition in Theorem 1.2.1) no vertical side not lying on the coordinate axis is involved. This is due to the fact (already mentioned above) that the corresponding monomials are minor.

The proof is a simple modification of that of Theorem 1.2.1. A slight change must be made only when deriving (III) from (II). Indeed, according to Theorem 1.1.3, the polynomial $P(\xi, \tau)$ differs from the polynomial $\tau^b P^{[1]} \dots P^{[\mu]}$ (where $P^{[j]}$ are constructed using formulas (1.1.15)) only in terms that are minor in the sense of Chapter 1 and, the more so, minor in the broader sense of the present chapter. If the polygon $N(P)$ contains no vertical sides, then all polynomials $P^{[j]}$ satisfy conditions (6). Moreover, if $N(P)$ does not contain horizontal sides, then all polynomials $P^{[j]}$ are $q^{(j)}$-parabolic in Petrovskiĭ's sense. If the side $\Gamma_j^{(1)}$ is horizontal, then the polynomial $P^{[1]}(\xi, \eta)$ does not depend on τ and is nonzero for all $\xi \in \mathbb{R}$.

Now let the side $\Gamma_m^{(1)}$ be vertical so that $P^{[m]}(\xi, \tau) \equiv P^{[m]}(\tau) = a\tau^b + O(\tau^{b-1})$. Here the polynomials $P^{[1]}(\xi, \tau) \dots P^{[m]}(\xi, \tau)$ and $\widehat{P}(\xi, \tau) = a\tau^b P^{[1]} \dots P^{[m-1]}(\xi, \tau)$ differ in terms minor in the sense of the present chapter. The theorem is proved.

2.2. Stability of polynomials admitting of estimate (4). Repeating the argument in Section 1.2.4 we can show that if a polynomial $P(\xi, \tau)$ satisfies the conditions of Theorem 2.1, then the "perturbed" polynomial

$$P_\delta(\xi, \tau) = \sum (a_{\alpha\beta} + \delta_{\alpha\beta}) \xi^\alpha \tau^\beta, \qquad |\delta_{\alpha\beta}| < \varepsilon \quad (P_0 \equiv P), \tag{8}$$

also satisfies these conditions for sufficiently small ε and, the more so, Petrovskiĭ's condition (1.5). Hence, the polynomials satisfying the conditions of Theorem 2.1 are interior points of the finite-dimensional set of polynomials satisfying condition (1.5) and having a given Newton polygon. We have the following

Theorem. *For a polynomial* $P(\xi, \eta)$ *conditions* (I), (II), *and* (III) *of Theorem* 2.1 *are equivalent to the following conditions.*

(IV) *There are* $\varepsilon_0 > 0$ *and* γ_0 *such that for* $|\varepsilon| < \varepsilon_0$ *all polynomials* (9) *satisfy Petrovskiĭ's condition, i.e.*

$$P_\delta(\xi, \tau) \neq 0 \quad \text{for} \quad \operatorname{Im} \tau \leqslant \gamma_0, \quad |\delta_{\alpha\beta}| < \varepsilon \leqslant \varepsilon_0. \tag{9}$$

(V) *There are constants* $c_1 > 0$ *and* γ_1 *such that*

$$d_\tau(\xi) > c_1|\xi| \quad \text{for} \quad \operatorname{Im}\tau < \gamma_1, \quad \xi \in \mathbb{R}, \tag{10}$$

where $d_\tau(\xi)$ *is the distance from the point* ξ *to the manifold of complex roots of the polynomial* $P(z,\tau)$ *for a fixed* τ.

Proof. As has already been said, the implication (I)$\Longrightarrow$(IV) is proved in the same way as the analogous assertion in Theorem 1.2.4.

(IV)$\Longrightarrow$(V). Each point z belonging to the circle $|z-\xi| < \varepsilon|\xi|$ can be represented as $z = \xi + h\xi$, $|h| < \varepsilon$. Therefore the polynomial $P(z,\tau)$ can be written in the form (8) (cf. the proof of Theorem 1.2.4) and use can be made of (9). The proof of (V)$\Longrightarrow$(I) will be given in §4 where a generalization of this theorem to the case $\xi \in \mathbb{R}^n$, $n > 1$, will be presented.

2.3. *N*-parabolic polynomials. In §1.3 we introduced the class of N quasi-elliptic polynomials that were hypoelliptic polynomials satisfying the two-sided estimate 1.2.1. We now study hypoelliptic polynomial satisfying the conditions of Theorem 2.1. We shall call them N-parabolic polynomials.

We begin with an analog of Proposition 1.3.2.

Proposition. *For a polynomial* $P(\xi,\tau)$ *satisfying the equivalent conditions in Theorems* 2.1 *and* 2.2 *the following conditions are equivalent.*

(i) P *is a hypoelliptic polynomial.*

(ii) *Newton's polygon* $N(P)$ *is regular.*

(iii) *There are* $\varkappa > 0$, $c > 0$, *and* γ_1 *such that*

$$(|\xi| + |\tau|)^\varkappa |\xi^\alpha \tau^\beta| < c_1|P(\xi,\tau)|, \quad \operatorname{Im}\tau < \gamma_1, \tag{11}$$

for any integral point $(\alpha,\beta) \in \delta_0(P)$[1].

Proof. (i)$\Longrightarrow$(ii) is true by virtue of Lemma 2 in Section 1.3.2. (ii)$\Longrightarrow$(iii) is proved like the analogous assertion in Proposition 1.3.2.

(iii)$\Longrightarrow$(i). From (11) it follows that $P_\gamma(\xi,\eta) = P(\xi,\eta+i\gamma)$ is a hypoelliptic polynomial for $\gamma < \gamma_0$. Since the polynomials $P_\gamma(\xi,\eta)$ and $P(\xi,\eta)$ differ by a linear combination of the derivatives of P_γ, the polynomial P is also hypoelliptic.

Theorem 2.1 and the above proposition imply

Theorem[2]**.** *The following definitions of an* N*-parabolic polynomial are equivalent.*

(I) *There are* $c > 0$, γ_0 *such that*

$$|\xi^\alpha \eta^\beta| < c|P(\xi,\tau)|, \qquad \operatorname{Im}\tau \leqslant \gamma_0,$$

[1] Recall (see Section 1.3.2) that a point (α,β) belongs to $\delta_0(P)$ if there is a point $(\alpha',\beta') \in N(P)$ such that $\alpha \leqslant \alpha'$ and $\beta \leqslant \beta'$, one of these inequalities being strict.

[2] Cf. Theorem 1.3.3.

for any points $(\alpha, \beta) \in N(P)$. Moreover, there is $\varkappa > 0$ such that for all integral points $(\alpha, \beta) \in \delta_0(P)$ inequality (11) holds.

(II) *$N(P)$ is a regular Newton polygon, and for its any side $\Gamma_j^{(1)}$ not lying on the coordinate axes the condition*

$$P_{q^{(j)}}(\xi, \eta) \neq 0 \quad \text{for} \quad \xi \neq 0, \quad |\tau| \neq 0, \quad \operatorname{Im} \tau \leqslant 0, \tag{12}$$

holds, where $q^{(j)}$ is the outer normal to $\Gamma_j^{(1)}$.

(III) *Let $\widehat{\tau}_j(\xi) = c_j \xi^{b_j}$ be the principal parts of Puiseux' expansions for $|\xi| \to \infty$ of the roots of the polynomial P. Then b_j is an even integer and $\operatorname{Im} c_j > 0$.*

(IV) *There exists a set of pairwise disproportionate vectors $q^{(j)}$, $j = 1, \dots, m$, with positive components and a set of $q^{(j)}$-parabolic polynomials $P^{[j]}(\xi, \tau)$ such that $P \sim P^{[1]} \dots P^{[m]}$, the polynomials $P^{[j]}$ being determined to within a constant factor.*

(V) *The polynomial P is hypoelliptic and satisfies estimate (4).*

The proof of the theorem is a modification of that of Theorem 1.3.3.

(I)$\Longrightarrow$(II) follows from Theorem 2.1 and the above proposition.

(II)$\Longrightarrow$(III). As in Theorem 1.3.3, denote by $\tau_{jl}(\xi)$ those roots of P whose principal parts are of the same degree, i.e. $\tau_{jl} = c_{jl}\xi^{b_j} + O(\xi^{b_j})$. According to Theorem 1.1.4, $c_{jl}\xi^{b_j}$ are the roots of the polynomial $P^{[j]}(\xi, \tau)$ related to $P_{q^{(j)}}$, $q^{(j)} = (1, b_j)$, by (1.1.15). The polynomials $P^{[j]}(\xi, \tau)$ essentially depend on both ξ and τ ($N(P)$ has neither vertical nor horizontal sides) and are solved with respect to the highest powers of both ξ and τ. It follows that the condition

$$P^{[j]}(\xi, \tau) \neq 0 \quad \text{for} \quad \xi \neq 0, \quad |\tau| \neq 0, \quad \operatorname{Im} \tau \leqslant 0, \tag{13}$$

remains valid for $|\xi| + |\tau| > 0$ and $\operatorname{Im} \tau \leqslant 0$. Therefore the polynomial $P^{[j]}$ is $q^{(j)}$-parabolic. Applying Proposition 1.2 we obtain (13).

(III)$\Longrightarrow$(IV)$\Longrightarrow$(V) is proved according to the same scheme as the analogous assertion in Theorem 1.3.3.

(V)$\Longrightarrow$(I) follows from the proposition.

2.4. *N*-stable correct polynomials. As was already shown in Section 2.2, the class of polynomials satisfying the conditions in Theorem 2.1 is stable relative to the addition of any monomials $c\xi^\alpha\tau^\beta$, $(\alpha, \beta) \in N(P)$ with sufficiently small coefficients c.

By N-stable correct polynomials we shall mean the ones that satisfy the conditions of Theorem 2.1 and are solved with respect to the highest power of τ.

Lemma. *For a polynomial $P(\xi, \tau)$ satisfying the conditions in Theorem 2.1 the following conditions are equivalent:*

(i) *the polynomial P is solved with respect to the highest power of τ (i.e. P is an N-stable correct polynomial);*

(ii) *the polynomial P is exponentially correct.*

Proof. (i)$\Longrightarrow$(ii). Generally speaking, the polynomial $P^{(\alpha)}(\xi,\tau)$ is a linear combination of monomials $\xi^{\varkappa}\tau^{\beta}$, $(\varkappa,\beta)\in\delta_0(P)$. If the polynomial P is solved with respect to the highest power of τ, then the polygon $N(P)$ has no horizontal side not lying on the coordinate axis. Repeating the argument of Lemma 4 in Section 1.3.2 it is easy to prove that in this case $\delta_0(P)=\delta(P)$, where $\delta(P)$ is the convex hull of the minor integral points (in the sense of Section 2.1) of $N(P)$. Applying Lemma 2.1 (ii) we conclude that for some $\varkappa>0$ we have

$$|P^{(\alpha)}(\xi,\tau)P^{-1}(\xi,\tau)|\leqslant \text{const}\,|\tau|^{-\varkappa}\to 0,\quad \operatorname{Im}\tau\to-\infty.$$

(ii)$\Longrightarrow$(i) is true by virtue of Lemma 1.1.

In view of the above results, we have the following

Theorem. *For a polynomial P the following conditions are equivalent.*

(I) *The polynomial P is N-stable correct.*

(II) *The polygon $N(P)$ has no horizontal side not belonging to the coordinate axis, and for any side $\Gamma_j^{(1)}$, that does not lie on the coordinate axes and is not vertical condition (12) is fulfilled.*

(III) *If $\hat{\tau}_j(\xi)=c_j\xi^{b_j}$ are the principal parts of Puiseux' expansions for $|\xi|\to\infty$ of the roots of the polynomial P, then either condition (13) is fulfilled or $\hat{\tau}_j(\xi)\equiv 0$ (i.e. $\tau_j(\xi)=O(1)$).*

(IV) *There is an N-parabolic polynomial $P_1(\xi,\tau)$ and an integer $b\geqslant 0$ such that $P\sim\tau^bP_1$.*

Proof. The proof of (I)$\Longrightarrow$(II)$\Longrightarrow$(I) is based on Theorem 2.1 and the equivalence of the conditions "the polynomial P is solved with respect to the highest power of τ" and "Newton's polygon $N(P)$ has no horizontal side not lying on the coordinate axis".

(I)$\Longrightarrow$(III). By virtue of the lemma, the polynomial P is exponentially correct, and, according to Proposition 2 in Section 1.4, either the principal parts $\hat{\tau}_j(\xi)$ of its roots have the form (13) or $\hat{\tau}_j(\xi)\equiv 0$ or $\hat{\tau}_j(\xi)=a_j\xi$ where a_j is real. Hence, it only remains to show that an N-stable correct polynomial has no roots of the form of

$$\hat{\tau}_j(\xi)=a_j\xi,\qquad \operatorname{Im}a_j=0. \tag{13'}$$

Indeed, assume that there is a root having the form (13$'$). Consider the polynomial $Q(\xi,\tau)=i\varepsilon\xi\prod_{l\neq j}(\tau-\hat{\tau}_l(\xi))$. This polynomial is a linear combination of monomials corresponding to the points of the polygon $N(P)$.

According to Theorem 2.2, the polynomial $P+Q$ does not satisfy (1.5) for a sufficiently small ε. And since the degree of Q relative to τ is less than the degree of P relative to τ, the polynomial $P+Q$ is correct in Petrovskiĭ's sense and is solved with respect to the highest power of τ. The structure of the polynomial $P+Q$ implies that one of its roots has Puiseux' expansion whose principal term is

$(a_j - i\varepsilon)\xi$.[1] If ε is real and $\varepsilon > 0$, we arrive at a contradiction to Proposition 1 in Section 1.4.

(III)$\Longrightarrow$(IV) is easily proved using Theorems 1.3 and 1.4.

(IV)$\Longrightarrow$(I) readily follows Theorem 2.3.

Corollary 1. *Let $P(\xi,\tau)$ be an N-stable correct polynomial and let (α,β) be a minor point of $N(P)$. Then the polynomial*

$$P_\varepsilon(\xi,\tau) = P(\xi,\tau) + \varepsilon\xi^\alpha\tau^\beta, \tag{14}$$

is N-stable correct for any (complex) ε.

Proof. According to the results in Section 1.1.4, if the monomial $\varepsilon\xi^\alpha\tau^\beta$ is added to the polynomial P, the principal parts of Puiseux' expansions of its roots do not change. Since β is less than the highest degree relative to τ, the polynomial P_ε remains being solved with respect to the highest power of τ. It now remains to use definition (III) of N-stable correct polynomials.

Thus, if the polynomial P is N-stable correct, then the polynomials P_ε are correct in Petrovskiĭ's sense for small $|\varepsilon|$ and any $(\alpha,\beta) \in N(P)$, and for arbitrary ε and $(\alpha,\beta) \in \delta(P)$.

Let a polynomial $P(\xi,\tau)$ be solved with respect to the highest power of τ and let $P^{[j]}(\xi,\tau)$ be the quasi-homogeneous polynomials related to P by (1.1.15). According to Section 1.1.4 the roots of these polynomials are the principal parts $\widehat{\tau}_j(\xi)$ of the expansions of the roots of P into Puiseux' series in decreasing powers of ξ. Therefore Proposition 1 in Section 1.4 implies that the conditions

$$P^{[j]}(\xi,\tau) \neq 0 \quad \text{for} \quad \operatorname{Im}\tau < 0 \tag{15}$$

are necessary for polynomial P to satisfy the correctness condition (1.5).

If condition (III) of the theorem holds, then (15) should be replaced by the stronger condition

$$P^{[j]}(\xi,\tau) \neq 0 \quad \text{for} \quad |\xi| + |\tau| > 0, \quad \operatorname{Im}\tau \leqslant 0. \tag{16}$$

Hence, we have

Corollary 2. *A polynomial $P(\xi,\tau)$ solved with respect to the highest power of τ is N-stable correct if and only if conditions* (16) *are fulfilled for $j = 1,\dots,n$.*

In view of what has been said, we state the following assertion that is of use for our further aims.

---

[1] Here we use the fact that the polynomials P and $\widehat{P} = \prod(\tau - \widehat{\tau}_l)$ differ only in minor monomials, and, in view of Section 1.1.4, when determining the principal parts of the roots, the polynomial P can be replaced by $\widehat{P}$.

Proposition. *Let a polynomial $P(\xi,\eta)$ be solved with respect to the highest power τ^m and let $\beta < m$ in polynomial (14). Then*

(i) *if polynomial (14) satisfies the correctness condition (1.5) for all sufficiently small $|\varepsilon|$ ($\varepsilon \in \mathbb{C}$), then $(\alpha,\beta) \in N(P)$;*

(ii) *if the polynomial remains correct for any $\varepsilon \in \mathbb{C}$, then (α,β) is a minor point of $N(P)$.*

Proof. (i) Let the point (α,β) lie outside $N(P)$. Then it is one of the vertices of $N(P_\varepsilon)$, say the vertex $\Gamma^{(0)}_{j+1}$, and let $\Gamma^0_j = (\alpha_j,\beta_j)$, $\beta_j > \beta$, be the neighboring vertex of $N(P_\varepsilon)$ (which simultaneously is a vertex of $N(P)$). According to what was said above, the quasi-homogeneous polynomial $P^{[j]}_\varepsilon$ corresponding to the side $\Gamma^{(0)}_j\Gamma^{(0)}_{j+1}$ must satisfy condition (15) for sufficiently small ε. However, as can easily be seen (see (1.1.15)), we have

$$P^{[j]}_\varepsilon = \tau^{\beta_j-\beta} + (\varepsilon/a_{\alpha_j\beta_j})\xi^{\alpha-\alpha_j},$$

and ε can be selected so that the roots of $P^{[j]}$ have negative imaginary parts, which contradicts (15),

(ii) By what has been proved in (i), it suffices to consider the case when the point (α,β) lies on the boundary of $N(P)$, say on the side joining the vertices $\Gamma^{(0)}_j$ and $\Gamma^{(0)}_{j+1}$. Then the quasi-homogeneous polynomial $P^{[j]}_\varepsilon = P^{[j]} + \varepsilon\xi^\alpha\tau^\beta$ must satisfy a condition of type (15) for any $\varepsilon \in \mathbb{C}$. However, this cannot hold since for arbitrary $\tau^0 \in \mathbb{C}$, $\xi^0 \in \mathbb{R}$ there is ε such that $P^{[j]}_\varepsilon(\xi^0,\tau^0) = 0$ (ε is determined from the linear equation $P^{[j]}(\xi^0,\tau^0) + \varepsilon(\xi^0)^\alpha(\tau^0)^\beta = 0$).

§3. Cauchy's problem for *N*-stable correct and *N*-parabolic differential operators in the case of one spatial variable

In this section we deal with differential operators

$$P(x,t,D_x,D_t) = \sum a_{\alpha\beta}(x,t)D^\alpha_x D^\beta_t = \sum_{j=0}^{m} P_j(x,t,D_x)D_t^{m-j} \tag{1}$$

on functions of the variables $(x,t) \in \mathbb{R}^2$, where x is called a spatial variable, t is time, $D_x = -i\,\partial/\partial x$, and $D_t = -i\,\partial/\partial t$. By $\tau = \sigma + i\gamma$ and ξ we shall denote the dual variables:

$$\langle(\tau,\xi),(t,x)\rangle = t\tau + x\xi.$$

In what follows we shall assume, without special stipulation, that operator (1) is solved with respect to D^m_t, i.e.

$$P_0(x,t;\xi) \equiv P_0(x,t).$$

Moreover, we shall assume that $P_0(x,t) \equiv 1$, i.e. operator (1) is written as

$$P(x,t;D_x,D_t) = D^m_t + \sum_{j=0}^{m} P_j(x,t,D_x)D_t^{m-j}. \tag{2}$$

3.1. Differential operators with constant coefficients. A differential operator $P(D_x, D_t)$ of the form of (1) with constant coefficients is said to be correct in Petrovskiĭ's sense or N-stable correct or N-parabolic if the polynomial $P(\xi, \tau)$ (the symbol of the operator P) possesses the corresponding property.

When dealing with operators correct in Petrovskiĭ's sense it is convenient to pass from the spaces $H^{(s)}$ (see Section 1.4.1) to the spaces $H^{(s)}_{[\gamma]}$ consisting of functions (distributions) $u(x,t)$ such that $\exp(\gamma t)u \in H^{(s)}$.

The definition of the Fourier transform of a decreasing functions implies that

$$(\widehat{\exp(\gamma t)}u)(\xi,\sigma) = (2\pi)^{-1}\iint \exp(-i\xi x - i(\sigma + i\gamma)t)u(x,t)\,dx\,dt = \widehat{u}(\xi,\tau).$$

By virtue of the definition of the spaces $H^{(s)}$, the inclusion $u \in H^{(s)}_{[\gamma]}$ means that the integral

$$\Big(\iint (1+\xi^2+\sigma^2)^s|\widehat{u}(\xi,\sigma+i\gamma)|^2 d\xi\,d\sigma\Big)^{1/2} \overset{\text{def}}{=} \|\exp(\gamma t)u\|_{(s)}$$

is convergent. As a norm in $H^{(s)}_{[\gamma]}$ it is more convenient to take the expression

$$\|u\|_{(s),\gamma} = \Big(\iint (1+|\tau|^2+\xi^2)^s|\widehat{u}(\xi,\tau)|^2\,d\xi\,d\sigma\Big)^{1/2}, \tag{3}$$

which is equivalent to the norm $\|\exp(\gamma t)u\|_{(s)}$, the equivalence constants depending on γ. For positive integral values of s, applying Newton's binomial formula to the integrand:

$$(1+\xi^2+\sigma^2+\gamma^2)^s = \sum_{\alpha+\beta\leqslant s} \frac{s!}{\alpha!\beta!(s-\alpha-\beta)!}\xi^{2\alpha}(\sigma^2+\gamma^2)^{2\beta},$$

we conclude that norm (3) is equivalent to

$$\Big(\sum_{\alpha+\beta\leqslant s} \|\exp(\gamma t)D_x^\alpha D_t^\beta u\|^2\Big)^{1/2}, \tag{3'}$$

the equivalence constant not depending on γ in this case. In the case of non-integral s, using the PDO with symbol $(1+\xi^2+\sigma^2+\gamma^2)^{s/2}$, we can rewrite norm (3) in the form

$$\|u\|_{(s),\gamma} = \|(1+D_x^2+D_t^2+\gamma^2)^{1/2}(\exp(\gamma t)u)\|. \tag{3''}$$

By analogy with (1.4.15), with polynomial (1) we associate the norm

$$\|u\|_{N(P),(s),\gamma} = \Big(\sum_{(\alpha,\beta)\in N(P)} \|D_x^\alpha D_t^\beta u\|^2_{(s),\gamma}\Big)^{1/2}. \tag{4}$$

Denote by $H^{N(P),s}_{[\gamma]}$ the space of functions with finite norm (4). Let $H^{(s)}_{[\gamma]+}$ and $H^{N(P),s}_{[\gamma]+}$ denote the subspaces of functions equal to zero for $t < 0$ (and, respectively, the subspaces of distributions with support belonging to the half-space $t \leqslant 0$).

Theorem 1. *For a differential operator (2) with constant coefficients the following conditions are equivalent.*

(I) *The polynomial $P(\xi,\tau)$ is N-stable correct.*

(II) $\exists\gamma_0,\ c>0$ *such that* $\forall s\in\mathbb{R}$ *the inequality*

$$\|u\|_{N(P),(s),\gamma}\leqslant c\|P(D_x,D_t)u\|_{(s),\gamma},\qquad \gamma\leqslant\gamma_0,\quad \forall u\in H^{(\infty)}_{[\gamma]} \tag{5}$$

holds.

(II_+) $\exists\gamma_0,\ c>0$ *such that* $\forall s\in\mathbb{R}$ *the inequality*

$$\|u\|_{N(P),(s),\gamma}\leqslant c\|P(D_x,D_t)u\|_{(s),\gamma},\qquad \gamma\leqslant\gamma_0,\quad \forall u\in H^{(\infty)}_{[\gamma_0]+},^{1)} \tag{5_+}$$

holds.

Proof. The equivalence of (I) and (II) follows from the definition of N-stable correct polynomials (see Section 2.4) and norms (3), (4) (cf. Theorem 1.4.2). Inequality (5_+) is a special case of (5). It remains to show that (II_+)$\Longrightarrow$(II). To prove this it should be noted that if $v(x,t)=u(x-x_0,t-t_0)$, then $\widehat{v}(\xi,\tau)=\exp(-i\xi x_0-i\sigma t_0+\gamma t_0)\widehat{u}(\xi,\tau)$, whence

$$\|v\|_{(s),\gamma}=\exp(\gamma t_0)\|u\|_{(s),\gamma},\ \|v\|_{N(P),(s),\gamma}=\exp(\gamma t_0)\|u\|_{N(P),(s),\gamma}.$$

By virtue of these relations, the function $u(x,t)$ in (5_+) can be replaced by the translated functions $u(x,t-t_0)$. Since the set of the translated functions is dense in $H^{(\infty)}_{[\gamma]}$, we obtain (5).

In an analogous way one proves

Theorem 2. *For a differential operator (2) the following conditions are equivalent:*

(I) *the polynomial $P(\xi,\tau)$ is N-parabolic;*

(II) $\exists\gamma_0,\ c>0,\ c'>0,\ \varkappa>0$ *such that* $\forall s\in\mathbb{R}$, *along with* (5), *the inequality*

$$\|u\|_{\delta(P),(s+\varkappa),\gamma}\leqslant c'\|P(D_x,D_t)u\|_{(s),\gamma}\quad \forall u\in H^{(\infty)}_{[\gamma]},\quad \gamma\leqslant\gamma_0, \tag{6}$$

holds.

(II_+) $\exists\gamma_0,\ c>0,\ c'>0,\ \varkappa>0$ *such that* $\forall s\in\mathbb{R}$, *along with (5_+), the inequality (6) is fulfilled on the space* $H^{(\infty)}_{[\gamma]+}$.

Remark (cf. Remark 1.4.2). In the left-hand norm in (6), in contrast to the analogous norm (1.3.4), the summation is extended over the integral points of $\delta(P)$ and not of $\delta_0(P)$. This is due to the fact at the very beginning we confined ourselves to symbols solved with respect to the highest power of τ. As was already said in the foregoing section, in this case Newton's polygon contains no horizontal side, and, consequently, $\delta_0(P)=\delta(P)$.

---

1) Here we use the fact that for the spaces $H^{(s)}_{[\gamma]+}$ the natural embeddings $H^{(s)}_{[\gamma']+}\subset H^{(s)}_{[\gamma'']+}$ hold for $\gamma''<\gamma'$.

3.2. *N*-stable correct and *N*-parabolic differential operators with variable coefficients. Consider operator (1), (2). At fixed (x,t) its symbol $P(x,t;\xi,\tau)$ is a polynomial in ξ and τ, and it is possible to define the polygon $N(P(x,t))$ of the polynomial. As in Section 1.4.3, denote by $N(P)$ the convex hull of the union of the polygons $N(P(x,t))$.

Definition (cf. Definition 1.4.3). Differential operator (2) is said to be N-stable correct (N-parabolic) if

(i) $N(P(x,t)) = N(P)\ \forall (x,t) \in \mathbb{R}^2$;

(ii) for any fixed $(x^0,t^0) \in \mathbb{R}^2$ the polynomial $P(x^0,t^0;\xi,\tau)$ is N-stable correct (N-parabolic).

Since the above-defined operators (in contrast to N quasi-elliptic operators in §1.4) will be considered globally, we shall have to make some additional assumptions concerning the behavior of the coefficients of operator (2) for $|x|+|t| \to 0$. For simplicity, we shall assume that the functions $a_{\alpha\beta}(x,t)$ in (2) are constant outside a sufficiently large circle $x^2+t^2=R^2$, i.e.

$$a_{\alpha\beta}(x,t) = a_{\alpha\beta} + a'_{\alpha\beta}, \qquad a'_{\alpha\beta}(x,t) \in \mathcal{D}. \tag{7}$$

Repeating almost literally the proof of Proposition 1.4.3[1)] one can prove the following

Proposition. *Let a symbol $P(x,t;\xi,\tau)$ satisfy conditions* (i) *and* (ii) *of the definition and let condition (7) hold for the coefficients $a_{\alpha\beta}(x,t)$. Then*

(i) *the coefficients of the symbols $P^{[j]}(x,t;\xi,\tau)$ defined for any (x,t) by formulas* (1.15) *satisfy conditions of type* (7);

(ii) *there are constants $c>0$ and γ_0 such that for all $(x,t) \in \mathbb{R}^2$ the inequality*

$$\Xi_{N(P)}(\xi,\tau) < c|P(x,t;\xi,\tau)|, \qquad \xi \in \mathbb{R}, \quad \operatorname{Im}\tau \leqslant \gamma_0, \tag{8}$$

holds.

Remarks. 1) The symbols $P(x,t;\xi,\tau)$ for which (8) is fulfilled are said to be uniformly N-stable correct. The uniform N-stable correctness of a symbol implies conditions (i) and (ii) of the definition. In general, the converse assertion is false without some additional assumptions concerning the behavior of the symbol at infinity.

However, all the results below on operators (2) remain valid if conditions (i) and (ii) of the definition and condition (7) are replaced by the condition of uniform N-stable correctness and the assumption of uniform boundedness on $\mathbb{R}^2$ of all derivatives of the functions $a_{\alpha\beta}(x,t)$.

[1)] Condition (7) allows us in fact to deal with functions on a compactum (as in the case of Proposition 1.4.3).

2) When $N(P)$ is a right triangle with vertices $(0,0)$, $(0,m)$, and $(mq,0)$, where q is even, our definition goes into the definition of operators with variable coefficients q-parabolic (or, simply, parabolic) in Petrovskiĭ's sense. For such symbols inequality (8) is equivalent to the following uniform estimate for the principal q-homogeneous part of $P_0(x,t;\xi,\tau)$:

$$c^{-1} \leqslant |P_0(x,t;\xi,\tau)|(|\xi|^q + |\tau|)^{-m} \leqslant c \qquad \forall(x,t) \in \mathbb{R}^2, \quad \forall(\xi, \operatorname{Re}\tau) \in \mathbb{R}^2, \quad \operatorname{Im}\tau \leqslant 0. \quad (9)$$

And this estimate is equivalent to the uniform positivity of the imaginary parts of the roots $\tau_{0j}(x,t)$ of the symbol P_0.

As in §1.4, with symbol (1) we associate the space $SL_{N(P)}$ of the symbols

$$Q(x,t;\xi,\tau) = \sum b_{\alpha\beta}(x,t)\xi^\alpha\tau^\beta, \quad (10)$$

for which $N(Q) \subset N(P)$ and the coefficients $b_{\alpha\beta}(x,t)$ satisfy conditions of type (7).

As in §1.4, we denote by $\delta(P)$ the convex polygon spanned to the minor integral points of $N(P)$. And let $S\mathcal{L}_{N(P)}$ denote the subspace of those $Q \in SL_{N(P)}$ for which $N(Q) \subset \delta(P)$. Modifying the proof of Theorem 1.4.3 we prove the following

Theorem. (i). *Let operator* (2) *be* N*-parabolic and let its coefficients satisfy condition* (7). *Then there are*

(a) *a set of even numbers* $q_1 > \cdots > q_\mu > 0$;
(b) *a set of operators* $P^{[j]}(x,t;D_x,D_t)$, $j = 1,\ldots,\mu$, q_j*-parabolic in Petrovskiĭ's sense, with coefficients satisfying a condition of type* (7);
(c) *and a symbol* $Q(x,t;\xi,\tau) \in S\mathcal{L}_{N(P)}$ *such that*

$$P(x,t;D_x,D_t) - P^{[1]}(x,t;D_x,D_t)\ldots P^{[\mu]}(x,t;D_x,D_t) = Q(x,t;D_x,D_t). \quad (11)$$

(ii) *Let operator* (2) *be* N*-stable correct and let its coefficients satisfy condition* (7). *Then there are*

(a) *an* N*-parabolic operator* $P_1(t,x;D_x,D_y)$ *with coefficients satisfying* (7);
(b) *an integer* $b > 0$;
(c) *and a symbol* $Q(x,t;\xi,\tau) \in S\mathcal{L}_{N(P)}$ *such that*

$$P(x,t;D_x,D_t) - P_1(x,t;D_x,D_t)D_t^b = Q(x,t;D_x,D_t). \quad (11')$$

3.3. A priory estimates for *N*-stable correct and *N*-parabolic operators with variable coefficients. In this section we shall prove analogs of Theorems 1 and 2 in Section 3.1 for the case of variable coefficients.

Theorem 1. *For an operator (2) with coefficients of the form of (7) the following conditions are equivalent.*

(I) *The operator is N-stable correct (in the sense of Definition 3.2).*

(II) *$\forall s \in \mathbb{R}$ there are $c = c(s) > 0$ and $\gamma_0 = \gamma_0(s)$ such that $\forall \gamma \leqslant \gamma_0$ the inequality*

$$\|u\|_{N(P),(s),\gamma} \leqslant c\|P(x,t;D_x,D_t)u\|_{(s),\gamma} \quad \forall u \in H^{(\infty)}_{[\gamma]} \tag{12}$$

holds.

(II_+) *$\forall s \in \mathbb{R}$ there are $c = c(s) > 0$ and $\gamma_0 = \gamma_0(s)$ such that $\forall \gamma \leqslant \gamma_0$ the inequality*

$$\|u\|_{N(P),(s),\gamma} \leqslant c\|P(x,t;D_x,D_t)u\|_{(s),\gamma} \quad \forall u \in H^{(\infty)}_{[\gamma]+} \tag{12_+}$$

is fulfilled.

The proof of the theorem is preceded by two lemmas.

Lemma 1. *Let $a(t,x) \in C^\infty$ and let*

$$\sup_{(t,x)\in\mathbb{R}^2} |D_x^\alpha D_t^\beta a(x,t)| < \varkappa_{\alpha\beta}.$$

Then $\forall s \in \mathbb{R}$ and $\forall \gamma < 0$ the inequality

$$\|au\|_{(s),\gamma} \leqslant (\varkappa_{00} + c_s(a)|\gamma|^{-1})\|u\|_{(s),\gamma}$$

holds, where the constant $c_s(a)$ depends only on the constants $\varkappa_{\alpha\beta}$, $\alpha + \beta = |s| + \nu$, where ν is a fixed number not depending on $a(x,t)$.

The proof is obvious for $s = 0$ (and in this case $c_0(a) \equiv 0$). In the case of arbitrary real s it is based on the technique of pseudodifferential operators. For this lemma in the indicated form see Volevich and Gindikin [6, Lemma 3.2].

Lemma 2. *Let (x^0,t^0) be a point in $\mathbb{R}^2$, let $P(D_x,D_t) = P(x^0,t^0;D_x,D_t)$, and let $S(x^0,t^0,r)$ be a circle of radius r with center at (x^0,t^0). Then for a sufficiently small r and a sufficiently large $-\gamma_0$ the inequality (12) on the functions $u \in \mathcal{D}$ with support in $S(x^0,t^0,r)$ is valid if and only if there holds an analogous inequality for the operator with constant coefficients $P(D_x,D_t)$.*

Proof (cf. the proof of Theorem 1.4.4.). Take a function $\psi(x,t) \in \mathcal{D}$ equal to 1 in the circle $S(x^0,t^0,r)$ and equal to zero outside the circle $S(x^0,t^0,2r)$. If $u(x,t)$ is a function with support lying in $S(x^0,t^0,r)$, then we have

$$P(x,t;D_x,D_t)u = P(D_x,D_t)u + P'(x,t;D_x,D_t)u,$$

where

$$P' = \sum a'_{\alpha\beta} D_x^\alpha D_t^\beta, \qquad a'_{\alpha\beta}(x,t) = \psi(x,t)(a_{\alpha\beta}(x,t) - a_{\alpha\beta}(x^0,t^0)).$$

According to Lemma 1, we have

$$\big| \|P(x,t;D_x,D_t)u\|_{(s),\gamma} - \|P(D_x,D_t)u\|_{(s),\gamma} \big| \leqslant \sum \|a'_{\alpha\beta}(x,t) D_x^\alpha D_t^\beta u\|_{(s),\gamma}$$
$$\leqslant \Big(\sum \max_{(x,t)\in S(x^0,t^0,2r)} |a'_{\alpha\beta}(x,t)| + c_1(a)|\gamma|^{-1} \Big) \|u\|_{N(P),(s)}$$

For any $\varepsilon < 1$ there is a sufficiently small r for which the sum on the right-hand side does not exceed $\varepsilon/2$. We then select γ_0 such that $c_1|\gamma|^{-1} < \varepsilon/2$ for $\gamma \leqslant \gamma_0$. Hence, the right-hand side does not exceed $\varepsilon\|\gamma\|_{N(P)}$; and therefore, if inequality (12) holds, then (4) is fulfilled, and if (4) takes place, then (12) also holds.

The proof of Theorem 1. (I)$\Longrightarrow$(II). Using Theorem 1 in Section 3.1 and Lemma 2 we prove inequality (12) for functions with sufficiently small supports. And we now "glue" together the local estimates by means of a partition of unity.

If the coefficients of operator (2) satisfy condition (7), then there is a circle $S(0,0,R)$ with center at the origin, outside which symbol (2) has constant coefficients. We select a set of functions $\chi_j(x,t)$ possessing the following properties:

(a) $\chi_j(x,t) \geqslant 0$, $\chi_j \in C^\infty$, $j = 0,1,\dots,J$;

(b) $\sum \chi_j(x,t) \equiv 1$ $\forall (x,t) \in \mathbb{R}^2$;

(c) $\forall j \in \{1,\dots,J\}$ there is $(x^j,t^j) \in \mathbb{R}^2$ such that the support of χ_j belongs to a sufficiently small neighborhood of (x^j,t^j); the function $\chi_0(x,t)$ is equal to 0 outside the circle $S(0,0,R)$ and to 1 inside a circle $S(0,0,R')$, $R' > R$.

As was mentioned above, by virtue of Lemma 2 and Theorem 1 in Section 3.1, we have the inequalities

$$\|\chi_j u\|_{N(P),(s),\gamma} \leqslant c\|P(x,t;D_x,D_t)(\chi_j u)\|_{(s),\gamma}, \quad \gamma \leqslant \gamma_0.$$

Adding together these estimates and applying Minkowski's inequality we obtain, in view of (b),

$$\|u\|_{N(P),(s),\gamma} = \|\sum \chi_j u\|_{N(P),(s),\gamma}$$
$$\leqslant \sum \|\chi_j u\|_{N(P),(s),\gamma} \leqslant c\sum \|P(x,t;D_x,D_t)(\chi_j u)\|_{(s),\gamma}. \tag{13}$$

According to Leibniz' formula, we have

$$P(\chi_j u) = \chi_j Pu + \sum_{(\alpha,\beta)\in\delta(P)} b_{j\alpha\beta} D_x^\alpha D_t^\beta u.$$

With regard to this inequality, estimate (13) takes the form

$$\|u\|_{N(P),(s),\gamma} \leqslant c_1 \|P(x,t;D_x,D_t)u\|_{(s),\gamma} + c_2 \sum_{(\alpha,\beta)\in\delta(P)} \|D_x^\alpha D_t^\beta u\|_{(s),\gamma}, \tag{14}$$

where the constants c_1 and c_2 do not depend on γ. Lemma 1.2 (ii) and the definition of the spaces $H^{(s)}_{[\gamma]}$ imply the inequality

$$\|D_x^\alpha D_t^\beta u\|_{(s),\gamma} \leqslant \varepsilon_{\alpha\beta}(\gamma)\|u\|_{N(P),(s),\gamma}, \quad (\alpha,\beta)\in\delta(P), \quad \varepsilon_{\alpha\beta}(\gamma)\to 0, \quad \gamma\to-\infty.$$

If $-\gamma_0$ is sufficiently large, then for $\gamma \leqslant \gamma_0$ the second term on the right-hand side of (14) does not exceed $(1/2)\|u\|_{N(P),(s),\gamma}$, whence follows (12).

(II)$\Longrightarrow$(II$_+$)$\Longrightarrow$(I). The first implication is trivial, and we proceed to the proof of the second one. We have to show that (12$_+$) implies conditions (i) and (ii) in Definition 3.2.

If inequality (12$_+$) holds, then, by virtue of Lemma 2, the operator $P(D_x,D_t) - P(x^0,t^0;D_x,D_t)$ ((x^0,t^0) is an arbitrary point in $\mathbb{R}^2$) satisfies an analogous inequality on the functions whose support is concentrated in the neighborhood of (x^0,t^0). Therefore (cf. Theorem 1 in Section 3.1) an analogous inequality is retained for the translated functions. Since the set of the translated functions is dense in $H^{(s)}_{[\gamma]}$, we arrive at (5$_+$) or (5), whence it follows that the symbol $P(\xi,\tau)$ is N-stable correct.

Let a point (α,β) belong to the polygon $N(P(x^1,t^1))$ corresponding to the polynomial $P(x^1,t^1;\xi,\tau)$. Then

$$\|D_x^\alpha D_t^\beta u\|_{(s),\gamma} \leqslant \|u\|_{N(P),(s),\gamma} \leqslant \text{const}\,\|P(D_x,D_t)\|_{(s),\gamma}$$

It follows (cf. Theorem 1 in Section 3.1) that

$$|\xi^\alpha\tau^\beta| \leqslant \text{const}\,|P(x^0,t^0;\xi,\tau)| \quad \text{for} \quad \operatorname{Im}\tau \leqslant \gamma_0.$$

A simple modification of the proof of Lemma 1.2.2 shows that $(\alpha,\beta)\in N(P(x^0,t^0))$, i.e. $N(P(x^1,t^1)) \subset N(P(x^0,t^0))$. Since the points (x^0,t^0) and (x^1,t^1) are arbitrary, all the polygons $N(P(x,t))$ coincide. The theorem is proved.

Theorem 2. *For an operator* (2) *with coefficients of the form of* (7) *the following conditions are equivalent.*

(I) *The operator is N-parabolic.*

(II) *$\forall s\in\mathbb{R}$ there are $c=c(s)>0$, $\varkappa>0$, and $\gamma_0=\gamma_0(s)$ such that, along with* (12), *the inequality*

$$\|u\|_{\delta(P),(s+\varkappa),\gamma} \leqslant c\|P(x,t;D_x,D_t)u\|_{(s),\gamma} \quad \forall u\in H^{(\infty)}_{[\gamma]}, \quad \gamma\leqslant\gamma_0 \tag{15}$$

holds.

(II$_+$) *Under conditions* (II), *along with* (12$_+$), *the inequality* (15) *is fulfilled for $u\in H^{(\infty)}_{[\gamma_0]+}$.*

Proof. (I)$\Longrightarrow$(II). According to Theorem 1, inequality (12) holds. The definition of parabolic polynomials and condition (i) in Definition 3.2 imply that

$$\|u\|_{\delta(P),(s+\varkappa),\gamma} \leqslant \text{const}\, \|u\|_{N(P),(s),\gamma}.$$

Using (12_+) we obtain (15). Since the implication (II)$\Longrightarrow$(II_+) is trivial, it remains to verify

(II_+)$\Longrightarrow$(I). From what was proved in Theorem 1, it follows that the operator (2) is N-stable correct, and therefore we have to verify only the regularity of the polygon $N(P)$. This property is proved by means of a literal repetition of the corresponding part of the proof of Theorem 1.4.4.

3.4. Cauchy's problem for *N*-stable correct differential operators with variable coefficients. By Cauchy's problem for operator (2) solved with respect to the highest derivative with respect to t is meant the problem of determining the function $u(x,t)$ satisfying the equation

$$P(x,t;D_x,D_t)u(x,t) = f(x,t), \qquad t > 0, \tag{16}$$

and the initial conditions

$$(D_t^{j-1}u)(x,0) = \varphi_j(x), \quad j = 1,\dots,m. \tag{17}$$

By means of the transformation

$$u = v + \sum_{j=1}^{m} t^{j-1}\varphi_j(x)/(j-1)!$$

problem (16), (17) can be reduced to an analogous problem with zero initial data:

$$(D_t^{j-1}u)(x,0) = 0, \quad j = 1,\dots,m. \tag{17'}$$

If conditions (17′) are fulfilled, then Equation (16) remains valid if the functions $u(x,t)$ and $f(x,t)$ are extended as zero to $t < 0$, i.e. if

$$u(x,t) = f(x,t) = 0 \quad \text{for} \quad t < 0. \tag{18}$$

Problem (16), (18) will be called the homogeneous Cauchy problem. It can be reformulated as the problem of inverting the continuous operator

$$(H_{[\gamma]}^{N(P),(s)})_+ \to H_{[\gamma]+}^{(s)} \qquad (u(x,t) \mapsto P(x,t;D_x,D_t)u(x,t)). \tag{19}$$

Along with operator (19), we shall consider the operator

$$H_{[\gamma]}^{N(P),(s)} \to H_{[\gamma]}^{(s)} \qquad (u \mapsto Pu) \tag{20}$$

defined throughout the space. The corresponding problem can be interpreted as Cauchy's problem with zero initial data for $t \to -\infty$.

Theorem. *For differential operator (2) with coefficients of the form of (7) the conditions below are equivalent.*

(I) *The operator is N-stable correct.*

(II) *For any $s \in \mathbb{R}$ there is $\gamma_0 = \gamma_0(s)$ such that for $\gamma \leqslant \gamma_0$ the mapping (20) is bijective.*[1]

[1]That is, operator (20) has no kernel and its image coincides with the right-hand space.

(II$_+$) *For any $s \in \mathbb{R}$ there is $\gamma_0 = \gamma_0(s)$ such that for $\gamma \leqslant \gamma_0$ the mapping* (19) *is bijective.*

Proof. (II)$\Longrightarrow$(I), (II$_+$)$\Longrightarrow$(I). If operators (19) and (20) are bijective, then, by Banach's theorem, they possess continuous inverse operators

$$R\colon H^{(s)}_{[\gamma]} \to H^{N(P),(s)}_{[\gamma]}, \qquad R_+\colon H^{(s)}_{[\gamma]+} \to (H^{N(P),(s)}_{[\gamma]})_+. \tag{21}$$

The boundedness of these operators implies a priori estimates (12) and (12$_+$). Applying Theorem 1 in Section 3.3 we prove that operator (2) is N-stable correct.

(I)$\Longrightarrow$(II), (II$_+$). The continuity of operators (19) and (20) follows from the definition of the space $H^{N(P),(s)}_{[\gamma]}$, and the absence of the kernel is implied by a priory estimates (12), (12$_+$) in Theorem 1 in the foregoing section. Thus, the theorem reduces to the proof of the surjectivity of the operators (19) and (20). Since both the assertions are proved in a similar way, we shall consider only the first of the operators.

So, we are going to prove that if $-\gamma_0$ is sufficiently large, then Equation (16) possesses a solution $u \in (H^{N(P),(s)}_{[\gamma]})_+$ for any right-hand side $f \in H^{(s)}_{[\gamma]+}$, $\gamma \leqslant \gamma_0$. To prove this we make use of Theorem 3.2, according to which

$$P(x,t;D_x,D_t) = P_0 + Q(x,t;D_x,D_t), \tag{22}$$

where the symbols of the operator Q belongs to $SL_{N(P)}$ and the operator P_0 is a composition of operators parabolic in Petrovskiĭ's sense and the operator of differentiation with respect to t:

$$P_0 = P^{[1]}(x,t;D_x,D_t)\dots P^{[\mu]}(x,t;D_x,D_t)D_t^b. \tag{22'}$$

Lemma. *There is γ_1 such that for $\gamma \leqslant \gamma_1$ there exists a continuous operator*

$$R_{0+}\colon H^{(s)}_{[\gamma]+} \to (H^{N(P),(s)}_{[\gamma]})_+ \tag{23}$$

serving as the right inverse operator of P_0, i.e. $P_0R_0f = f \quad \forall f \in H^{(s)}_{[\gamma]+}$.

Proof. According to the well-known results for operators parabolic in Petrovskiĭ's sense (e.g. see Agranovich and Vishik [1][1)]), the operators

$$(H^{N(P^{[i]}),(s)}_{[\gamma]+})_+ \to H^{(s)}_{[\gamma]+} \qquad (w \mapsto P^{[i]}w)$$

possess continuous inverse operators

$$R^{[i]}\colon H^{(s)}_{[\gamma]+} \to (H^{N(P^{[i]})(s)}_{[\gamma]})_+, \qquad P^{[i]}R^{[i]} \equiv 1,$$

[1)] An independent proof of the solvability of Cauchy's problem for operators parabolic in Petrovskiĭ's sense will be in fact given in §5.

for a sufficiently large $-\gamma$. The ordinary differential operator $P^{[0]} = D_t^b$ possesses the same property, and we denote its right inverse operator as $R^{[0]}$. We select $N > \deg P$ and consider the operator

$$R_0 = R^{[0]} R^{[\mu]} \dots R^{[1]} \tag{24}$$

on the space $H_{[\gamma]+}^{(s+N)}$, $\gamma < \gamma_2$. This operator transforms $H_{[\gamma]+}^{(s+N)}$ into itself and is the right inverse operator of P_0 on this space. We shall show that it can be extended to continuous operator (23).

Indeed, the operator P_0 is N-stable correct and $N(P_0) = N(P)$. Therefore there is γ_0 such that for $\gamma < \gamma_0$ the inequality

$$\|v\|_{N(P),(s),\gamma} \leqslant c\|P_0 v\|_{(s),\gamma}$$

is fulfilled. Replacing v by $R_0 g$, $g \in H_{[\gamma]+}^{(s+N)}$, and using the property that $P_0 R_0 g = g$ we obtain

$$\|R_0 g\|_{N(P),(s),\gamma} \leqslant c\|g\|_{(s),\gamma},$$

i.e. operator (24) that was originally defined on $H_{[\gamma]+}^{(s+N)}$ is continued by continuity to operator (23).

We now prove the surjectivity of operator (19). To this end the solution $u \in (H_{[\gamma]}^{N(P),(s)})_+$ to Equation (16) is sought in the form $u = R_0 g$, where $g \in H_{[\gamma]+}^{(s)}$ and R_0 is the operator in the lemma. Then for g the equation

$$g + Q(x,t;D_x,D_t)R_0 g = f, \tag{25}$$

is obtained. Since $Q \in S\mathcal{L}_{N(P)}$, we have

$$\begin{aligned}\|QR_0 g\|_{(s),\gamma} &\leqslant \text{const}\,\|R_0 g\|_{\delta(P),(s),\gamma}\\ &\leqslant \varepsilon(\gamma)\|R_0 g\|_{N(P),(s),\gamma} \leqslant \varepsilon_1(\gamma)\|g\|_{(s),\gamma}, \quad \varepsilon_1(\gamma) \to 0, \quad \gamma \to -\infty.\end{aligned}$$

Hence, $\forall \varepsilon < 1$ there is γ such that

$$\|QR_0 g\|_{(s),\gamma} \leqslant \varepsilon\|g\|_{(s),\gamma}.$$

By virtue of this inequality, Equation (25) possesses a solution $g \in H_{[\gamma]+}^{(s)}$ which is determined by Neumann's series

$$g = \sum_{k=0}^{\infty} (-QR_0)^k f.$$

The theorem is proved.

§4. Stable-correct and parabolic polynomials in several variables

In this section we extend the results of §2 to the case of polynomials in the variables $\xi = (\xi_1, \ldots, \xi_n) \in \mathbb{R}^n$, $\tau \in \mathbb{C}$:

$$P(\xi, \tau) = \sum a_{\alpha_1,\ldots,\alpha_n,\beta} \xi_1^{\alpha_1} \ldots \xi_n^{\alpha_n} \tau^\beta. \tag{1}$$

In the Introduction we associated with each polynomial (1) a polygon $\Delta(P)$. We now present necessary and sufficient conditions on polynomial (1) under which the inequality

$$\sum_{(\alpha,\beta)\in\Delta(P)} |\xi|^\alpha |\tau|^\beta \leqslant c|P(\xi,\tau)|, \quad \operatorname{Im}\tau \leqslant \gamma_0, \quad (\xi, \operatorname{Re}\tau) \in \mathbb{R}^{n+1}, \tag{2}$$

is fulfilled. Among the polynomials satisfying (2) we separate out analogs of N-stable correct and N-parabolic polynomials.

In the multidimensional case the main idea is that the polynomial $P(\xi, \tau)$ is regarded as a polynomial in the two variables τ and $\rho = |\xi|$, depending on the parameter $\omega = \xi/|\xi|$. Accordingly, we embed the class of polynomials in the class of pseudo-polynomials, i.e. polynomials in τ and ρ depending smoothly on ω. In this broader class we manage to obtain an analog of the factorization in §2, in which the role of polynomials parabolic in Petrovskiĭ's sense is played by pseudo-polynomials parabolic in Petrovskiĭ's sense.

4.1. Polynomials and pseudo-polynomials. We introduce in $\mathbb{R}^n$ the polar coordinates $\xi = \rho\omega$, $\rho = |\xi|$, $\omega = \xi/|\xi|$; the angular variable ω runs over the unit sphere S^{n-1}. By a pseudo-polynomial will be meant a function

$$Q(\xi, \tau) \equiv Q_\omega(|\xi|, \tau) = \sum_{(\varkappa,\beta)\in\nu_Q} b_{\varkappa\beta}(\omega)|\xi|^\varkappa \tau^\beta,$$

where ν_Q is a finite set of nonnegative integral points and $b_{\varkappa\beta}$ are functions belonging to $C^\infty(S^{n-1})$.

Every polynomial (1) can be rewritten as

$$P(\xi, \tau) = P_\omega(\rho, \tau) = \sum a_{\varkappa\beta}(\omega)\rho^\varkappa \tau^\beta, \tag{3}$$

where

$$a_{\varkappa\beta}(\omega) = \sum_{\alpha_1+\cdots+\alpha_n=\varkappa} a_{\alpha_1,\ldots,\alpha_n,\beta}\omega_1^{\alpha_1} \ldots \omega_n^{\alpha_n} \in C^\infty(S^{n-1}),$$

i.e. polynomial (1) is a pseudo-polynomial.

For a fixed ω the pseudo-polynomial Q_ω is a polynomial in two variables, and therefore we can define Newton's polygon $N(Q_\omega)$ and the polygon $\delta(Q_\omega)$ for Q. We set

$$\Delta(Q) = \bigcup_{\omega\in S^{n-1}} N(Q_\omega), \qquad \delta(Q) = \bigcup_{\omega\in S^{n-1}} \delta(Q_\omega).$$

In the case of polynomial (1) these definitions coincide with the definitions of the polygons $\Delta(P)$ and $\delta(P)$ presented earlier. Obviously, we have

$$N(P_\omega) \subset \Delta(P), \qquad \delta(P_\omega) \subset \delta(P). \tag{4}$$

Let P be a polynomial (1). We retain the notation $L_{\Delta(P)}$ and $\mathcal{L}_{\Delta(P)}$ for the linear spaces of pseudo-polynomials Q satisfying (respectively) the conditions $\Delta(Q) \subset \Delta(P)$ and $\delta(Q) \subset \delta(P)$.

All definitions of the foregoing sections are extended trivially to the pseudo-polynomials, and therefore we shall speak, without any special stipulation, about pseudo-polynomials correct in Petrovskiĭ's sense, hyperbolic, parabolic in Petrovskiĭ's sense, N-stable correct, and N-parabolic.

4.2. Condition for the existence of estimate (2). The multidimensional analog of Theorem 2.1 is the following

Theorem. *For polynomial* (1) *the conditions below are equivalent.*

(I) *There are $c > 0$ and γ_0 such that inequality* (2) *holds.*

(II) *For each $\omega \in S^{n-1}$ the pseudo-polynomial $P_\omega(\rho, \tau)$ satisfies the equivalent conditions of Theorem* 2.1, *and we have*

$$N(P_\omega) = \Delta(P) \quad \forall \omega \in S^{n-1}. \tag{5}$$

(III) *There exists a set of vectors $q^{(j)} = (q_1^{(j)}, q_2^{(j)})$, $j = 1, \dots, \mu$, $q_1^{(j)} \geqslant 0$, $q_2^{(j)} > 0$, and a set of $q^{(j)}$-homogeneous quasi-polynomials $P_\omega^{[j]}(\rho, \tau)$ such that*

$$P_\omega^{[j]}(\rho, \tau) \neq 0 \quad \text{for} \quad \rho^2 + |\tau|^2 > 0, \qquad \operatorname{Im} \tau \leqslant 0, \tag{6}$$

and an integer $b \geqslant 0$ such that

$$P(\xi, \tau) - \tau^b \prod_{j=1}^{\mu} P_\omega^{[j]}(|\xi|, \tau) = Q_\omega(|\xi|, \tau) \in \mathcal{L}_{\Delta(P)}. \tag{7}$$

Proof. (I)$\Longrightarrow$(II). Replacing ξ in (2) by $|\xi|\omega$ we obtain the inequality

$$\sum_{(\varkappa,\beta)\in\Delta(P)} |\xi|^{\varkappa} |\tau|^{\beta} \leqslant c|P_\omega(|\xi|, \tau)|, \quad \operatorname{Im} \tau \leqslant \gamma_0. \tag{2'}$$

By virtue of (4), we have $N(P_\omega) \subset \Delta(P)$, whence it follows that the polynomial $P_\omega(\rho, \tau)$ with fixed ω satisfies the conditions of Theorem 2.1. On the other hand, Lemma 1.2.2 remains valid in the case of estimates of type (2′) (the trivial modification of the proof of the lemma is left to the reader). And therefore $\Delta(P) \subset N(P_\omega)$. Comparing this inclusion with (4) we obtain (5).

(II)$\Longrightarrow$(III) (cf. Section 3.2). Since for each ω the polynomial $P_\omega(\rho,\tau)$ satisfies the equivalent conditions of Theorem 2.1, for each ω we can write relation (7) where, in general, the number b and the orders of homogeneity $q^{(j)}$ of the polynomials $P^{[j]}$ may depend on ω while the coefficients of the polynomial Q_ω even may not be continuous functions. However, the vectors $q^{(j)}$ and the number b are uniquely determined by the polygon $N(P_\omega)$, which, according to (5), depends on ω. To prove that the coefficients of the polynomials $P^{[j]}_\omega$ are smooth functions (of ω) we use explicit formulas (cf. Section 1.1.3). Let $(\varkappa_j,\beta_j)$ be the vertices of the polygon $\Delta(P)$. Then

$$P^{[j]}_\omega(|\xi|,\tau) = \sum_{\langle q^{(j)},(|\alpha|,\beta)\rangle = m_j} a_{\alpha_1\dots\alpha_n\beta}\xi_1^{\alpha_1}\dots\xi_n^{\alpha_n}\tau^\beta \times\Big[\sum_{\alpha_1+\dots\alpha_n=\varkappa_j} a_{\alpha_1\dots\alpha_n\beta_j}\omega_1^{\alpha_1}\dots\omega_n^{\alpha_n}|\xi|^{\varkappa_j}\tau^{\beta_j+1}\Big]^{-1}. \tag{8}$$

Since (5) holds, we have

$$\sum_{|\alpha|=\varkappa_j} a_{\alpha_1\dots\alpha_n\beta_j}\omega^{\alpha_1}\dots\omega_n^{\alpha_n}\neq 0 \quad \forall\omega\in S^{n-1}.$$

Consequently, the modulus of this function has a positive infinium on the unit sphere, whence follows that the coefficients of pseudo-polynomial (8) belong to $C^(S^{n-1})$. And hence the right-hand side of (7) possesses this property.

(III)$\Longrightarrow$(I). To prove this implication it is in fact necessary to repeat the argument in Theorem 1.2.1 (see the derivation of (III)$\Longrightarrow$(I)). Then for each ω we obtain

$$\sum_{(\varkappa,\beta)\in N(P_\omega)} |\rho^\varkappa\tau^\beta| \leqslant c|P_\omega(\rho,\tau)|, \quad \rho\in\mathbb{R}, \quad \operatorname{Im}\tau\leqslant\gamma_0.$$

Since the polygon $N(P_\omega)$ is uniquely determined by the vectors $q^{(j)}$ and the number b (not depending on ω), the polygon does not depend on ω. Further, for at least one $\omega_0\in S^{n-1}$ the polygon $N(P_{\omega_0})$ coincides with $\Delta(P)$, and therefore $N(P_\omega)=\Delta(P)$ $\forall\omega\in S^{n-1}$. If one examines carefully the proof of Theorem 1.2.1 (this was in fact done in the proof of Proposition 1.4.3), one sees that it is possible to select unified constants c and γ_0 for which these inequalities hold for all ω. Replacing ρ in these inequalities by $|\zeta|$ and recalling that $P_\omega(|\xi|,\tau)=P(\xi,\tau)$ we obtain (2). The theorem is proved.

4.3. Conditions for fulfilment of inequality (2) in terms of complex zeros of polynomial (1). We now state a multidimensional analog of Theorem 2.2.

Let with each $\tau\in\mathbb{C}$ a surface

$$\Sigma_\tau = \{\zeta=(\zeta_1,\dots,\zeta_n)\in\mathbb{C}^n, P(\tau,\zeta)=0\}$$

be associated and let $d_\tau(\xi)$ denote the distance from the point $\xi \in \mathbb{R}^n$ to the surface Σ_τ:

$$d_\tau(\xi) = \inf_{\zeta\in\Sigma_\tau} |\xi - \zeta|.^{1)}$$

Theorem. *For polynomial* (1) *the conditions* (I), (II), *and* (III) *in Theorem* 4.1 *are equivalent to the following conditions:*

(IV) *There are* ε_0 *and* γ_0 *such that all the polynomials*

$$P_\delta(\xi,\tau) = \sum (a_{\alpha_1\ldots\alpha_n\beta} + \delta_{\alpha_1\ldots\alpha_n\beta})\xi_1^{\alpha_1}\ldots\xi_n^{\alpha_n}\tau^\beta, \quad |\delta_{\alpha_1\ldots\alpha_n\beta}| < \varepsilon, \tag{9}$$

satisfy simultaneously Petrovskiĭ's condition; more precisely,

$$P_\delta(\xi,\tau) \neq 0, \quad \operatorname{Im}\tau < \gamma_0, \quad \varepsilon < \varepsilon_0, \quad (\xi, \operatorname{Re}\tau) \in \mathbb{R}^{n+1}. \tag{10}$$

(V) *There are constants* $c_1 > 0$ *and* γ_0 *such that*

$$d_\tau(\xi) > c_1|\xi| \quad \textit{for} \quad \operatorname{Im}\tau < \gamma_1, \quad \xi \in \mathbb{R}^n. \tag{11}$$

Remark. In the absence of the parameter τ the inequality (11) is a condition for ellipticity. In the present context (11) can be interpreted as a condition for ellipticity with a parameter (cf. Agranovich and Vishik [1]). We note that in condition (11) Newton's polygon is no longer involved.

The proof of the theorem. (I)$\Longrightarrow$(IV). The derivation of (9) from (2) is carried out in the same way as in the case $n = 1$

(IV)$\Longrightarrow$(V). Let (10) be fulfilled and let T be a complex $(n \times n)$ matrix, the sum of the moduli of its elements not exceeding ε. Then for the polynomial $P(\xi + T\xi, \tau)$ Petrovskiĭ's condition is fulfilled for $\operatorname{Im}\tau \leqslant \gamma_0$ and $\varepsilon < \varepsilon_0$. Since each point of the complex ball $|\zeta - \xi| < \varepsilon|\xi|$ can be represented as $\xi + T\xi$, we have $P(\xi,\tau) \neq 0$ for the points ζ belonging to this ball and $\operatorname{Im}\tau \leqslant \gamma_0$. Hence, inequality (11) with $c_1 = \varepsilon$ holds.

The proof of the implication (V)$\Longrightarrow$(I) is based on the two lemmas below.

Lemma 1. *Let* $\mathcal{D}$ *be a bounded set in* $\mathbb{R}^n$ *possessing the property that any polynomial of degree no higher that* μ *vanishing on* $\mathcal{D}$ *is identically equal to zero. Then there is a constant* $K = K(\mathcal{D}, n, \mu)$ *such that for any polynomial* $Q(\zeta)$ *of degree no higher than* μ *and for all* $\zeta \in \mathbb{C}^n$ *the inequality*

$$|Q^{(\alpha)}(\zeta)| \leqslant K \sup_{\theta\in\mathcal{D}} |Q(\zeta + \theta)|,$$

is fulfilled.

---

1) If the polynomial $P(\xi,\tau)$ does not depend on ξ (e.g. $P \equiv \tau^b$), then we formally put $d_\tau(\xi) = +\infty$.

Lemma 2. *Let Q be a polynomial of degree μ and let $d(\xi)$ be the distance from the point ξ to the (complex) surface of zeros $\{Q(\zeta) = 0\}$. Then*

$$|Q(\xi + \zeta)| \leqslant 2^{\mu}|Q(\xi)| \quad \text{for} \quad |\zeta| \leqslant d(\xi).$$

For the proof of these lemmas see Hörmander [3(Lemma 3.1.5 and Lemma 4.1.1)]. We now complete the proof of the theorem, i.e. show that (11) implies (2).

Set $\rho = |\xi|$ and write

$$P(\xi, \tau) = P(\rho\omega, \tau) = \sum_{\alpha} P_{\alpha}(\tau)\omega^{\alpha}\rho^{|\alpha|}. \tag{12}$$

(recall that $\omega^{\alpha} = \omega_1^{\alpha_1} \dots \omega_n^{\alpha_n}$ and $|\alpha| = \alpha_1 + \dots + \alpha_n$). Consider an auxiliary polynomial in $\mathbb{C}^n$:

$$Q_{\tau,\rho}(\eta) = P(\tau, \rho\eta). \tag{13}$$

Then (12) implies the following expressions for the coefficients of $P_{\alpha}(\tau)$:

$$\alpha!\, P_{\alpha}(\tau)\rho^{|\alpha|} = Q_{\tau,\rho}^{(\alpha)}(0).$$

We now apply Lemma 1 to polynomial (13) taking the spherical zone

$$\mathcal{D} = \{\eta \in \mathbb{C}^n, 1 - \varepsilon < |\eta| < 1 + \varepsilon\}$$

as the set $\mathcal{D}$. This yields

$$\alpha!\, |P_{\alpha}(\tau)|\rho^{|\alpha|} \leqslant K \sup_{1-\varepsilon<|\eta|<1+\varepsilon} |P(\tau, \rho\eta)|. \tag{14}$$

If $\eta \in \mathcal{D}$, $\xi = \rho\omega$, and $|\omega| = 1$, then

$$|\xi - \rho\eta| = |\xi||\omega - \eta| < \varepsilon|\xi|.$$

If ε does not exceed the constant c_1 in (11) and $\operatorname{Im}\tau < \gamma_0$, then, according to (11), we have

$$|\xi - \rho\eta| < d_{\tau}(\xi).$$

And therefore we can apply Lemma 2 to the right-hand side of (14), which results in

$$\alpha!\, |P_{\alpha}(\tau)|\rho^{|\alpha|} \leqslant 2^{\mu}K|P(\xi, \tau)|, \quad \xi \in \mathbb{R}^n, \quad \operatorname{Im}\tau \leqslant \gamma_0. \tag{15}$$

If $|\tau|$ is sufficiently large, then $|P_{\alpha}(\tau)| > \text{const}\, |\tau|^{m_{\alpha}}$, where m_{α} is the degree of the polynomial $P_{\alpha}(\tau)$. Hence, inequality (15) is equivalent to (2). The theorem is proved.

4.4. Parabolic polynomials. A hypoelliptic polynomial satisfying the equivalent conditions of Theorems 4.1 and 4.2 is said to be parabolic. According to Lemma 1 in Section 1.3.2, a parabolic polynomial is solved with respect to the highest power of any of the variables $\xi_1, \dots, \xi_n, \tau$. It follows that the polygon $\Delta(P)$ is regular. Proposition 2.3 remains valid if $\xi \in \mathbb{R}$ in its formulation is replaced by $\xi \in \mathbb{R}^n$ and $N(P)$ is replaced by $\Delta(P)$. Comparing this assertion with Theorems 4.1 and 4.2 we arrive at the following description of parabolic polynomials.

Theorem. *For polynomial* (1) *the conditions below are equivalent.*

(I) *The polynomial P is parabolic (i.e. it is hypoelliptic and there are $c > 0$ and γ_0 such that* (2) *holds).*

(II) *The pseudo-polynomials $P_\omega(\rho, \tau)$ are $N(P_\omega)$-parabolic $\forall \omega \in S^{n-1}$ (Theorem* 2.3) *and condition* (5) *is fulfilled.*

(III) *There are even numbers $q_1, \dots, q_\mu$ and pseudo-polynomials $P_\omega^{[j]}(\rho, \tau)$, $j = 1, \dots, \mu$, q_j-parabolic in Petrovskiĭ's sense such that*

$$P(\xi, \tau) - \prod_{j=1}^{\mu} P_\omega^{[j]}(|\xi|, \tau) = Q_\omega(|\xi|, \tau) \in \mathcal{L}_{\Delta(P)}.$$

Remark. The condition of hypoellipticity in (I) can be replaced by the existence of $\varkappa > 0$ and $c_1 > 0$ such that

$$(|\tau| + |\xi|)^{\varkappa} \sum_{(\alpha,\beta)\in\delta(P)} |\xi|^\alpha |\tau|^\beta \leqslant c_1 |P(\xi, \tau)| \quad \text{for} \quad \operatorname{Im}\tau \leqslant \gamma_0. \tag{16}$$

4.5. Stable-correct polynomials. A polynomial (1) solved with respect to the highest power of τ and satisfying the equivalent conditions of Theorems 4.1 and 4.2 is said to be stable-correct. Proposition 2.4 remains valid in the multidimensional case as well, i.e. for polynomials satisfying the conditions of Theorems 4.1 and 4.2 the requirements that the polynomials should be solved with respect to the highest power of τ and that they should be exponentially correct are equivalent. With regard to Theorem 4.1, there holds an analog Theorem 2.4.

Theorem. *For polynomial* (1) *the following conditions are equivalent.*

(I) *The polynomial P is stable-correct (i.e. P is an exponentially correct polynomial and there are $c > 0$ and γ_0 such that* (2) *holds).*

(II) *The pseudo-polynomials $P_\omega(\rho, \tau)$ are $N(P_\omega)$-stable correct (Theorem* 2.4) *$\forall \omega \in S^{n-1}$ and condition* (5) *is fulfilled.*

(III) *There is an integer b and a parabolic polynomial $P_1(\xi, \tau)$ such that $P(\xi, \tau) - \tau^b P_1(\xi, \tau) \in \mathcal{L}_{\Delta(P)}$.*

§5. Cauchy's problem for stable-correct differential operators with variable coefficients

5.1. We now extend the results obtained in §3 to the case of differential operators on functions of $n+1$ variables, $n > 1$. Here attention is focused on the facts where the multidimensional situation differs from the one-dimensional case. As to the assertions that do not differ from those in §3 in either the statement or the proof, we shall not dwell on them.

The spaces $H_{[\gamma]}^{(s)}(\mathbb{R}^n)$, $n > 1$, are defined in a natural way, and an analog of the norm (3.3) is introduced for them:

$$\|u\|_{(s),\gamma} = \left(\int_{\mathbb{R}^{n+1}} (1 + |\tau|^2 + |\xi|^2)^s |\widehat{u}(\xi,\tau)|^2 \, d\xi \, d\sigma \right)^{1/2}, \tag{1}$$

where $\sigma = \operatorname{Re}\tau$ and $x = (x_1, \dots, x_n)$ and $\xi = (\xi_1, \dots, \xi_n)$ are the primal and dual variables relative to the form $\langle x, \xi\rangle = x_1\xi_1 + \dots + x_n\xi_n$. With a polynomial $P(\xi,\tau)$ or, more precisely, with the polygon $\Delta(P)$ (or $\delta(P)$) the Banach norm

$$\|u\|_{N(P),(s),\gamma} = \left(\sum_{(\alpha,\beta)\in\Delta(P)} \|D_x^\alpha D_t^\beta u\|_{(s),\gamma}^2 \right)^{1/2} \tag{2}$$

is associated, and let $H_{[\gamma]}^{\Delta(P),(s)}$ denote the space of functions with finite norm (2). As in §3, we denote, respectively, by $H_{[\gamma]+}^{(s)}$ and $(H_{[\gamma]}^{\Delta(P),(s)})_+$ the spaces of functions and distributions vanishing for $t < 0$. In the case under consideration Theorems 1 and 2 in Section 3.3 remain true.

5.2. We now consider polynomial symbols (in ξ and τ) with coefficients depending on the variables x and t. These symbols have the form (3.1) where $\alpha = (\alpha_1, \dots, \alpha_n)$ are multiindices, $D_x^\alpha = D_1^{\alpha_1} \dots D_n^{\alpha_n}$, $D_j = -i\,\partial/\partial x_j$, $j = 1, \dots, n$, and $D_t = -i\,\partial/\partial t$. Symbols (3.1) are said to be stable-correct (parabolic) if conditions (i) and (ii) in Definition 3.2 are fulfilled.

We shall assume that the coefficients of the symbols (3.1) are stabilized at infinity, i.e. have the form (3.7).

Along with pseudo-polynomials in the foregoing section, one can consider pseudo-polynomial symbols, i.e. expressions of the form of

$$Q(x,t;\xi,\tau) = \sum q_{\varkappa\beta}(x,t,\omega)|\xi|^\varkappa \tau^\beta, \qquad \omega = \xi/|\xi|, \tag{3}$$

where $q_{\varkappa\beta}(x,t,\omega)$ are $\mathbb{C}^\infty$ functions of all the variables x, t, and ω and do not depend on x and t for sufficiently large $|x| + |t|$. Noting that

$$|\xi| = \sum \xi_j(\xi_j/|\xi|) = \sum \xi_j\omega_j$$

we can rewrite the pseudo-polynomial symbol (3) in the form

$$Q(x,t;\xi,\tau) = \sum b_{\alpha\beta}(x,t,\omega)\xi^\alpha\tau^\beta, \tag{3'}$$

where the functions $b_{\alpha\beta}(x,t,\omega)$ possess the same properties as the functions $q_{\varkappa\beta}$ in (3).

Let $SL_{\Delta(P)}$ and $S\mathcal{L}_{\Delta(P)}$ denote the spaces of pseudo-polynomial symbols (3′) such that the summation index $(|\alpha|,\beta)$ in them runs over $\Delta(P)$ and $\delta(P)$, respectively. The definitions of stable correctness, parabolicity, parabolicity in Petrovskiĭ's sense, etc. are readily extended to pseudo-polynomial symbols.

Proposition. *Let symbol* (3.1) *satisfy conditions* (i) *and* (ii) *in Definition* 3.2 *(in which $N(P)$ should be replaces by $\Delta(P)$). Then the following assertions hold.*

(a) *The pseudo-polynomials $P^{[j]}(x,t;\xi,\tau)$ corresponding to the non-vertical sides of $\Delta(P)$ and determined by formulas of type* (4.8) *are pseudo-polynomial symbols uniformly parabolic in Petrovskiĭ's sense, i.e. there are q_j, m_j, and $c_j > 0$ such that*

$$c_j^{-1} \leqslant \left|P^{[j]}(x,t;\xi,\tau)\right|\left(|\xi|^{q_j} + |\tau|\right)^{-m_j} \leqslant c_j. \tag{4}$$

(b) *The polynomial symbol* (3.1) *is uniformly stable-correct, i.e. there are constants $c > 0$ and γ_0 such that*

$$\Xi_{\Delta(P)}(\xi,\tau) < c|P(x,t;\xi,\tau)| \quad \text{for} \quad \xi \in \mathbb{R}^n, \quad \operatorname{Im}\tau \leqslant \gamma_0. \tag{5}$$

(c) *For the parabolic polynomial symbol* (3.1) *there are pseudo-polynomial symbols $P^{[j]}(x,t;\xi,\tau)$, $j = 1,\dots,\mu$, parabolic in Petrovskiĭ's sense, satisfying estimates of type* (4), *and such that*

$$P(x,t;\xi,\tau) - \prod_{j=1}^{\mu} P^{[j]}(x,t;\xi,\tau) = Q(x,t;\xi,\tau) \in S\mathcal{L}_{\Delta(P)}. \tag{6}$$

(d) *For the stable-correct polynomial symbol* (3.1) *there is a number $b \geqslant 0$ and a parabolic polynomial symbol $P_1(x,t;\xi,\tau)$ such that*

$$P(x,t;\xi,\tau) - \tau^b P_1(x,t;\xi,\tau) = Q(x,t;\xi,\tau) \in S\mathcal{L}_{\Delta(P)}. \tag{7}$$

Proof. (a). This assertion follows from explicit formulas of type (4.8) (cf. the proof of Theorem 4.2). Inequalities (4) make it possible in fact to repeat the argument of Theorem 1.2.1 and to obtain a uniform estimate (5). Assertions (c) and (d) are proved in the same way as their analogs for polynomials (Theorem 4.2).

Assertion (b) in the proposition allows one to repeat literally the argument of Section 3.3 for the situation under consideration and to prove analogs of Theorems 1 and 2 of Section 3.3.

5.3. We now discuss the factorization of stable-correct differential operators with variable coefficients in the case of $n+1$ variables. As was shown in §3, for $n = 1$ the factorization (3.10), (3.11) takes place for the class of *differential* operators. We now show that for $n > 1$ to the factorization (6), (7) of the symbol there corresponds factorization in the class of *pseudo-differential* operators.

We remind the reader of some well-known definitions. With each function $a(x;\omega)$ belonging to $C^\infty(\mathbb{R}^n \times S^{n-1})$ and not depending on x for large $|x|$ the classical pseudodifferential operator or, briefly, PDO (or, in a different terminology, singular integral operator) is associated:

$$a(x;D_x)f(x) = (2\pi)^{-n/2}\int \exp(i\langle x,\xi\rangle)a(x;\xi/|\xi|)\widehat{f}(\xi)\,d\xi. \tag{8}$$

The function $a(x;\omega)$ is called the symbol of operator (8).

We note some properties of operator (8) important to our further aims.

(i) For any $s \in \mathbb{R}$ the operator

$$H^{(s)}(\mathbb{R}^n) \to H^{(s)}(\mathbb{R}^n) \qquad (f(x) \mapsto a(x;D)f(x))$$

is continuous.

(ii) If $a_j(x,\omega)$, $j = 1,2$, are two symbols of the above-mentioned type, then $\forall s \in \mathbb{R}$ and $\forall k \in \{1,\dots,n\}$ the operator

$$[(a_1a_2)(x;D) - a_1(x,D)a_2(x,D)]D_k\colon H^{(s)}(\mathbb{R}^n) \to H^{(s)}(\mathbb{R}^n)$$

is bounded.

(iii) $\forall j \in \{1,\dots,n\}$ the relation

$$D_ja(x;D) = a(x;D)D_j + a_{(j)}(x;D), \qquad a_{(j)}(x,\omega) = D_j(x;u),$$

holds.

For the proofs of properties (i) and (ii) see any textbook on PDO (for instance, Taylor [1]), and property (iii) follows directly from definition (8).

With each symbol $Q(x,t;\xi,\tau) \in SL_{\Delta(P)}$ of the form of (3′) we associate the PDO

$$Q(x,t;D_x,D_t) = \sum b_{\alpha\beta}(x,t;D_x)D_x^\alpha D_t^\beta, \tag{9}$$

where $b_{\alpha\beta}(x,t,D_x)$ are operators of type (8) acting on the variables x and depending smoothly on the parameter t. Then, according to (i), $\forall s \in \mathbb{R}$ the operator

$$H_{[\gamma]}^{\Delta(P),(s)} \to H_{[\gamma]}^{(s)} \quad (u \mapsto Q(x,t;D_x,D_t)u), \quad Q \in SL_{\Delta(P)},$$

is continuous. In case $Q \in S\mathcal{L}_{\Delta(P)}$, the operator

$$H_{[\gamma]}^{\delta(P),(s)} \to H_{[\gamma]}^{(s)} \quad (u \mapsto Qu), \quad Q \in S\mathcal{L}_{\Delta(P)}, \tag{10}$$

is continuous.

Theorem. (i) *Let $P(x,t;D_x,D_t)$ be a parabolic differential operator. Then there are PDO $P^{[j]}(x,t;D_x,D_t)$ parabolic in Petrovskiĭ's sense such that*

$$T \stackrel{\text{def}}{=} P(x,t;D_x,D_t) - P^{[1]}(x,t;D_x,D_t)\dots P^{[\mu]}(x,t;D_x,D_t) \tag{11}$$

is continuous operator from $H^{\delta(P),(s)}_{[\gamma]}$ into $H^{(s)}_{[\gamma]}$ $\forall s \in \mathbb{R}$ and $\forall \gamma < \gamma_0$.

(ii) *For each stable-correct differential operator $P(x,t;D_x,D_t)$ there is an integer $b \geqslant 0$ and a parabolic differential operator $P_1(x,t;D_x,D_t)$ such that the symbol of the operator $P(x,t;D_x,D_t) - P_1(x,t;D_x,D_t)D_t^b$ belongs to $S\mathcal{L}_{\Delta(P)}$.*

Proof. (ii) follows trivially from Proposition 5.2 (d), and therefore we consider the proof of (i).

Setting $\widehat{P}(x,t;\xi,\tau) = \prod P^{[j]}(x,t;\xi,\tau)$ we rewrite operator (11) in the form

$$T = \left[P(x,t;D_x,D_t) - \widehat{P}(x,t;D_x,D_t)\right] + \left[\widehat{P}(x,t;D_x,D_t) - P^{[1]}(x,t;D_x,D_t)\dots P^{[\mu]}(x,t;D_x,D_t)\right] = Q(x,t;D_x,D_t) + T_1.$$

According to Proposition 5.2 (c), the symbol of the operator $Q(x,t;D_x,D_t)$ belongs to $S\mathcal{L}_{\Delta(P)}$, whence follows the continuity of operator (10).

We now analyze the structure of T_1. To simplify the notation, we assume that $\mu = 2$. We write

$$P^{[i]}(t,x;\xi,\tau) = \sum a_{i\alpha\beta}(t,z,\omega)\xi^\alpha\tau^\beta.$$

The symbol $P^{[j]}$ is solved with respect to the highest power τ^{m_i}. Without loss of generality, we can assume that the monomial τ^{m_i} enters $P^{[j]}$ with a coefficient identically equal to 1. In view of (iii), the operator T_1 can be rewritten

$$T_1 = \sum\{(a_{1\alpha\beta}a_{2\alpha'\beta'})(x,t,D_x) - a_{1\alpha\beta}(x,t,D_x)a_{2\alpha'\beta'}(x,t,D_x)\} \times D_x^{\alpha+\alpha'}D_t^{\beta+\beta'} + \sum B_{\alpha''\beta''}D_x^{\alpha''}D_t^{\beta''} = T_1' + T_1'',$$

the summation in T_1'' extending over (α'',β'') such that $(|\alpha''|,\beta'') \in \delta(P)$ and the operators $B_{\alpha''\beta''}$ being products of classical PDO with respect to the variables x, whose symbols are the functions $a_{i\alpha\beta}(x,t;\xi)$ and their derivatives with respect to x and t. In view of (i), to T_1'' there corresponds a continuous operator from $H^{\delta(P),(s)}_{[\gamma]}$ into $H^{(s)}_{[\gamma]}$.

We now consider a typical term in the sum of operators T_1'. Since the highest derivatives with respect to t are involved in the operators $P^{[i]}$ with constant coefficients, the sum contains only those terms for which $|\alpha| + |\alpha'| > 0$. Therefore for some k we have

$$D_x^{\alpha+\alpha'}D_t^{\beta+\beta'} = D_{x_k}D_x^{\alpha''}D_t^{\beta+\beta'}, \quad (|\alpha''|,\beta+\beta') \in \delta(P).$$

According to (ii), the composition of the operator in curly brackets and the operator D_{x_k} is a bounded operator in $H^{(s)}_{[\gamma]}$ $\forall s \in \mathbb{R}$. It follows that the operator $T_1' \colon H^{\delta(P),(s)}_{[\gamma]} \to H^{(s)}_{[\gamma]}$ is continuous. The theorem is proved.

5.4. We now pass to the problem of solvability of the (homogeneous) Cauchy problem. As in §3, we consider the operators (cf. (3.19) and (3.20))

$$H_{[\gamma]}^{\Delta(P),(s)} \to H_{[\gamma]}^{(s)} \quad (u(x,t) \mapsto P(x,t;D_x,D_t)u(x,t)), \tag{12}$$

and

$$(H_{[\gamma]}^{\Delta(P),(s)})_+ \to H_{[\gamma]+}^{(s)} \quad (u(x,t) \mapsto P(x,t;D_x,D_t)u). \tag{12$_+$}$$

In the case under consideration Theorem 3.4 remains true.

Theorem. *For a differential operator whose coefficients satisfy conditions of type* (3.7) *the following conditions are equivalent.*

(I) *The operator is stable-correct.*

(II) $\forall s \in \mathbb{R}$ *there is* $\gamma_0 = \gamma_0(s)$ *such that for* $\gamma \leqslant \gamma_0$ *the mapping* (12) *is bijective.*

(II$_+$) $\forall s \in \mathbb{R}$ *there is* $\gamma_0 = \gamma_0(s)$ *such that for* $\gamma \leqslant \gamma_0$ *the operator* (12$_+$) *is bijective.*

The proof of this theorem differs from the proof of its one-dimensional analog (Theorem 3.4) in the part where the surjectivity of operators (12) and (12$_+$) is proved. Indeed, owing to Theorem 5.3, this assertion can also be proved following the scheme of Theorem 3.4. Recall that in this theorem we resorted to the assertion that the differential operator $P^{[j]}(x,t;D_x,D_t)$ parabolic in Petrovskiĭ's sense generates the bijective mappings

$$H_{[\gamma]}^{\Delta(P^{[j]}),(s)} \to H_{[\gamma]}^{(s)}, \qquad (H_{[\gamma]}^{\Delta(P^{[j]}),(s)})_+ \to H_{[\gamma]+}^{(s)}.$$

This result can be extended comparatively simply to differential operators whose coefficients are singular integral operators with respect to x. We thus arrive at the proof of the theorem in question. We now present another and more natural approach to the proof of the bijectivity of (12) and (12$_+$) that does not use the factorization of the corresponding operator and is based on the apparatus of PDO.

5.5.

Definition. A symbol $P(x,t;\xi,\tau)$ with smooth stabilized coefficients (satisfying (3.7)) is called an *exponentially correct symbol of constant strength* if the following conditions are fulfilled:

(i) $\forall (x^0,t^0) \in \mathbb{R}^{n+1}$ the polynomial $P(\xi,\tau) = P(x^0,t^0;\xi,\tau)$ is exponentially correct (see Section 1.1);

(ii) there are constants $A > 0$ and γ_0 such that $\forall (x',t'), (x'',t'') \in \mathbb{R}^{n+1}$ the inequality

$$|P(x',t';\xi,\tau)P^{-1}(x'',t'';\xi,\tau)| < A, \qquad (\xi, \operatorname{Re}\tau) \in \mathbb{R}^{n+1}, \qquad \operatorname{Im}\tau \leqslant \gamma_0$$

holds.

Obviously, by virtue of Proposition 5.2 (b), stable-correct symbols satisfy the conditions of this definition.

Fix an arbitrary point $(x^0, t^0) \in \mathbb{R}^{n+1}$ and set $P(\xi, \tau) = P(x^0, t^0; \xi, \tau)$. We define the space

$$H_{[\gamma]}^{P,(s)} = \{u \in H_{[\gamma]}^{(s)}, P(D_x, D_t)u \in H_{[\gamma]}^{(s)}\} \tag{13}$$

and endow it with the natural norm

$$\|u\|_{(s),\gamma} + \|P(D_x, D_t)u\|_{(s),\gamma}.$$

Replacing $H_{[\gamma]}^{(s)}$ in (13) by $H_{[\gamma]+}^{(s)}$ we define the space $(H_{[\gamma]}^{P,(s)})_+$.

According to (ii), the space (13) does not depend on the choice of the point $(x^0, , t^0)$. Its replacement by another point (x', t') results in an equivalent norm but the space continues to have the same set of elements. Further, in view of Proposition 5.2 (b), in case of stable-correct symbols the space (13) coincides with the space of functions (distributions) having a finite norm (2).

Theorem. *Let $P(x, t; \xi, \tau)$ be exponentially correct and have a constant strength. Then $\forall s \in \mathbb{R}$ there is $\gamma_0 = \gamma_0(s)$ such that* (12) *and* (12_+) *are bijective mappings.*

We do not present the detailed proof and confine ourselves giving the main points of the scheme. For detailed proofs of this and more general theorems see, for instance, the monograph by Volevich and Gindikin [1, Chapter 5]. The proof of isomorphism (12_+) is based on the calculus of PDO with holomorphic symbols.

Denote by $\mathfrak{S}^m$ the class of functions $b(y; \xi, \tau)$, $y = (x, t) \in \mathbb{R}^{n+1}$, belonging to C^∞ as functions of all their variables, holomorphic with respect to τ for $\operatorname{Im} \tau \leqslant \gamma_0$, not depending on y for sufficiently large $|y|$, and satisfying inequalities of the form of

$$|D_y^\alpha b(y; \xi, \tau)| \leqslant B_\alpha(\gamma)(1 + |\xi| + |\tau|)^m \quad \text{for} \quad \operatorname{Im} \tau \leqslant \gamma \leqslant \gamma_0. \tag{14}$$

Let $\mathfrak{S}$ be the union of all $\mathfrak{S}^m$, $-\infty < m < \infty$.

With each function $b(y; \xi, \tau) \in \mathfrak{S}$ the PDO

$$b(y, D) = (2\pi)^{-(n+1)/2} \int_{\substack{\mathbb{R}^{n+1} \\ \operatorname{Im} \tau = \gamma}} \exp(i\langle x, \xi\rangle + it\tau) b(y; \xi, \tau) \widehat{f}(\xi, \tau)\, d\xi\, d\tau, \tag{15}$$

can be associated.

This operator possesses the following properties:

1) the value of the integral on the right-hand side of (15) does not depend on the choice of $\gamma < \gamma_0$;

2) if the support of $f \in H_{[\gamma]+}^{(s)}$ lies in the half-space $t \geqslant T > 0$, then the support of $b(y; D)f$ also lies in that same half-space;

3) the inequality

$$\|b(y; D)f\|_{(s-m),\gamma} \leqslant \varkappa B(\gamma)\|f\|_{(s),\gamma} \tag{16}$$

holds, where $\varkappa$ does not depend on the symbol b, and

$$B(\gamma) = \max_{|\alpha| \leqslant n+2+|s-m|} B_\alpha(\gamma). \tag{17}$$

We have to prove that the differential equation

$$P(y; D_x, D_y)u(y) = f(y) \in H^{(s)}_{[\gamma]+} \tag{18}$$

has at least one solution $u \in (H^{P,(s)}_{[\gamma]})_+$. We shall seek the solution in the form

$$u = R(y; D_x, D_t)g, \tag{19}$$

where $g \in H^{(s)}_{[\gamma]+}$ is a function to be determined and $R(y; \xi, \tau) = P^{-1}(y; \xi, \tau)$.

Lemma 1. *Let $P(y; \xi, \tau)$ be exponentially correct and have a constant strength. Then $P^{-1}(y; \xi, \tau) \in \mathfrak{S}^m$ for $m = 0$.*

Assuming that the lemma has been proved, we come back to Equation (18). Since $R \in \mathfrak{S}$, expression (19) makes sense. Apply the differential operator $P(y; D_x, D_t)$ to the right-hand side of (19) and insert it under the sign of the integral determining the PDO R. On differentiating

$$P[R(y; \xi, \tau) \exp(i\langle x, \xi\rangle + it\tau)]$$

according to the Leibniz-Hörmander rule we derive for g the pseudodifferential equation

$$g + H(y; D_x, D_t)g = f, \tag{20}$$

where

$$H(y; \xi, \tau) = \sum_{\alpha > 0} \frac{1}{\alpha!} P^{(\alpha)}(y; \xi, \tau) D^\alpha_y P^{-1}(y; \xi, \tau).$$

Lemma 2. *If $P(y; \xi, \tau)$ is an exponentially correct symbol of constant strength, then $H(y; \xi, \tau) \in \mathfrak{S}^m$ for $m = 0$ and, moreover,*

$$|D^\beta_y H(y; \xi, \tau)| < \varepsilon_\beta(\operatorname{Im}\tau), \quad \varepsilon_\beta(\operatorname{Im}\tau) \to 0, \quad \operatorname{Im}\tau \to -\infty. \tag{21}$$

Hence, if $-\gamma_0$ is sufficiently large, then, according to (16), (17), and (21), for $\gamma \leqslant \gamma_0$ the norm of the operator

$$H(y; D_x, D_t) \colon H^{(s)}_{[\gamma]+} \to H^{(s)}_{[\gamma]+}$$

is strictly less than 1, whence it follows that Equation (20) possesses a solution $g \in H^{(s)}_{[\gamma]+}$ determining by Neumann's series

$$g = \sum_{j=0}^{\infty} (-H(y; D_x, D_t))^j f.$$

It now remains to show that $u \in (H^{P,(s)}_{[\gamma]})_+$. To this end we apply the differential operator $P(D_x, D_t) = P(x^0, t^0; \xi, \tau)$ to the right-hand side of (19). This results in

$$P(D_x, D_t)u = H_0(x, t; D_x, D_t)g,$$

where

$$H_0(x, t; \xi, \tau) = \sum_{\alpha \geqslant 0} \frac{1}{\alpha!} P^{(\alpha)}(y^0, \xi, \tau) D_y^\alpha P^{-1}(y; \xi, \tau). \tag{22}$$

Lemma 3. *If $P(y; \xi, \tau)$ is an exponentially correct symbol of constant strength, then $H_0(x, t; \xi, \tau) \in \mathfrak{S}^m$ for $m = 0$.*

Lemma 3 and the properties of PDO with symbol belonging to $\mathfrak{S}$ imply that the operator H_0 transforms $H^{(s)}_{[\gamma]+}$ into itself, i.e. $Pu \in H^{(s)}_{[\gamma]+}$, and, consequently, $u \in (H^{P,(s)}_{[\gamma]})_+$.

It now remains to outline the proofs of Lemmas 1, 2 and 3.

1) We first of all show that there are $c > 0$ and γ_0 such that

$$|P(y; \xi, \tau)| > c \quad \text{for} \quad \operatorname{Im} \tau \leqslant \gamma_0. \tag{23}$$

Indeed, in view of the condition of constant strength, it suffices to verify inequality (23) at a fixed point $y = y_0$, i.e. for the polynomial $P(\xi, \tau) = P(y^0; \xi, \tau)$. The condition of exponential correctness (Lemma 1.1) implies that the polynomial $P(\xi, \tau)$ is solved with respect to the highest power of τ, i.e. $P(\xi, \tau) = a\tau^m + O(\tau^{m-1})$. Therefore $\partial_\tau^m P(\xi, \tau) = am!$. Since $(\partial_\tau^m P)P^{-1} \to 0$ as $\operatorname{Im} \tau \to -\infty$ (uniformly relative to the variables ξ and $\operatorname{Re} \tau$), we arrive at inequality (23).

2) Consider the family of polynomials $P(y; \xi, \tau)$ depending on the parameter $y \in \mathbb{R}^{n+1}$. Since, by virtue of the condition of constant strength, the degrees of all polynomials $P(y; \xi, \tau)$ are bounded by a unified constant, the whole family belongs to a finite-dimensional space of polynomials. Therefore a finite number of points $y^1, \ldots, y^J$ can be chosen so that each of the polynomials $P(y; \xi, \tau)$ is representable as a linear combination

$$P(y; \xi, \tau) = \sum_{j=1}^{J} c_j(y) P(y^j, \xi, \tau). \tag{24}$$

Differentiating (24) with respect to y and using the condition of constant strength be obtain the inequality

$$|D_y^\beta P(y; \xi, \tau)| < c_\beta |P(\xi, \tau)|. \tag{25}$$

3) We now prove Lemma 3. The symbol $P(y; \xi, \tau)$ is obviously a smooth function of all its arguments and is holomorphic with respect to τ for $\operatorname{Im} \tau \leqslant \gamma_0$, where γ_0 is taken from inequality (23). According to (23), for $\alpha = 0$ the symbol $b = R$

satisfies estimate (14) with $m = 0$. Let us show that a similar estimate also holds for the derivatives of R with respect to y. By Leibniz' formula, $D_y^\alpha P^{-1}$ is a linear combination of expressions of the form of

$$(D_y^{\alpha^1} P/P)\dots(D_y^{\alpha^k} P/P)P^{-1}.$$

By virtue of (23) and (25), these expressions do not exceed a constant.

4) We now prove Lemma 2. To this end we rewrite the symbol H in the form

$$H(y;\xi,\tau) = \sum_{\alpha>0} \frac{P^{(\alpha)}(y;\xi,\tau)}{P(\xi,\tau)} h_\alpha(y;\xi,\tau),$$

where

$$h_\alpha(y;\xi,\tau) = P(\xi,\tau)D_y^\alpha P^{-1}(y;\xi,\tau)/\alpha!.$$

What has been said in 3) implies that the functions h_α and their derivatives with respect to y do not exceed a constant. Further, differentiating H according to Leibniz' formula we find

$$D_y^\beta H(y;\xi,\tau) = \sum_{\alpha>0}\sum_{\delta} \binom{\beta}{\delta} D_y^\delta \frac{P^{(\alpha)}(y;\xi,\tau)}{P(\xi,\tau)} D_y^{\beta-\delta} h_\alpha(y;\xi,\tau).$$

Using (24) we derive

$$D_y^\delta P^{(\alpha)}(y;\xi,\tau)P^{-1}(\xi,\tau) = \sum D_y^\delta c_j(y) P^{(\alpha)}(y^j;\xi,\tau)P^{-1}(\xi,\tau).$$

Thus, finally, with regard to the estimates for the derivatives of h_α and the condition of constant strength, we have

$$|D^\beta H(y;\xi,\tau)| < \text{const} \sum |P^{(\alpha)}(y^j;\xi,\tau)P^{-1}(y^j;\xi,\tau)|.$$

Using the definition of exponentially correct polynomials we prove the assertion of Lemma 2.

Lemma 3 is verified in a similar way.

Hence, we have presented the proof of the continuity and surjectivity of the operator

$$H_{[\gamma]}^{P,(s)} \to H_{[\gamma]}^{(s)} \quad (u \mapsto P(y;D_x,D_t)u). \tag{26}$$

The proof of the injectivity is based on the same ideas. Applying the PDO with symbol $R(y;\xi,\tau) = P^{-1}(y;\xi,\tau)$ to relation (18) on the left we rewrite it as

$$u + H_1 u = R(y;D_x,D_t)f,$$

whence

$$\|u\|_{P,(s),\gamma} \leqslant \|Rf\|_{P,(s),\gamma} + \|H_1 u\|_{P,(s),\gamma}. \tag{27}$$

We have already proved in fact that

$$\|Rf\|_{P,(s),\gamma} \leqslant \text{const}\, \|f\|_{(s),\gamma} = \text{const}\, \|P(y; D_x, D_t)u\|_{(s),\gamma}. \tag{28}$$

Combining the estimates for exponentially correct symbols of constant strength (Lemmas 1, 2, and 3) and the standard PDO technique (for more detail see Volevich and Gindikin [1], Chapter 5) we can prove the inequality

$$\begin{gathered} \|H_1 u\|_{P,(s),\gamma} \leqslant \varepsilon_s(\gamma)\|u\|_{P,(s),\gamma} \quad \forall u \in H^{(\infty)}_{[\gamma]} \\ \gamma \leqslant \gamma_0, \quad \varepsilon_1(\gamma) \to 0, \quad \gamma \to -\infty. \end{gathered} \tag{29}$$

Substituting (28) and (29) into (27) we obtain

$$\|u\|_{P,(s),\gamma} \leqslant c\|P(y; D_x, D_t)u\|_{(s),\gamma} + \varepsilon_s(\gamma)\|u\|_{P,(s),\gamma}.$$

Now, taking γ_1 such that $\varepsilon_s(\gamma) < 1/2$ for $\gamma \leqslant \gamma_1$ we arrive at the inequality

$$\|u\|_{P,(s),\gamma} \leqslant 2c\|P(y; D_x, D_t)u\|_{(s),\gamma} \quad \forall u \in H^{(\infty)}_{[\gamma]}, \quad \gamma \leqslant \gamma_1, \tag{30}$$

which implies the injectivity of the operator (26).

5.6. For our further aims it is advisable to note some a priori estimates contained in the above argument.

Theorem. *Let $P(y; \xi, \tau)$, $y = (x, t)$, be an exponentially correct symbol of constant strength, whose coefficients satisfy conditions of type* (3.7). *Then*

(i) *$\forall \alpha > 0$ and $\forall s \in \mathbb{R}$ there is γ_0 and a constant $\varepsilon_{s\alpha}(\gamma) \to 0$, $\gamma \to -\infty$, such that the inequality*

$$\|P^{(\alpha)}(y; D_x, D_t)u\|_{(s),\gamma} \leqslant \varepsilon_{s\alpha}(\gamma)\|P(y; D_x, D_t)u\|_{(s),\gamma}, \qquad \gamma \leqslant \gamma_0, \quad \forall u \in H^{(\infty)}_{[\gamma]} \tag{31}$$

is fulfilled;

(ii) *$\forall \alpha > 0$ and $\forall r \in \mathbb{R}$ there is ρ_0 and a constant $\varepsilon^*_{r\alpha}(\rho) \to 0$, $\rho \to \infty$, such that the inequality*

$$\|\overline{P}^{(\alpha)}(y; D_x, D_t)v\|_{(r),\rho} \leqslant \varepsilon^*_{r\alpha}(\rho)\|P^*(y; D_x, D_t)v\|_{(r),\rho}, \qquad \rho \geqslant \rho_0, \quad \forall v \in H^{(\infty)}_{[\rho]} \tag{32}$$

holds.

Proof. (i) By virtue of (30), we must first estimate the left-hand side of (31) with a small constant $\varepsilon_{s\alpha}(\gamma)$ by means of the left-hand side of (30). Using (24) we obtain

$$\begin{aligned} \|P^{(\alpha)}(y; D_x, D_t)u\|_{(s),\gamma} &\leqslant \sum_{j=1}^{J} \|c_j(y)P^{(\alpha)}(y^j; D_x, D_t)u\|_{(s),\gamma} \\ &\leqslant \text{const} \sum \|P^{(\alpha)}(y^j; D_x, D_t)u\|_{(s),\gamma}. \end{aligned} \tag{33}$$

Applying Fourier's transformation and the definition of exponentially correct polynomials we find

$$\|P^{(\alpha)}(y^j; D_x, D_t)u\|_{(s),\gamma} \leqslant \varepsilon^j(\gamma)\|P(y^j; D_x, D_t)u\|_{(s),\gamma},$$

where $\varepsilon^j(\gamma) \to 0$, $\gamma \to -\infty$. Substituting this into (33) and using the condition of constant strength we arrive at (31).

(ii) We first of all note that if the polynomial $P(\xi, \tau)$ is exponentially correct, then the polynomial $\overline{P}(\xi, -\tau)$ possesses the same property. Recalling that $P^*(y; \xi, \tau) = \overline{P}(y; \xi, \tau) + \Sigma\overline{P}^{(\alpha)}_{(\alpha)}(y; \xi, \tau)/\alpha!$ we conclude that $P^*(y; \xi, -\tau)$ is an exponentially correct symbol of constant strength. By what has been proved in (i), the operator $P^*(y; D_y, D_t)$ satisfies an inequality of type (31). Making change of variable $t \to -t$ in this inequality we obtain (32).

5.7. Estimates (31) and (32) can be extended to spaces of functions of finite algebraic growth or decrease with respect to the variables $y = (x, t)$. Denote by $H^{(s)}_{(l)}$ the space of functions $u \in S'$ such that $(1 + |y|^2)^{1/2}u \in H^{(s)}$. Let $\|\quad\|^{(s)}$ be the norm in $H^{(s)}$. We define the $H^{(s)}_{(l)}$ norm

$$\|u\|^{(s)}_{(l)} = \|(1 + |y|^2)^{l/2}u\|^{(s)} = \|(1 + |D|^2)^{s/2}(1 + |y|^2)^{l/2}u\|, \tag{34}$$

which is equivalent to

$$'\|u\|^{(s)}_{(l)} = \|(1 + |y|^2)^{l/2}(1 + |D|^2)^{s/2}u\|. \tag{34'}$$

We can also consider the spaces $H^{(s)}_{(l),\gamma+}$ and define the norms $\|\quad\|^{(s)}_{(l),\gamma}$ and $'\|\quad\|^{(s)}_{(l),\gamma}$ in them. Here we do not discuss this in detail referring the reader to Chapter 5 in the monograph by Volevich and Gindikin [1].

Theorem. *Under the conditions of Theorem 5.6 $\forall s, l \in \mathbb{R}$ there is $\gamma_0(s, l)$ such that the inequality*

$$\sum_{\alpha>0} \|P^{(\alpha)}(y, D_y)u\|^{(s)}_{(l),\gamma} \leqslant \varepsilon_{sl}(\gamma)\|P(y, D_y)u\|^{(s)}_{(l),\gamma} \tag{35}$$

holds and an analogous inequality is fulfilled for P^*v in the norms $\|\quad\|^{(-s)}_{(-l),-\gamma}$.

Proof. Replacing u in (31) by $(1 + |y|^2)^{1/2}u$ and performing the summation of these inequalities over $\alpha > 0$ we obtain

$$\sum_{\alpha>0} \|P^{(\alpha)}(y, D_y)((1 + |y|^2)^{l/2}u)\|^{(s)}_{\gamma} \leqslant \varepsilon_s(\gamma)\|P(y, D_y)(1 + |y|^2)^{l/2}u\|^{(s)}_{\gamma}. \tag{36}$$

By Leibniz's formula, we have

$$P[(1+|y|^2)^{l/2}u] = (1+|y|^2)^{l/2}Pu + \sum_{\beta>0}(D^\beta(1+|y|^2)^{l/2}/\beta!)P^{(\beta)}u.$$

Noting that the operator of multiplication by the function

$$(1+|y|^2)^{-l/2}D^\beta(1+|y|^2)^{l/2}/\beta!$$

is bounded operator in $H_\gamma^{(s)}$ (see Volevich and Gindikin [1]) we conclude that

$$\|P((1+|y|^2)^{l/2}u)\|_\gamma^{(s)} \leqslant \|Pu\|_{(l),\gamma}^{(s)} + K_1\sum_{\beta>0}\|P^{(\beta)}u\|_{(l),\gamma}^{(s)}. \tag{37}$$

If we show that

$$\sum_{\beta>0}\|P^{(\beta)}u\|_{(l),\gamma}^{(s)} \leqslant K_2\sum_{\delta>0}\|P^{(\delta)}(1+|y|^2)^{l/2}u\|_\gamma^{(s)}, \tag{38}$$

where K_2 does not depend on γ, then substituting (37) and (38) into (36) we derive (35). To verify (38) it suffices to note that

$$\begin{aligned}(1+|y|^2)^{l/2}P^{(\beta)}u &= (1+|y|^2)^{l/2}P^{(\beta)}\left[(1+|y|^2)^{-l/2}(1+|y|^2)^{l/2}u\right]\\ &= \sum\left[(1+|y|^2)^{l/2}D^\delta(1+|y|^2)^{-l/2}/\delta!\right]P^{(\beta+\delta)}\left[(1+|y|^2)^{l/2}u\right].\end{aligned}$$

Remark. In the book by Volevich and Gindikin [1] estimates are also presented for exponentially correct operators of constant strength in spaces of functions of exponential decrease or growth.

CHAPTER III

DOMINANTLY CORRECT OPERATORS

Introduction

In the foregoing chapter we studied stable-correct differential operators which are a natural generalization of operators parabolic in Petrovskiĭ's sense. Their symbols are polynomials $P(\xi,\tau)$, $\xi\in\mathbb{R}^n$, $\tau\in\mathbb{C}$, solved with respect to the highest power of τ and admitting of an adequate estimate from below by means of the sum of the moduli of all its monomials:

$$\sum_{(\varkappa,\beta)\in\Delta(P)} |\xi|^{\varkappa}|\tau|^{\beta} \leqslant c|P(\xi,\tau)|, \qquad \operatorname{Im}\tau \leqslant \gamma_0, \quad (\xi, \operatorname{Re}\tau)\in\mathbb{R}^{n+1}. \tag{1}$$

By virtue of this inequality, all derivatives involved in the corresponding differential operator (even in the case of variable coefficients) are estimated by means of that operator in the norms of the spaces $H^{(s)}_{[\gamma]}$.

Inequality (1) implies a stronger estimate for minor monomials:

$$\sum_{(\varkappa,\beta)\in\delta(P)} |\xi|^{\varkappa}|\tau|^{\beta} \leqslant \varepsilon(\operatorname{Im}\tau)|P(\xi,\tau)|, \qquad \varepsilon(\operatorname{Im}\tau)\to 0, \quad \operatorname{Im}\tau\to-\infty, \tag{2}$$

i.e. the minor monomials of the polynomial P are estimated by means of the polynomial itself with an arbitrary large constant when the imaginary part of $-\tau$ is sufficiently large. In the language of differential operators, inequality (2) means that in the spaces $H^{(s)}_{[\gamma]}$ the corresponding differential operator with an arbitrarily large constant majorizes the $H^{(s)}_{[\gamma]}$ norms of all lower derivatives provided that $-\gamma$ is sufficiently large.

As is known from the classical theory, along with parabolic operators, strictly hyperbolic operators (their definition is given in §1) possess this property. It turns out that the property is retained for the composition of hyperbolic and parabolic operators. Therefore it seems natural to try to unify all these operators as a class of operators with a dominant (relative to Newton's polygon) principal part.

Definition. A polynomial $P(\xi,\tau)$ solved with respect to the highest power of τ is said to be dominantly correct if it satisfies inequality (2).

Remark. In this chapter we use the same definition of minor monomials of a polynomial as in the foregoing chapter. Since dominantly correct polynomials are solved with respect to the highest power of τ, the polygon $\delta(P)$ coincides with $\delta_0(P)$, and (2) implies that dominantly correct polynomials are exponentially correct.

The aim of the present chapter is to give an algebraic description of dominantly correct polynomials and to prove the correctness of Cauchy's problem for dominantly correct operators with variable coefficients.

The presentation of the material is organized in the following way. §1 is of auxiliary character and compiles some known facts relating to strictly hyperbolic polynomials and differential operators that are necessary for the further presentation. §2 provides the description of dominantly correct polynomials for the case of two variables. The main result consist in that, to within minor monomials, such a polynomial is a product of a strictly hyperbolic polynomial by a stable-correct polynomial. In §3 we consider dominantly correct differential operator with variable coefficients. An a priori estimate corresponding to (2) is obtained for them and an existence and uniqueness theorem is proved for the solution of Cauchy's problem in the spaces $H^{(s)}_{[\gamma]}$. §4 is devoted to the extension of the results of §§3 and 4 to the case $n > 1$.

It should be stressed that, in contrast to stable-correct operators, dominantly correct differential operators with variable coefficients are not operators of constant strength. Therefore the derivation of a priori estimates and the proof of the solvability of Cauchy's problem are substantially more complex than the derivation of the analogous assertions in the foregoing chapter. In Chapter 7 we shall return to the analytical problems presented in this chapter.

§1. Strictly hyperbolic operators

As was already mentioned in the Introduction, stable-correct polynomials are dominantly correct. We now consider another class of polynomials for which inequality (2) is fulfilled.

1.1. A homogeneous polynomial $H_0(\xi, \tau)$, $\xi \in \mathbb{R}^n$, $\tau \in \mathbb{C}$, is said to be *strictly hyperbolic* if

(i) the polynomial H_0 is solved with respect to the highest power of τ, i.e.

$$H_0(\xi, \tau) = \tau^m + \sum_{j=1}^{m} \sum_{|\alpha|=m-j} a_{\alpha_j} \xi^\alpha \tau^{m-j}; \tag{1}$$

(ii) the roots τ_{0j}, $j = 1, \ldots, m$, of polynomial (1) are real;

(iii) for $|\xi| \neq 0$ all the roots $\tau_{0j}(\xi)$ are distinct.

As was already mentioned in Section 2.1.2, homogeneous polynomials satisfying only conditions (i) and (ii) are said to be hyperbolic.

Proposition. *For a homogeneous polynomial H_0 of the mth degree the conditions below are equivalent.*

(I) *The polynomial H_0 is strictly hyperbolic.*

(II) *There is $c > 0$ such that*

$$|\operatorname{Im}\tau|\big(|\tau| + |\xi|\big)^{m-1} \leqslant c|H_0(\xi, \tau)|, \qquad \operatorname{Im}\tau \leqslant 0, \quad (\xi, \operatorname{Re}\tau) \in \mathbb{R}^{n+1}. \tag{2}$$

Proof. (II)$\Longrightarrow$(I). Let us show that (2) implies conditions (i) to (iii). Putting $\xi = 0$ in (2) we find

$$c^{-1}|\operatorname{Im}\tau|\,|\tau|^{m-1} \leqslant |H_0(0,\tau)|,$$

whence follows (i). Further, according to (2), $H_0(\xi,\tau) \neq 0$ for $\operatorname{Im}\tau < 0$. By virtue of the homogeneity, we have $H_0(\xi,-\tau) = (-1)^m H_0(-\xi,\tau)$, whence $H_0(\xi,\tau) \neq 0$ for $\operatorname{Im}\tau > 0$, i.e. the roots of the polynomial H_0 can only be real, i.e. (ii) is fulfilled.

We now prove that for $\xi \neq 0$ all roots are distinct. Assume the contrary, i.e. let there be $\xi^0 \neq 0$ such that $\tau_{01}(\xi^0) = \tau_{02}(\xi^0)$. Consider inequality (2) along the ray

$$\xi(t) = t\xi^0, \quad t > 0, \quad \tau(t) = \tau_{01}(\xi(t)) - i\gamma, \quad \gamma > 0.$$

On the ray the left-hand side of (2) is estimated from below by means of

$$\operatorname{const}\gamma(\gamma + t)^{m-1}.$$

Noting that $\tau_{0j}(\xi(t)) = t\tau_{0j}(\xi^0)$ we obtain the inequality

$$\begin{aligned} |H_0(\xi(\tau),\tau(t))| &= \prod_{j=1}^{m} |\tau_{01}(\xi(t)) - i\gamma - \tau_{0j}(\xi(t))| \\ &= \gamma^2 \prod_{j=3}^{m} |t(\tau_{01}(\xi^0) - \tau_{0j}(\xi^0)) - i\gamma| \leqslant \operatorname{const}\gamma^2(\gamma+t)^{m-2} \end{aligned}$$

for the right-hand side. For $t \to \infty$ we arrive at a contradiction.

(I)$\Longrightarrow$(II). The condition that the roots are not multiple means that there is $\varepsilon > 0$ and a covering $\{\Omega_l\}$ of the unit sphere $\{(\sigma,\xi) \in \mathbb{R}^{n+1}, \sigma^2 + |\xi|^2 = 1\}$ such that in every neighborhood Ω_l the inequality $|\sigma - \tau_{0j}(\xi)| < \varepsilon$ can hold for at most one j.

To the covering $\{\Omega_l\}$ the covering of $\mathbb{R}^{n+1} \setminus \{0\}$ by the cones

$$V_l = \{(\sigma,\xi) \in \mathbb{R}^{n+1}, (\sigma^2 + |\xi|^2)^{-1/2}(\sigma,\xi) \in \Omega_l\}$$

corresponds. Since the polynomial H_0 is solved with respect to τ^m, inequality (2) is obvious for $\operatorname{Re}\tau = 0$, $\xi = 0$, and it suffices to establish (2) in each cone V_l. By what has been said and the homogeneity of the roots (i.e. $\tau_{0j}(t\xi) = t\tau_{0j}(\xi)$), for a given l there can be only a single j for which $|\sigma - \tau_{0j}(\xi)| < \varepsilon(\sigma^2 + |\xi|^2)^{1/2}$. If such j exists, we obtain the estimate

$$|\tau - \tau_{0j}(\xi)| \geqslant |\operatorname{Im}\tau| \quad \text{for} \quad (\sigma,\xi) \in V_l \quad (\sigma = \operatorname{Re}\tau). \tag{3}$$

If $k \neq j$, then $|\sigma - \tau_{0k}(\xi)| > \varepsilon(\sigma^2 + |\xi|^2)^{1/2}$, whence

$$|\tau - \tau_{0k}(\xi)| > \varepsilon(|\operatorname{Im}\tau|^2 + |\operatorname{Re}\tau|^2 + |\xi|^2)^{1/2}. \tag{3'}$$

Multiplying (3) and (3′) we obtain (2).

1.2.

Definition. A polynomial $H(\xi,\tau)$ is said to be *strictly hyperbolic* (hyperbolic in Petrovskiĭ's sense) if its principal homogeneous part possesses this property.

Theorem. *A polynomial $H(\xi,\tau)$ of degree m is strictly hyperbolic if and only if there are $c>0$ and $\gamma_0<0$ such that*

$$|\operatorname{Im}\tau|\big(|\tau|+|\xi|\big)^{m-1}\leqslant c|H(\xi,\tau)|,\qquad \operatorname{Im}\tau\leqslant\gamma_0,\quad (\operatorname{Re}\tau,\xi)\in\mathbb{R}^{n+1}. \tag{4}$$

Corollary. *Every strictly hyperbolic polynomial is dominantly correct.*

The proof of the theorem. Let (4) hold. If (ξ,τ) in (4) is replaced by $(t\xi,t\tau)$, the two sides of (4) are divided by t^m, and the passage to the limit is performed as $t\to+\infty$, this results in (2), whence (Proposition 1.1) it follows that H is strictly hyperbolic.

Conversely, if H_0 is a strictly hyperbolic polynomial, then, with account of (2), we have

$$\begin{aligned}|H(\xi,\tau)|&>|H_0(\xi,\tau)|-|H(\xi,\tau)-H_0(\xi,\tau)|\\&\geqslant c^{-1}|\operatorname{Im}\tau|\big(|\tau|+|\xi|\big)^{m-1}-c_1(1+|\tau|+|\xi|)^{m-1}.\end{aligned}$$

If $|\operatorname{Im}\tau|$ is sufficiently large, we arrive at inequality (5).

1.3. We now consider a differential operator with variable coefficients

$$H(x,t;D_x,D_t)=\sum_{\alpha_1+\cdots+\alpha_n+\beta\leqslant m}a_{\alpha_1\ldots\alpha_n\beta}(x,t)D_1^{\alpha_1}\ldots D_n^{\alpha_n}D_t^{\beta}, \tag{5}$$

and let $H(x,t;\xi,\tau)$ be its symbol. Operator (5) is said to be strictly hyperbolic, if the principal homogeneous part $H_0(x,t;\xi,\tau)$ of its symbol possesses the following properties:

(i) the symbol H_0 is solved with respect to τ^m; subsequently we shall assume that the coefficient in τ^m is identically equal to 1, i.e.

$$H(x,t;\xi,\tau)=\tau^m+\sum_{j=1}^{m}h_j(x,t,\xi)\tau^{m-j};$$

(ii) the roots $\tau_{0j}(x,t,\xi)$ are real and uniformly non-multiple in the sense that there is $\lambda>0$ such that

$$|\tau_{0j}(x,t,\xi)-\tau_{0k}(x,t,\xi)|>\lambda\quad\forall(x,t),\quad|\xi|=1,\quad j\neq k. \tag{6}$$

In the well-known works by Petrovskiĭ [2], Leray [1], etc. a priory estimates were obtained for strictly hyperbolic operators and the solvability of Cauchy's problem was proved. We have the following

Theorem. *Let operator* (5) *be strictly hyperbolic and let (for simplicity) its coefficients satisfy conditions of type* (2.3.7):

(i) $\forall s \in \mathbb{R}\ \exists \gamma_0(s)$ *such that the inequality*

$$|\gamma| \|u\|_{(s+m-1),\gamma} \leqslant c\|H(x,t;D_x,D_t)u\|_{(s),\gamma} \quad \forall u \in H^{(\infty)}_{[\gamma]}, \quad \gamma \leqslant \gamma_0(s), \tag{7}$$

is fulfilled.

(ii) $\forall s \in \mathbb{R}\ \exists \gamma_0(s)$ *such that* $\forall f \in H^{(s)}_{[\gamma]}$ *there is a single function* $u \in H^{(s+m-1)}_{[\gamma]}$ *satisfying the equation*

$$H(x,t;D_x,D_t)u = f$$

in the sense of distribution theory.

(ii$_+$) *All assertions in* (ii) *remain valid if the spaces* $H^{(s)}_{[\gamma]}$ *are replaced by* $H^{(s)}_{[\gamma]+}$.

For this theorem in the form given here see the survey paper by Volevich and Gindikin [6]; it is a special case of the more general results obtained in Chapter 7.

§2. Dominantly correct polynomials in two variables

In this section we shall describe the structure of polynomials $P(\xi,\tau)$, $\xi \in \mathbb{R}$, satisfying inequality (0.2). The main result consists in that, to within minor monomials, each polynomial of this kind is a product of a strictly hyperbolic polynomial by a stable-correct polynomial.

The definition of dominantly correct polynomials involves the polygon $\delta(P)$. We begin with a detailed description of this polygon for polynomials in two variables correct in Petrovskiĭ's sense and, in particular, consider the problem of reconstructing $N(P)$ from $\delta(P)$, which plays an important role in studying dominantly correct polynomials of several variables (§4) and dominantly correct differential operators with variable coefficients (§§3 and 4).

2.1. The polygon $\delta(P)$. Let $P(\xi,\tau)$ be a polynomial in two variables, let

$$b_1 > b_2 > \cdots > b_m \geqslant 0 \tag{1}$$

be the degrees of its roots $\tau_j(\xi)$, and let μ_j, $j = 1,\ldots,m$ be the number of the roots of degree b_j. As was shown in §1.1.4, the vertices (α_j,β_j), $j = 0,\ldots,m+1$, $(\alpha_0,\beta_0) = (0,0)$, are uniquely determined by the numbers b_j and μ_j (see formulas (1.1.23) and (1.1.23′)) while the numbers b_j and μ_j are reconstructed uniquely from the numbers (α_j,β_j) (see (1.1.25)). Thus the polygon $N(P)$ is determined by the system of inequalities

$$\alpha + b_j\beta \leqslant \alpha_j + b_j\beta_j, \quad j = 1,\ldots,m, \qquad \alpha \geqslant 0, \quad \beta \geqslant 0. \tag{2}$$

If a polynomial P satisfies Petrovskiĭ's correctness condition, the numbers (1) are integers (see Proposition 1 in Section 2.1.4), and consequently the right-hand sides of (2) involve integers.

Lemma. *Let $P(\xi, \tau)$ be a polynomial correct in Petrovskiĭ's sense and let (α, β) be an integral minor point of $N(P)$. Then*

(i) *if $b_m > 0$ (i.e. $N(P)$ has no vertical side not lying on the coordinate axis), then $(\alpha + 1, \beta) \in N(P)$;*

(ii) *in the general case either $(\alpha + 1, \beta) \in N(P)$ or $(\alpha, \beta + 1) \in N(P)$.*

Proof. (i) If $b_m > 0$, then the minor points cannot belong to the sides of $N(P)$ not lying on the coordinate axes. Consequently, $\alpha + b_j\beta < \alpha_j + b_j\beta_j$, and, since we deal with integers, $\alpha + b_j\beta \leqslant \alpha_j + b_j\beta_j - 1$, i.e. $(\alpha + 1, \beta) \in N(P)$.

(ii) If (α, β) belongs to the interior of $N(P)$ or lies on a coordinate axis, then $(\alpha+1, \beta) \in N(P)$. In case the point (α, β) belongs to the vertical side $\Gamma_m^{(1)}$, it cannot coincide with the vertex (α_m, β_m). And therefore $\beta \leqslant \beta_m - 1$, i.e. $(\alpha, \beta+1) \in N(P)$.

The above lemma allows one to describe completely the correspondence between the polygons $N(P)$ and $\delta(P)$.

1) If $N(P)$ has no vertical side not lying on the coordinate axis (i.e. $b_m > 0$), then $\delta(P)$ is determined by the system of inequalities

$$\alpha + b_j\beta \leqslant \alpha_j + b_j\beta_j - 1, \qquad j = 1, \ldots, m. \tag{2'}$$

We note that the straight line $\alpha + b_1\beta = \alpha_1 + b_1\beta_1 - 1 = b_1\beta_1 - 1 = b_1(\beta_1 - 1/b_1)$ intersects the axis $\{\beta\}$ at the point $(0, \beta_1 - 1/b_1)$ which is not integral when $b_1 > 1$. In this case inequality (2′) should be supplemented with the inequality

$$\beta \leqslant \beta_1 - 1, \tag{2''}$$

and the vertices of the polygon $\delta(P)$ are the points

$$(0,0), (0, \beta_1 - 1), (b_1 - 1, \beta_1 - 1), (\alpha_2 - 1, \beta_2), \ldots, (\alpha_{m+1} - 1, 0). \tag{3}$$

2) If $N(P)$ contains a vertical side (not lying on the coordinate axis), then $\delta(P)$ contains the points belonging to the line segment $\{\alpha = \alpha_m, 0 \leqslant \beta \leqslant \beta_m - 1\}$. Hence, in this case $\delta(P)$ is the convex hull of the points

$$(0,0), (0, \beta_1 - 1), (b_1 - 1, \beta_1 - 1), (\alpha_2 - 1, \beta_2), \ldots,$$
$$(\alpha_m - 1, \beta_m), (\alpha_m, \beta_m - 1), (\alpha_m, 0). \tag{3'}$$

All the points (3′) except, possibly, $(\alpha_{m-1} - 1, \beta_m)$ are vertices of $\delta(P)$. The latter point is a vertex in the case $b_{m-1} > 1$ and is an interior point when $b_{m-1} = 1$.

3) Our aim is to reconstruct $N(P)$ from $\delta(P)$. Before indicating the method for reconstructing $N(P)$ form $\delta(P)$ it is advisable to find out whether this procedure leads to a unique result.

Let a polygon N be determined by a set of numbers

$$(b_1, \mu_1), \ldots, (b_{m-1}, \mu_{m-1}), (1, 1), \tag{4}$$

the case $b_{m-1} = 1$ not being excluded here. Let a polygon N' be determined by the set of numbers

$$(b_1, \mu_1), \ldots, (b_{m-1}, \mu_{m-1}), (0, 1), \tag{4'}$$

Then the polygons N and N' coincide to the left of the line $\alpha = \alpha_m$ (see Figure 2), where the numbers α_j, β_j, are found from the numbers $(b_1, \ldots, b_{m-1}, 0)$,

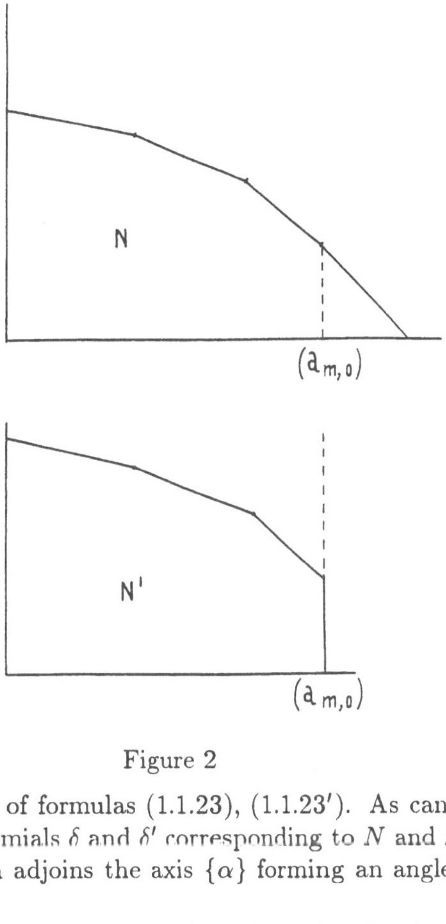

Figure 2

$(\mu_1, \ldots, \mu_{m-1}, 1)$ by means of formulas (1.1.23), (1.1.23$'$). As can easily be seen, the polygons of minor monomials δ and δ' corresponding to N and N' coincide, the side of the polygon δ which adjoins the axis $\{\alpha\}$ forming an angle with that axis equal to $\pi/4$.

So, let (γ_j, δ_j), $j = 0, \ldots, k$, $\gamma_0 = \gamma_1 = 0$, $\delta_k = \delta_0 = 0$ be the given vertices of the polygon $\delta(P)$ corresponding to the original polygon $N(P)$. In the reconstruction of $N(P)$ from $\delta(P)$ one should distinguish between the following three cases.

If $\delta(P)$ has a vertical side not lying on the coordinate axis, then $N(P)$ is sure to have a vertical side, not lying on the coordinate axis, with height no less than 2. The polygon $N(P)$ is reconstructed uniquely, and its vertices are the points (cf. (4′))

$$(0,0),(0,\delta_1+1),(\gamma_2+1,\delta_2),\ldots,(\gamma_{k-2}+1,\delta_{k-2}),(\gamma_{k-1},\delta_{k-1}+1),(\gamma_{k-1},0). \quad (5')$$

If the angle between the side of $\delta(P)$ adjoining the axis $\{\alpha\}$ and that axis does not exceed $\pi/4$, then $N(P)$ is sure to have no vertical side not lying on the coordinate axis. The polygon $N(P)$ is reconstructed uniquely and its vertices are the points

$$(0,0),(0,\delta_1+1),(\gamma_2+1,\delta_2),\ldots,(\gamma_k+1,\delta_k). \quad (5)$$

In case $\delta(P)$ has an angle equal to $\pi/4$, the two above-mentioned reconstruction methods result in two different polygons corresponding to the sets (4) and (4′). We have thus proved the following

Theorem. *Let $P(\xi,\tau)$ be a polynomial correct in Petrovskiĭ's sense. If P has no roots of the form of $\tau(\xi)=O(1)$ (we call them "zero" roots) or if the multiplicity of such roots exceeds 1, then the polygon $N(P)$ and, hence, the numbers b_j, μ_j are reconstructed uniquely from the polygon $\delta(P)$. In the general case all exponents $b_j>1$ and the corresponding multicities μ_j are reconstructed uniquely from $\delta(P)$. The character of the possible non-uniqueness is indicated in (4) and (4′).*

2.2. Product of a strictly hyperbolic polynomial by a stable-correct polynomial. As was proved above, stable-correct and strictly hyperbolic polynomials are dominantly correct. We now prove that in the case of two variables the product of such polynomials is also a dominantly correct polynomial.

Proposition. *Let $H(\xi,\tau)$ be a homogeneous strictly hyperbolic polynomial of degree h and let $S(\xi,\tau)$ be a stable-correct polynomial, the polynomial $H(\xi,\tau)$ having no zero roots in the case when $S(\xi,\tau)$ has them. Then the polynomial*

$$P(\xi,\tau)=H(\xi,\tau)S(\xi,\tau) \quad (6)$$

is dominantly correct.

The proof is based on

Lemma 1. *Let polynomial (6) satisfy the conditions of the proposition and let $(\alpha,\beta)\in\delta(P)$. Then there are points $(\alpha',\beta')\in N(S)$ and $(\alpha'',\beta'')\in\delta(S)$, a polynomial $q(\xi,\tau)$ of degree no higher than $h-1$, and a constant c such that the representation*

$$\xi^\alpha\tau^\beta=q(\xi,\tau)\xi^{\alpha'}\tau^{\beta'}+cH(\xi,\tau)\xi^{\alpha''}\tau^{\beta''} \quad (7)$$

takes place.

Assuming that the lemma is proved we prove the proposition.

We have

$$\left|\frac{\xi^\alpha\tau^\beta}{P(\xi,\tau)}\right|\leqslant\left|\frac{q(\xi,\tau)}{H(\xi,\tau)}\right|\left|\frac{\xi^{\alpha'}\tau^{\beta'}}{S(\xi,\tau)}\right|+\left|\frac{c\xi^{\alpha''}\tau^{\beta''}}{S(\xi,\tau)}\right|.$$

2.3. Description on dominantly correct polynomials. In the foregoing section it was shown that the compositions of strictly hyperbolic and stable-correct polynomials are dominantly correct polynomials. We now show that the polynomials of that kind exhaust (to within minor monomials) the class of dominantly correct polynomials in two variables. We have the following

Theorem. *For a polynomial $P(\xi,\tau)$ the following conditions are equivalent:*

(I) *the polynomial P is dominantly correct, i.e. it is solved with respect to the highest power of τ and inequality (0.2) is fulfilled;*

(II) *the polynomial P is correct in Petrovskiĭ's sense; moreover, for any polynomial $Q(\xi,\tau) \in \mathcal{L}_{N(P)}$ there is $\gamma_0 = \gamma_0(Q)$ such that*

$$|P(\xi,\tau)+Q(\xi,\tau)| \neq 0 \quad \textit{for} \quad \operatorname{Im}\tau \leqslant \gamma_0, \quad (\xi, \operatorname{Re}\tau) \in \mathbb{R}^2,$$

i.e. the polynomial $P+Q$ is correct in Petrovskiĭ's sense;

(III) *the principal parts $\widehat{\tau}_j(\xi)$ of Puiseux' expansions of the roots of P belong to one of the following types (see Proposition 2 in Section 2.1.4):*

(a′) $\widehat{\tau}_j(\xi) = c_{j1}\xi^{\varkappa_{j1}}$, $\operatorname{Im} c_{j1} > 0$, *$\varkappa_{j1}$ is even;*

(b′) $\widehat{\tau}_j(\xi) = c_{j1}\xi$, $\operatorname{Im} c_{j1} = 0$;

(c) $\widehat{\tau}(\xi) \equiv 0$,

the roots of type (b′) being distinct, i.e. $c_{jk} = c_{k1}$, $j \neq k$;

(IV) *there is a strictly hyperbolic polynomial $H(\xi,\tau)$ having no zero roots and a stable-correct polynomial $S(\xi,\tau)$ such that*

$$P(\xi,\tau) - H(\xi,\tau)S(\xi,\tau) \in \mathcal{L}_{N(P)}. \tag{10}$$

Proof. For the proof of (IV)⟹(I) see Proposition 2.2.

(I)⟹(II). If $Q \in \mathcal{L}_{N(P)}$, then, according to (0.2), there is $\gamma(Q)$ such that

$$|Q(\xi,\tau)| < |P(\xi,\tau)|/2 \quad \text{for} \quad \operatorname{Im}\tau < \gamma(Q),$$

whence

$$|P(\xi,\tau)+Q(\xi,\tau)| > \frac{1}{2}|P(\xi,\tau)| \quad \text{for} \quad \operatorname{Im}\tau \leqslant \gamma_0, \gamma(Q).$$

(II)⟹(III). Without loss of generality we can assume that $\tau_j(\xi) = c_j\xi^{b_j}$ since discarding the lower terms in Puiseux' expansions of the roots we change only minor monomials of the polynomial (see Section 1.1.4). By Proposition 1 in Section 2.1.4, the numbers b_j are integers, and we have $\operatorname{Im} c_j = 0$ if b_j is odd and $\operatorname{Im} c_j \geqslant 0$ if b_j is even. We have to show that there are no roots belonging to the following type:

$$\tau_j(\xi) = c_j\xi^{b_j}, \quad b_j > 1, \quad b_j \text{ is odd}, \quad \operatorname{Im} c_j = 0, \quad c_j \neq 0. \tag{11}$$

Indeed, if the polynomial P has a root of the form of (11), then the polynomial $Q(\xi,\tau) = i\xi \prod(\tau - \tau_k(\xi))$ belongs to $\mathcal{L}_{N(P)}$, i.e. the polynomial

$$P(\xi,\tau) + Q(\xi,\tau) = (\tau - c_j\xi^{q_j} + i\xi) \prod_{k\neq j}(\tau - \tau_k(\xi))$$

is correct in Petrovskiĭ's sense. However, this contradicts Proposition 1 in Section 2.1.4 according to which a polynomial correct in Petrovskiĭ's sense cannot have roots of the form $\tau = c_j\xi^{q_j} + i\xi$, $q_j > 1$, where c_j is real. Hence, with account of Proposition 1 in Section 2.1.4, we have proved that the polynomial under consideration can have only roots of types (a'), (b') and (c). It remains to show that the roots of type (b') must necessarily be distinct.

Assume the contrary, i.e. let the polynomial P have the form $(\tau - a\xi)^2 P'$, where a is real, and let the polynomial P' be correct in Petrovskiĭ's sense. Then $Q(\xi,\tau) = \xi P'$ belongs to $\mathcal{L}_{N(P)}$, and consequently the polynomial $P + Q$ is correct in Petrovskiĭ's sense, which contradicts Proposition 1 in Section 2.1.4 since $P + Q$ possesses roots $\tau = a\xi + i\sqrt{\xi}$.

(III)$\Longrightarrow$(IV). Let $c_j\xi^{b_j}$ be the principal parts of the roots of P. Then the polynomial $\widehat{P}(\xi,\tau) = \prod(\tau - c_j\xi^{b_j})$ differs from P only in minor monomials. Denoting by $H(\xi,\tau)$ the product of all factors of the form of $\tau - c_j\xi$ and by $S(\xi,\tau)$ the product of the other factors we obtain, in view of Theorem 2.2.4 and 1.1, relation (10).

§3. Dominantly correct differential operators with variable coefficients (the case of two variables)

In this section we consider differential operators

$$P(x,t;D_x,D_t) = \sum a_{\alpha\beta}(x,t) D_x^\alpha D_t^\beta, \tag{1}$$

whose symbols $P(x,t;\xi,\tau)$ are dominantly correct polynomials for all x and t. Under the additional assumption that the polygon of minor monomials $\delta(P)$ does not depend on x, t we shall prove that, to within lower terms, operator (1) is a product of a strictly hyperbolic differential operator by a stable-correct differential operator. Combining the results of §2.3 with the results on the solvability of Cauchy's problem for strictly hyperbolic operators (Theorem 1.3) we prove the unique solvability of Cauchy's problem for dominantly correct differential operators with variable coefficients.

3.1. Structure of dominantly correct differential operators. A differential operator (1) is said to be dominantly correct if its symbol satisfies the following conditions:

(i) the polygons of minor monomials $\delta(P(x,t))$ of the symbols $P(x,t;\xi,\tau)$ do not depend on (x,t), i.e.

$$\delta(P(x,t)) = \delta(P) \quad \forall (x,t);$$

(ii) for any fixed (x^0, t^0) the polynomial $P(\xi,\tau) = P(x^0,t^0;\xi,\tau)$ is dominantly correct.

Since dominantly correct polynomials are solved with respect to the highest power of τ, the symbol $P(x,t;\xi,\tau)$ possesses the same property. In what follows we shall additionally assume that the coefficient in the highest power of τ is identically equal to 1, i.e.

$$P(x,t;\xi,\tau) = \tau^M + \sum_{j\geqslant 1} P_j(x,t;\xi)\tau^{M-j}. \tag{2}$$

As in §2.3, we shall assume that the coefficients of operator (1) belong to C^∞ and do not depend on (x,t) for sufficiently large (x,t), i.e. condition (2.3.7) holds. As a rule, in what follows we shall not stipulate this condition.

Proposition. *Let symbol* (2) *be dominantly correct. Then it is representable as*

$$P(x,t;\xi,\tau) = H(x,t;\xi,\tau)S(x,t;\xi,\tau) + Q(x,t;\xi,\tau), \tag{3}$$

where the symbols H, S, and Q possess the following properties:

1) *$S(x,t;\xi,\tau)$ is a stable-correct symbol (in the sense of Section 2.3.2), its coefficient in the highest power of τ is equal to 1, and the coefficients of S satisfy condition (2.3.7);*

2) *For any (x,t) the symbol $H(x,t;\xi,\tau)$ is a homogeneous strictly hyperbolic polynomial, its coefficient in the highest power of τ is equal to 1, and the coefficients of H satisfy condition (2.3.7). If the symbol $S(x,t;\xi,\tau)$ has a zero root of multiplicity k for some $x = x^0$, $t = t^0$, then it has a zero root of multiplicity k for all (x,t), and in this case the symbol $H(x,t;\xi,\tau)$ can be solved with respect to the highest power of ξ;*

3) *$Q(x,t;\xi,\tau) \in S\mathcal{L}_{N(P)}$.*

Proof. Let $b_j(x,t)$, $j = 1,\dots,m$, be the degrees of the roots $\tau_k(x,t;\xi)$ of polynomial (2) indexed in the decreasing order and let $\mu_j(x,t)$ be the number of roots of degree $b_j(x,t)$. We first assume that one of the three conditions below is fulfilled:

(a) symbol (2) has no zero roots, i.e. $b_m(x^0,t^0) > 0$ $\forall (x^0,t^0)$;

(b) $b_m(x^0,t^0) = 0$, $\mu_m(x^0,t^0) = 1$ $\forall (x^0,t^0)$;

(c) $\exists (x^0,t^0)$ such that $b_m(x^0,t^0) = 0$, $\mu_m(x^0,t^0) > 1$.

As is seen from Theorem 2.1, under these assumptions the polygons $N(P(x^0,t^0))$ are reconstructed uniquely from the polygons $\delta(P(x^0,t^0))$. Since, by the definition of dominantly correct symbols, the latter polygons do not depend on (x,t), Newton's polygons $N(P(x^0,t^0))$ and, consequently, the numbers $b_j(x^0,t^0)$ and $\mu_j(x^0,t^0)$ do not depend on (x^0,t^0) either. In particular, the quantifier $\exists$ in condition (c) should be replaced by $\forall$.

We now follow the same argument as in the proof of Theorem 1.4.3. According to Theorem 2.3, for each (x,t) there is a number b, a set of polynomials $P^{[j]}(x,t;\xi,\tau)$ parabolic in Petrovskiĭ's sense, a homogeneous strictly hyperbolic polynomial $H(x,t;\xi,\tau)$, $H(x,t;\xi,0) \neq 0$, and a polynomial $Q(x,t;\xi,\tau) \in \mathcal{L}_{N(P)}$ such that

$$P(x,t;\xi,\tau) - \tau^b(x,t;\xi,\tau)\prod P^{[j]}(x,t;\xi,\tau) = Q(x,t;\xi,\tau), \qquad (3')$$

where $b = 0$ if the polygon $N(P)$ has no vertical side not lying on the coordinate axis. We now show that the coefficients of all polynomials involved in (3′) satisfy conditions of type (2.3.7). Indeed, according to formulas (1.1.15), the coefficients of the polynomials H and $P^{[j]}$ are products of coefficients of the polynomial P and the functions $a^{-1}_{\alpha_j\beta_j}(x,t)$ where (α_j,β_j) are the vertices of $N(P(x,t))$ distinct from $(0,0)$ and from the vertex $(\alpha_{m+1},0)$ lying on the vertical side (if $b_m = 0$). Since, by what has been proved, the polygon $N(P(x,t))$ does not depend on x and t, we have

$$a_{\alpha_j\beta_j}(x,t) \neq 0.$$

Since, according to condition (2.3.7), the functions $a_{\alpha_j\beta_j}(x,t)$ are in fact defined on a compactum, there are $\lambda_j > 0$ such that $|a_{\alpha_j\beta_j}(x,t)| > \lambda_j$, whence it follows that the functions $a^{-1}_{\alpha_j\beta_j}(x,t)$ satisfy condition (2.3.7). Therefore the coefficients of $Q(x,t;\xi,\tau)$ also satisfy this condition, i.e. $Q \in S\mathcal{L}_{N(P)}$. We also note that, in view of the compactness, the polynomials $P^{[j]}$ uniformly satisfy Petrovskiĭ's parabolicity condition, and H has distinct roots uniformly with respect to $(x,t) \in \mathbb{R}^2$. Thus, under above assumptions (a), (b), and (c), the proposition is completely proved.

Assume now that at some separate points x^0, t^0 the symbol (2) has a zero root of multiplicity 1, i.e. $b_m(x^0,t^0) = 0$, $\mu(x^0,t^0) = 1$. This relates to the property that the polynomial $P(x,t;\xi,\tau)$ possesses roots of the first degree,one of them vanishing at some separate points. In this case factorization (3′) is not unique at the point $x = x^0$, $t = t^0$. To eliminate this ambiguity, we include the zero root in the hyperbolic symbol, i.e. we write (3′) with $b = 0$. Repeating literally the above argument we prove the proposition.

Arguing in the same way as in Theorem 1.4.3 we deduce from (3) the following

Theorem. *A dominantly correct differential operator with symbol* (2) *is represented in one of the following two forms:*

$$P(x,t;D_x,D_t) = H(x,t;D_x,D_t)S(x,t;D_x,D_t) + Q_1(x,t;D_x,D_t), \qquad (4)$$

$$P(x,t;D_x,D_t) = S(x,t;D_x,D_t)H(x,t;D_x,D_t) + Q_2(x,t;D_x,D_t), \qquad (4')$$

where the symbols H and S satisfy the conditions of the proposition, and we have $Q_j(x,t;\xi,\tau) \in S\mathcal{L}_{N(P)}$, $j = 1,2$.

3.2. A priori estimate. We have the following

Theorem. *Let $P(x,t;\xi,\tau)$ be a dominantly correct symbol with coefficients satisfying condition* (2.3.7). *Then $\forall s \in \mathbb{R}\ \exists\gamma_0 = \gamma_0(s)$ such that the inequality*

$$\|u\|_{\delta(P),(s),\gamma} \leqslant \varepsilon_s(\gamma)\|Pu\|_{(s),\gamma} \quad \forall u \in H^{(\infty)}_{[\gamma]} \qquad (5)$$

holds, where

$$\varepsilon_s(\gamma) \to 0 \quad \text{as} \quad \gamma \to \infty. \tag{5'}$$

Proof. 1) By virtue of (5′), for large $-\gamma$ the inequality (5) remains valid when an expression $\varepsilon_s(\gamma)K\|u\|_{\delta(P),(s),\gamma}$ is added to the right-hand side, where K does not depend on γ. Replacing P on the right-hand side by expressions (4) and (4′) we reduce (5) to the system of inequalities

$$\|D_x^\alpha D_t^\beta u\|_{(s),\gamma} \leqslant \varepsilon_s(\gamma)(\|H \cdot Su\|_{(s),\gamma} + \|S \cdot Hu\|_{(s),\gamma} + \|u\|_{\delta(P),(s),\gamma}),$$
$$\forall(\alpha,\beta) \in \delta(P), \quad \forall u \in H^{(\infty)}_{[\gamma]}. \tag{6}$$

2) As in Section 2.2, denote by $\delta^0(P)$ the set of points $(\alpha,\beta) \in \delta(P)$ representable in the form (2.8). With the set $\delta^0(P)$ we associate in a natural way the norm $\|\ \|_{\delta^0(P),(s),\gamma}$ and the class of symbols $S\mathcal{L}^0_{N(P)}$ that are linear combinations of monomials $\xi^\alpha\tau^\beta$, $(\alpha,\beta) \in \delta^0(P)$. We estimate $\|\ \|_{\delta^0(P),(s),\gamma}$ by means of the right-hand side of (6). By (2.8), we have

$$\|u\|^2_{\delta^0(P),(s),\gamma} = \sum_{\substack{\alpha'+\beta' \leqslant h-1\\ (\alpha'',\beta'')\in N(S)}} \|D_x^{\alpha'} D_t^{\beta'} D_x^{\alpha''} D_t^{\beta''} u\|^2_{(s),\gamma}$$
$$\leqslant \sum_{(\alpha'',\beta'')\in N(S)} \|D_x^{\alpha''} D_t^{\beta''}\|^2_{(s+h-1),\gamma} = \|u\|^2_{N(S),(s+h-1),\gamma}.$$

Using Theorem 2.3.3 for large $-\gamma$ we can estimate $\|u\|_{N(S),(s+h-1),\gamma}$ from above by means of $\|Su\|_{(s+h-1),\gamma}$, i.e. we obtain the inequality

$$\|u\|_{\delta^0(P),(s),\gamma} \leqslant c\|Su\|_{(s+h-1),\gamma}. \tag{7}$$

We now apply the estimate in Theorem 1.3 (i) to the right-hand side of (7). This finally results in

$$\|u\|_{\delta^0(P),(s),\gamma} \leqslant c_1|\gamma|^{-1}\|H \cdot Su\|_{(s),\gamma} \quad \forall u \in H^{(\infty)}[\gamma]. \tag{8}$$

3) We now derive inequality (6) for $(\alpha,\beta) \in \delta(P) \setminus \delta^0(P)$ i.e. for (α,β) representable as (2.8′) or (2.8″).

4) Let (2.8′) hold. Then the operator $D_x^\alpha D_t^\beta$ can be rewritten in the form

$$D_x^\alpha D_t^\beta = D_x^{\alpha''} D_t^{\beta''} H(x,t;D_x,D_t) + D_x^{\alpha''} D_t^{\beta''}(D_t^h - H(x,t;D_x,D_t)).$$

The second operator on the right-hand side has a symbol belonging to $S\mathcal{L}^0_{N(P)}$, i.e. the norm of its value on the function u can be estimated by means of the right-hand side of (8). Let us proceed to the estimation of the first operator on the right-hand side. According to Lemma 2.2.1, there is $\varkappa > 0$ such that

$$|\xi^{\alpha''}\tau^{\beta''}| \leqslant \text{const}\,|\tau|^{-\varkappa} \sum_{(\alpha,\beta)\in N(S)} |\xi^{\alpha}\tau^{\beta}|,$$

whence it follows that

$$\|D_x^{\alpha''}D_t^{\beta''}w\|_{(s),\gamma} \leqslant \text{const}\,|\gamma|^{-\varkappa}\|w\|_{N(S),(s),\gamma}.$$

Applying Theorem 2.3.3 once again we coincide that

$$\|D_x^{\alpha''}D_t^{\beta''}w\|_{(s),\gamma} \leqslant \text{const}\,|\gamma|^{-\varkappa}\|Sw\|_{(s),\gamma}.$$

Replacing w in this inequality by Hu we finally obtain the estimate

$$\|D_x^{\alpha''}D_t^{\beta''}Hu\|_{(s),\gamma} \leqslant \text{const}\,|\gamma|^{-\varkappa}\|S\cdot Hu\|_{(s),\gamma}.$$

5) It now remains to consider the case (2.8″). It can take place only when $N(S)$ has a vertical side and, consequently the symbol H can be solved with respect to the highest power of ξ, i.e.

$$H(x,t;\xi,\tau) = q_h(x,t)\xi^h + O(|\xi|^{h-1}).$$

Then the operator $D_x^\alpha D_t^\beta$ can be rewritten as

$$D_x^\alpha D_t^\beta = D_x^{\alpha''}D_t^{\beta''}q_h^{-1}H + D_x^{\alpha''}D_t^{\beta''}(D_x^h - q_h^{-1}H).$$

The symbol of the second operator on the right-hand side belongs to $S\mathcal{L}^0_{N(P)}$. The first operator is estimated in just the same way as in the case $q_h = 1$.

3.3. Cauchy's problem. With a dominantly correct differential operator $P(x,t;D_x,D_t)$, along with the space

$$H_{[\gamma]}^{N(P),(s)} = \{u \in H_{[\gamma]}^{(s)}, D_x^\alpha D_t^\beta u \in H_{[\gamma]}, \forall(\alpha,\beta) \in N(P)\},$$

a broader space will also be associated:

$$\mathcal{H}_{[\gamma]}^{N(P),(s)} = \{u \in H_{[\gamma]}^{(s)}, D_x^\alpha D_t^\beta u \in H_{[\gamma]}^{(s)}, \forall(\alpha,\beta) \in \delta(P)\}$$

with the natural norm $\|\ \ \|_{\delta(P),(s),\gamma}$. Denote by $(H_{[\gamma]}^{N(P),(s)})_+$ and $(\mathcal{H}_{[\gamma]}^{N(P),(s)})_+$ the subspaces consisting, respectively, of functions and distributions vanishing for $t > 0$.

By the generalized solution to Cauchy's problem for the differential equation

$$P(x,t;D_x,D_t)u(x,t) = f(x,t) \in H_{[\gamma]+}^{(s)} \tag{9}$$

will be meant a function (distribution) $u(x,t) \in (\mathcal{H}_{[\gamma]}^{N(P),(s)})_+$ satisfying (9) in the sense of distribution theory:

$$(u, {}^tP\varphi) = (f,\varphi) \quad \forall\varphi \in \mathcal{D}. \tag{9'}$$

Theorem. *If $P(x,t;D_x,D_t)$ is a dominantly correct operator, then $\forall s \in \mathbb{R}$ $\exists \gamma_0(s)$ such that for $\gamma < \gamma_0(s)$ there exists a unique solution $u \in (\mathcal{H}_{[\gamma]}^{N(P),(s)})_+$ for any right-hand side $f \in H_{[\gamma]+}^{(s)}$.*

Proof. We associate unbounded closed operator with Equation (9). By virtue of Theorem 3.2, it will be injective. And its surjectivity is proved using the theorem on the invertibility of stable-correct and strictly hyperbolic operators. We split the proof into several stages.

1) We associate with Equation (9) the unbounded operator

$$\mathcal{P}\colon (\mathcal{H}_{[\gamma]}^{N(P),(s)})_+ \to H_{[\gamma]+}^{(s)} \quad (u \mapsto Pu) \tag{10}$$

with domain $D_{\mathcal{P}} = (H_{[\gamma]}^{N(P),(s)})_+$. Operator (10) can be closed. Indeed, let a sequence u_j be given such that $u_j \to 0$, $j \to \infty$, in the norm of the space $\mathcal{H}_{[\gamma]}^{N(P),(s)}$. Let $\mathcal{P}u_j \to f$ in the norm of $H_{[\gamma]}^{(s)}$. We have to prove that $f = 0$.

Indeed, by virtue of the obvious embedding

$$H_{[\gamma]}^{N(P),(s)} \subset \mathcal{H}_{[\gamma]}^{N(P),(s)} \subset H_{[\gamma]}^{N(P),(s-1)},$$

the operator (10) can be regarded as a restriction of the continuous operator

$$(H_{[\gamma]}^{N(P),(s-1)})_+ \to H_{[\gamma]+}^{(s-1)} \quad (u \mapsto Pu).$$

Therefore $\mathcal{P}u_j \to 0$ in $H_{[\gamma]}^{(s-1)}$, and $f = 0$.

In what follows we assume that operator (10) has already been closed, i.e. (10) is a closed operator.

2) The operator (10) is injective. Indeed, let $u \in \mathcal{D}_{\mathcal{P}}$ and let $\mathcal{P}u = 0$. By the definition of an unbounded closed operator, there is a sequence $u_j \in (H_{[\gamma]}^{N(P),(s)})_+$ such that $u_j \to u$ in $(\mathcal{H}_{[\gamma]}^{N(P),(s)})_+$ and $Pu_j \to 0$ in $H_{[\gamma]}^{(s)}$. The a priori estimate (5) is continued by continuity to the space $H_{[\gamma]}^{N(P),(s)}$, whence

$$\|u_j\|_{\delta(P),(s),\gamma} \leqslant \text{const}\,\|Pu_j\|_{(s),\gamma} \to 0 \quad \text{as} \quad j \to \infty.$$

Since $u_j \to u$, we have $u \equiv 0$.

3) The image of operator (10) is closed. Indeed, let $f \in H_{[\gamma]+}^{(s)}$ be a limit point in the range of operator (10), i.e. there be a sequence $u_j \in (H_{[\gamma]}^{N(P),(s)})_+$ such that $Pu_j \to f$ in $H_{[\gamma]}^{(s)}$. By virtue of inequality (5), $\{u_j\}$ is a Cauchy sequence in the space $\mathcal{H}_{[\gamma]}^{N(P),(s)}$ and, consequently, it converges in this space to an element $u \in (\mathcal{H}_{[\gamma]}^{N(P),(s)})_+$. The closedness of operator (10) implies that $\mathcal{P}u = f$.

4) To prove the theorem it remains to show that the image of operator (10) coincides with the entire space $H^{(s)}_{[\gamma]+}$. To this end (see Section 2.3.4) we consider the operator

$$\widehat{\mathcal{P}}: (\mathcal{H}^{N(P),(s)}_{[\gamma]})_+ \to (H^{(s)}_{[\gamma]})_+ \quad (v \mapsto H \cdot Sv). \tag{11}$$

Since the operator $H \cdot S$ differs from P by an operator with symbol belonging to $S\mathcal{L}_{N(P)}$, Theorem 3.2 holds for this operator as well. Therefore we can repeat literally the above argument and prove that $\widehat{\mathcal{P}}$ is a closed injective operator with a closed image.

5) We now prove that the image of (11) coincides with $H^{(s)}_{[\gamma]+}$. In view of the closedness of the image, it suffices to show that the image of the operator contains a dense set in $H^{(s)}_{[\gamma]+}$, say $H^{(s+N)}_{[\gamma]}$, where N is sufficiently large. To this end we consider the equation

$$\widehat{\mathcal{P}}u \stackrel{\text{def}}{=} H(x,t;D_x,D_t) \cdot S(x,t;D_x,D_t)u = f \in H^{(s+N)}_{[\gamma]+}. \tag{12}$$

By Theorem 1.3, if $-\gamma$ is sufficiently large, then the equation $Hg = f \in H^{(s+N)}_{[\gamma]}$ possesses a unique solution $g \in H^{(s+N+h-1)}_{[\gamma]+}$. Further, according to the results of §2.3, the equation

$$S(x,t;D_x,D_t)u = g \in H^{(s+N+h-1)}_{[\gamma]+},$$

is sure to have a solution $u \in H^{(s+N+h-1)}_{[\gamma]+}$. If N is sufficiently large, then

$$H^{(s+N+h-1)}_{[\gamma]+} \subset (H^{N(P),(s)}_{[\gamma]})_+,$$

i.e. Equation (12) possesses a solution belonging to the domain of operator $\mathcal{P}$.

6) We now complete the proof of the theorem (cf. the end of the proof of Theorem 2.3.4). Since the closed operator (11) is injective and surjective, by Banach's theorem it has a bounded inverse operator $\widehat{R}: H^{(s)}_{[\gamma]+} \to (\mathcal{H}^{N(P),(s)})_+$, and, by virtue of (5), the norm of the operator tends to zero as $\gamma \to -\infty$. With regard to expansion (4), the operator equation $\mathcal{P}u = f$ can be rewritten as

$$u + \widehat{R}Q_1(x,t;D_x,D_t)u = \widehat{R}f \in H^{(s)}_{[\gamma]+}. \tag{13}$$

Since the operator $Q_1 : (\mathcal{H}^{N(P),(s)})_+ \to H^{(s)}_{[\gamma]+}$ is continuous and its norm does not depend on γ, for large $|\gamma|$ the norm of the operator $\widehat{R}Q_1$ is less than unity, and hence Equation (13) has a unique solution $u \in (\mathcal{H}^{N(P),(s)}_{[\gamma]})_+$. The theorem is proved.

§4. Dominantly correct polynomials and the corresponding differential operators (the case of several spatial variables)

4.1. Description of dominantly correct polynomials. As in §2.4, with each polynomial $P(\xi,\tau)$, $\xi \in \mathbb{R}^n$, we associate a family of pseudo-polynomials $P_\omega(|\xi|,\tau) \overset{\text{def}}{=} P(|\xi|\omega,\tau)$, i.e. polynomials on two variables depending on the parameter $\omega \in S^{n-1}$. We shall describe dominantly correct polynomials in several variables in terms of such families; the definition of these polynomials was given in the Introduction to the present chapter.

Theorem. *For a polynomial*

$$P(\xi,\tau) = \tau^m + \sum_{\beta<m} a_{\alpha_1\ldots\alpha_n\beta}\xi_1^{\alpha_1}\cdots\xi_n^{\alpha_n}\tau^\beta, \tag{1}$$

the following conditions are equivalent:

(I) *polynomial* (1) *is dominantly correct, i.e. inequality* (2) *in the Introduction holds;*

(II) *polynomial* (1) *is correct in Petrovskiĭ's sense, and for every polynomial* $Q(\xi,\tau)$, $\Delta(Q) \subset \delta(P)$, *there is* $\gamma_0 = \gamma_0(Q)$ *such that*

$$|P(\xi,\tau)+Q(\xi,\tau)| \neq 0 \quad \textit{for} \quad \operatorname{Im}\tau \leqslant \gamma_0, \quad (\xi, \operatorname{Re}\tau) \in \mathbb{R}^{n+1};$$

(III) *for all* $\omega \in S^{n-1}$ *the polynomials* $P_\omega(\rho,\tau)$ *satisfy the equivalent conditions of Theorem 2.3, the polygons* $\delta(P_\omega)$ *not depending on* ω *and hence coinciding with* $\delta(P)$;

(IV) *there are an integer* $b \geqslant 0$, *a parabolic polynomial* $R(\xi,\tau)$, *and a homogeneous strictly hyperbolic pseudo-polynomial* $H_\omega(|\xi|,\tau)$ *such that*

$$P(\xi,\tau) - \tau^b R(\xi,\tau) H_\omega(|\xi|,\tau) \in \mathcal{L}_{N(P)},^{1)} \tag{2}$$

and if $b > 0$, *then the pseudo-polynomial* H_ω *has no zero roots, i.e.*

$$H_\omega(1,0) \neq 0 \quad \forall\omega \in S^{n-1} \quad (b \geqslant 1). \tag{3}$$

Proof. (I)$\Longrightarrow$(II) is proved in the same way as the analogous implication in the case $n = 1$ (see Theorem 2.3).

(II)$\Longrightarrow$(III). Consider the polynomial $P_\omega(\rho,\tau) = P(\rho\omega,\tau)$. We have $\delta(P_\omega) \subset \delta(P)$, and therefore the polynomial P_ω satisfies the equivalent conditions in Theorem 2.3. Moreover, this polynomial remains correct under the addition of monomials $\varepsilon\rho^\alpha\tau^\beta$, $(\alpha,\beta) \in \delta(P)$, with an arbitrary coefficient $\varepsilon \in \mathbb{C}$. Therefore, according

---

1) Recall that, as in Section 2.4.1, $\mathcal{L}_{N(P)}$ denotes the space of pseudo-polynomials $Q_\omega(|\xi|,\tau)$ such that $N(Q_\omega) \subset \delta(P)$.

to Proposition 2.2.4 (ii), we have $\delta(P) \subset \delta(P_\omega)$, whence $\delta(P_\omega) = \delta(P)$ for any $\omega \in S^{n-1}$.

(III)$\Longrightarrow$(IV). Since, by the hypothesis, each polynomial $P_\omega(\rho,\tau) = P(\rho\omega,\tau)$ is dominantly correct, Theorem 2.3 implies that for any $\omega \in S^{n-1}$ there is an integer $b(\omega)$, an N-parabolic polynomial $R^0_\omega(\rho,\tau)$, and a strictly hyperbolic polynomial $H^0_\omega(\rho,\tau)$, $H^0_\omega(1,0) \neq 0$, such that

$$P_\omega(\rho,\tau) - \tau^{b(\omega)} R^0_\omega(\rho,\tau) H^0_\omega(\rho,\tau) \in \mathcal{L}_{N(P_\omega)} \quad \forall \omega \in S^{n-1}, \tag{2'}$$

where R^0_ω is the product of polynomials $P^{[j]}_\omega(\rho,\tau)$, $j = 1,\dots,k(\omega)$, $b_j(\omega)$-parabolic in Petrovskiĭ's sense. By Theorem 2.1, the number $k(\omega)$, the (even) numbers $b_j(\omega)$, and the numbers $\varkappa_i(\omega)$ (the degrees of $P^{[j]}_\omega$ with respect to τ) are uniquely determined by the polygon $\delta(P_\omega)$ of minor monomials. Since, by the hypothesis, these polygons do not depend on ω (and, consequently, coincide with $\delta(P)$), the numbers k, b_j, $\varkappa_j$, $j = 1,\dots,k$, do not depend on ω either. Further, if the vertices of the polygon $N(P_\omega)$ are denoted as $(0,0)$, $(\alpha_{(j)}(\omega), \beta_{(j)}(\omega))$, $j = 1,\dots,m(\omega) \geqslant k$, then the vertices $(\alpha_{(j)}, \beta_{(j)})$, $j = 1,\dots,k+1$, do not depend on ω and coincide with the vertices of the polygon $\Delta(P)$.

Denote by $R(\xi,\tau)$ the sum of the monomials of the form of

$$a_{\alpha_1\dots\alpha_n\beta}\xi_1^{\alpha_1}\dots\xi_n^{\alpha_n}\tau^{\beta-\beta_{(k+1)}},$$

where the points $(\alpha_1 + \dots + \alpha_n, \beta)$ belong to the broken line joining the vertices $(\alpha_{(1)}, \beta_{(1)}),\dots,(\alpha_{(k+1)}, \beta_{(k+1)})$. The method of defining the polynomials $P^{[j]}_\omega$ (see Section 1.1.3) implies that $R^0_\omega(\rho,\tau)$ in (2′) can be replaced by $R_\omega(\rho,\tau) = R(\rho\omega,\tau)$.[1)] To this end, in order to examine the factors $\tau^{b(\omega)}H^0_\omega$ we consider the three cases below.

1) Assume first that there is $\omega = \omega^0$ such that $b(\omega^0) > 1$, i.e. the polygon $N(P_\omega)$ contains a vertical side of height no less than 2 not lying on the coordinate axis. By Theorem 2.1, in this case the polygon $N(P_\omega)$ is determined uniquely by the polygon $\delta(P_\omega)$, and, consequently, the numbers $b(\omega)$ do not depend on ω. Thus, all the polygons $N(P_\omega)$ and, hence, $\Delta(P)$ have the vertices $(\alpha_{(j)}, \beta_{(j)})$, $j = 1,\dots,k+3$, the side joining $(\alpha_{(k+2)}, \beta_{k+2})$ and $(\alpha_{(k+3)}, \beta_{(k+3)})$ being vertical, i.e. $\alpha_{(k+2)} = \alpha_{(k+3)}$, $\beta_{(k+3)} = 0$. Denote by $P_0(\xi,\tau)$ the sum of the monomials corresponding to the side joining $(\alpha_{k+1}, \beta_{k+1})$ and $(\alpha_{(k+2)}, \beta_{k+2})$ (this side relates to hyperbolic roots of the polynomial P), and let

$$a_{(k+1)}(\xi) = \sum_{\alpha_1+\dots+\alpha_n=\alpha_{(k+1)}} a_{\alpha_1\dots\alpha_n}\xi_1^{\alpha_1}\dots\xi_n^{\alpha_n} = |\xi|^{\alpha_{(k+1)}} a_{(k+1)}(\omega)$$

---

1) We stress that, generally speaking, the polynomials $R^0_\omega(\rho,\tau)$ depending on ω in an uncontrollable way. As to the polynomial $R_\omega(\rho,\tau)$, its coefficients are smooth functions of ω.

be the sum of the monomials corresponding to the vertex $(\alpha_{(k+1)}, \beta_{(k+1)})$. Since this a common vertex of all $N(P_\omega)$, we have $a_{(m+1)}(\omega) \neq 0\ \forall \omega \in S^{n-1}$. It follows that the function

$$H_\omega(|\xi|, \tau) = P_0(\xi, \tau) a^{-1}_{(k+1)}(\xi) \tag{4}$$

is a pseudo-polynomial, and we have $H_\omega = H^0_\omega$ for each $\omega \in S^{n-1}$. We thus arrive at (2) with b, $R(\xi, \tau)$, and $H_\omega(|\xi|, \tau)$ determined above, and since we have $H_\omega = H^0_\omega$, condition (3) is fulfilled.

2) Now let $b(\omega) \leqslant 1$ and let there be $\omega = \omega^0$ such that $b(\omega^0) = 0$. In this case we have $\tau^{b(\omega)} H^0_\omega = H_\omega$, where the function H_ω is determined by means of (5). We thus obtain factorization (2) with $b = 0$.

3) Finally, if $b(\omega) \equiv 1$, then we can again write factorization (2) with $b = 1$ and H_ω in (4). The function H_ω must satisfy condition (3) since otherwise we would arrive at a contradiction to the condition that $\delta(P_\omega)$ does not depend on ω.

(IV)$\Longrightarrow$(I). The proof of this implication is analogous to the proof of Proposition 2.2. It is only necessary to establish, instead of expansion (2.7), its multidimensional analog

$$|\xi|^\alpha \tau^\beta = q(\omega, |\xi|, \tau)|\xi|^{\alpha'} \tau^{\beta'} + c(\omega) H_\omega(|\xi|, \tau)|\xi|^{\alpha''} \tau^{\beta''}, \tag{5}$$

where the degree of the pseudo-polynomial q is less than that of the pseudo-polynomial H_ω, and $(\alpha', \beta') \in \Delta(P)$, $(\alpha'', \beta'') \in \delta(P)$. Expansion (5) is proved following the plan of the proof of Lemma 1 in Section 2.3, and we leave this to the reader.

4.2. Dominantly correct differential operators. A differential operator

$$D_t^m + \sum_{\beta < m} a_{\alpha_1 \dots \alpha_m \beta}(x, t) D_1^{\alpha_1} \dots D_n^{\alpha_n} D_t^\beta,$$

where $x \in \mathbb{R}^n$, is said to be dominantly correct if its symbol

$$P(x, t; \xi, \tau) = \tau^m + \sum_{\beta < m} a_{\alpha_1 \dots \alpha_n \beta}(x, t) \xi_1^{\alpha_1} \dots \xi_n^{\alpha_n}, \tag{6}$$

satisfies the following conditions (cf. Section 3.1):

(i) the polygons of minor monomials $\delta(P(x,t))$ do not depend on (x, t), i.e. $\delta(P(x,t)) = \delta(P) = \cup \delta(P(x,t))$;
(ii) the polynomial $P(\xi, \tau) = P(x^0, t^0, \xi, \tau)$ is dominantly correct $\forall (x^0, t^0) \in \mathbb{R}^{n+1}$.

The multidimensional analog of Proposition 3.1 is proved by means of a literal repetition of the proof of the implication (III)$\Longrightarrow$(IV) in Theorem 3.1.

Proposition. *A dominantly correct symbol* (6) *is representable as*

$$P(x,t;\xi,\tau) = H_\omega(x,t;|\xi|,\tau)S(x,t;\xi,\tau) + Q_\omega(x,t;|\xi|,\tau), \tag{7}$$

where the polynomial symbol $S(x,t;\xi,\tau)$ *and the pseudo-polynomial symbols* H_ω *and* Q_ω *possess the following properties:*

1) $S(x,t;\xi,\tau)$ *is a stable-correct symbol (in the sense of Section* 3.1 *where* $N(P)$ *should be replaced by* $\Delta(P)$*), the coefficient in the highest power of* τ *is equal to* 1, *and the coefficients satisfy condition* (2.3.7);

2) $H_\omega(x,t;|\xi|,\tau)$ *is a strictly hyperbolic pseudo-polynomial, its coefficient in the highest power of* τ *is equal to* 1, *and the other coefficients satisfy condition* (2.3.7);

3) *if the symbol* S *is not parabolic (i.e.* $S \sim \tau^b P_1$, $b > 0$ *where* P_1 *is a parabolic symbol), then the symbol* H_ω *has no zero roots, i.e.*

$$H_\omega(x,t;1,0) \neq 0 \quad \forall (x,t) \in \mathbb{R};$$

4) *the pseudo-polynomial* Q_ω *belongs to* $S\mathcal{L}_{\Delta(P)}$ *(see* §2.5*), i.e. is a linear combination of monomials of the form of* $a(x,t;\omega)|\xi|^\alpha\tau^\beta$, $(\alpha,\beta) \in \delta(P)$.

Theorem. *A dominantly correct differential operator with symbol* (6) *is represented in the form*

$$P(x,t;D_x,D_t) = S(x,t;D_x,D_t)\cdot H(x,t;D_x,D_t) + Q_1, \tag{8}$$
$$P(x,t;D_x,D_t) = H(x,t;D_x,D_t)\cdot S(x,t;D_x,D_t) + Q_2, \tag{8$'$}$$

where Q_1 *and* Q_2 *are bounded operator from* $\mathcal{H}_{[\gamma]}^{N(P),(s)}$ [1] *into* $H_{[\gamma]}^{(s)}$ *for any* $s \in \mathbb{R}$ *and* $\gamma \leqslant \gamma_0$.

Proof. The composition of the differential operator $S(x,t;D_x,D_t)$ and the pseudodifferential operator $H(x,t;D_x,D_t)$ is a pseudodifferential operator with symbol

$$S(x,t;\xi,\tau)H(x,t;\xi,\tau) + \sum_{\delta>0} S^{(\delta)}(x,t;\xi,\tau)H_{(\delta)}(x,t;\xi,\tau)/\delta!.$$

With regard to relation (7), Q_1 is a pseudodifferential operator with symbol

$$Q_1(x,t;\xi,\tau) = -\sum_{\delta>0}(S^{(\delta)}H_{(\delta)})(x,t;\xi,\tau)/\delta! + Q(x,t;\xi,\tau).$$

From the description of the polygon $\Delta(P)$ that was in fact presented in §2 it follows that the symbol Q_1 is a linear combination of expressions of the form of

$$q_{\alpha\beta}(x,t;\omega)D_x^\alpha D_t^\beta, \qquad (|\alpha|,\beta) \in \delta(P),$$

[1] For the definition of this space see Section 3.3.

i.e. $Q_1 \in S\mathcal{L}_{\Delta(P)}$, and the corresponding PDO transforms $\mathcal{H}_{[\gamma]}^{N(P),(s)}$ into $H_{[\gamma]}^{(s)}$, i.e. (8) is proved.

Comparing (8) and (8′) we see that

$$Q_2 = Q_1 + [S, H].$$

We now show that the commutator on the right-hand side is a linear combination of operators of the form of

$$Q_{\alpha\beta} D_x^\alpha D_t^\beta, \qquad (|\alpha|, \beta) \in \delta(P), \tag{9}$$

where $Q_{\alpha\beta}$ are bounded operators on $H_{[\gamma]}^{(s)}$ for any s and γ. Writing

$$S = \sum_{(|\alpha|,\beta)\in\Delta(S)} S_{\alpha\beta}(x,t) D_x^\alpha D_t^\beta, \qquad H = \sum_{|\rho|+\delta\leqslant h} H_{\rho\delta}(x,t;D_x) D_x^\beta D_t^\delta$$

we represent $S \cdot H - H \cdot S$ as a linear combination of operators

$$[S_{\alpha\beta} D_x^\alpha D_t^\beta, H_{\rho\delta} D_x^\beta D_t^\delta] = [S_{\alpha\beta}(x,t), H_{\rho\delta}(x,t)] D_x^{\alpha+\rho} D_t^{\beta+\delta} + \cdots, \tag{10}$$

where the dots symbolize the terms appearing under the commutation of the operators of differentiation with the operator of multiplication by the function $S_{\alpha\beta}(x,t)$ and the PDO $H_{\rho\delta}(x,t;D_x)$; these operators are obviously written in the form (9). Since the coefficients in the highest powers of D_t in the operators S and H are equal to 1, the terms in (10) corresponding to the senior points $(|\alpha| + |\rho|, \beta + \delta)$ of the polygon $\Delta(P)$ involve differentiation with respect to the variables $x_1, \ldots, x_n$. Since $[S_{\alpha\beta}, H_{\rho\delta}]D_j$ are bounded operators on $H_{[\gamma]}^{(s)}$, the right-hand operators in (10) are readily represented in the form (9). The theorem is proved.

4.3. Cauchy's problem for dominantly correct differential operators. We now extend Theorems 3.2 and 3.3 to the case of several variables

Theorem. *Let $P(x,t;D_x,D_t)$ be a dominantly correct operator. Then $\forall s \in \mathbb{R}$ $\exists \gamma_0 = \gamma_0(s)$ such that the assertions below hold:*

(i) *the estimate*

$$\|u\|_{\delta(P),(s),\gamma} \leqslant \varepsilon_s(\gamma) \|P(x,t;D_x,D_t)u\|_{(s),\gamma}, \tag{11}$$

where $\varepsilon_s(\gamma) \to 0$, $\gamma \to -\infty$, is fulfilled;

(ii) *$\forall f \in H_{[\gamma]+}^{(s)}$ there is a function $u \in (H_{[\gamma]}^{\delta(P),(s)})_+$ satisfying the equation*

$$P(x,t;D_x,D_t)u = f, \tag{12}$$

in sense of distribution theory.

The proof of (i). This assertion is proved following the scheme for the one-dimensional case (Theorem 3.2) based on analogs of expansions (2.8), (2.8′), and (2.8″); if $(\alpha, \beta) \in \delta(P)$, then

$$(\alpha,\beta) = (\alpha',\beta') + (\alpha'',\beta''), \quad (|\alpha'|,\beta') \in \Delta(S),\ |\alpha''| + \beta'' \leqslant h-1, \tag{13}$$

$$(\alpha,\beta) = (\alpha',\beta') + (0,h), \quad (|\alpha'|+1,\beta') \in \Delta(S),\ (|\alpha'|,\beta'+1) \notin \Delta(S), \tag{13'}$$

$$(\alpha,\beta) = (\alpha',\beta') + (\alpha'',0), \quad |\alpha''| = h, \quad (|\alpha'|,\beta'+1) \in \Delta(S), \quad (|\alpha'|+1,\beta') \notin \Delta(S). \tag{13''}$$

As in Theorem 3.2, denote by $\delta^0(P)$ the set of points $(\alpha,\beta) \in \delta(P)$ representable as (13) and associate with this space the class of symbols $S\mathcal{L}^0_{\Delta(P)}$ and the corresponding norm. Combining the estimates for stable-correct operators with the Petrovskiĭ-Leray inequality (see Theorem 1.3) we obtain, as in Theorem 3.2, the inequality

$$\|u\|_{\delta^0(P),(s),\gamma} \leqslant \text{const}\, |\gamma|^{-1} \|H \cdot Su\|_{(s),\gamma}. \tag{14}$$

In the case (13′) the norm of $D_x^\alpha D_t^\beta$ is estimated as in Theorem 3.2. The operator $D_x^\alpha D_t^\beta$ is rewritten in the form

$$D_x^\alpha D_t^\beta = D_x^\alpha D_t^{\beta-h} H(x,t;D_x,D_t) + D_x^\alpha D_t^\beta (D_t^h - H).$$

The second operator on the right-hand side is a PDO with symbol belonging to $S\mathcal{L}_{\Delta(P)}$, and therefore the norm of its value on the function u can be estimated by means of the right-hand side of (14). As in the case $n=1$, we have

$$\|D_x^\alpha D_t^{\beta-h} Hu\|_{(s),\gamma} \leqslant \varepsilon_s(\gamma) \|S \cdot Hu\|_{(s),\gamma}.$$

A more intricate problem is the estimation of the norm $\|D_x^\alpha D_t^\beta u\|_{(s),\gamma}$ in the case (13″) since it is this point where the specificity of the multidimensional case manifests itself. The case (13″) is possible only when the symbol S possesses zero roots, i.e. (Proposition 4.2) the symbol $H_\omega(x,t;|\xi|,0) \overset{\text{def}}{=} H(x,t;\xi,0)$ is nonzero for $|\xi| \neq 0$ and is an elliptic symbol with respect to (x,ξ) of order h depending on the parameter t. We make use of the well-known estimate for elliptic PDO (e.g. see Eskin [1]):

$$\left|(1+|D_x|^2)^{h/2} w, \mathbb{R}^n_x\right| \leqslant c\big(|H(x,t;D_x,0)w, \mathbb{R}^n_x| + |w, \mathbb{R}^n_x|\big),$$

where $|\ \ |$ is the ordinary L_2 norm in $\mathbb{R}^n$. Replacing w in this estimate by $D_x^\alpha D_t^\beta u(x,t)$ we find

$$\begin{aligned}\|D_x^\alpha D_t^\beta u\| &= \|D_x^{\alpha'}(D_x^{\alpha''} D_t^\beta u)\|_{(s),\gamma} \\ &\leqslant \text{const}\,\big(\|H(x,t;D_x,0) D_x^{\alpha''} D_t^\beta u\|_{(s),\gamma} + \|D_x^{\alpha''} D_t^\beta u\|_{(s),\gamma}\big).\end{aligned} \tag{15}$$

Since we have $\xi^{\alpha''}\tau^{\beta} \in \mathcal{L}^0_{\Delta(P)}$, the second term on the right-hand side has already been estimated. As to the first term, we have

$$\begin{aligned} H(x,t;D_x,0)D_x^{\alpha''}D_t^{\beta} &= D_x^{\alpha''}D_t^{\beta}H(x,t;D_x,D_t) \\ &+ D_x^{\alpha''}D_t^{\beta}(H(x,t;D_z,0) - H(x,t;D_x,D_t)) \\ &+ [H(x,t;D_x,D_t), D_x^{\alpha''}D_t^{\beta}] = I_1 + I_2 + I_3. \end{aligned}$$

Expressions of the type I_1 were already estimated, I_2 is a PDO with a symbol belonging to $S\mathcal{L}^0_{\Delta(P)}$, and the commutator I_3 is a linear combination of PDO with symbols of the form of

$$(D_x^{\alpha''-\mu}D_t^{\beta-\nu}H)(x,t;\xi,\tau)\xi^{\mu}\tau^{\nu}, \qquad |\mu| + \nu > 0. \tag{16}$$

If $|\mu| > 0$, then symbol (16) belongs to $S\mathcal{L}^0_{\Delta(P)}$, i.e. we have to consider the case $\mu = 0$, $\nu > 0$. Then (16) has the form

$$(D_t^{\beta-\nu}H)(x,t;\xi,0)\xi^{\alpha''}\tau^{\nu} + \cdots, \tag{16'}$$

where the dots designate symbols belonging to $S\mathcal{L}^0_{\Delta(P)}$. Symbol (16′) is a linear combination of polynomial symbols $\xi^{\mu}\tau^{\nu}$ with coefficients depending on x, t, and ω, and μ and ν satisfy conditions of type (13′), (15′).

We now summarize the results. We have shown that if conditions (13″) are fulfilled, then $\|D_x^{\alpha}D_t^{\beta}u\|_{(s),\gamma}$ can be estimated by means of the right-hand side of (11) and a finite set of norms $\|D^{\mu}D_t^{\nu}u\|_{(s),\gamma}$, where μ and γ satisfy conditions analogous to those for α and β with the additional requirement that $\nu < \beta$. Repeating these estimates a finite number of times we prove inequality (11).

The proof of (ii). The existence theorem for the solution to Cauchy's problem for strictly hyperbolic differential operators is extended to strictly hyperbolic PDO (e.g. see Eskin [2]). In view of this, we can literally repeat, based on inequality (11), the argument in Theorem 3.3 and prove the solvability of Equation (12). A direct proof of this assertion will be given in Chapter 7.

CHAPTER IV

OPERATORS OF PRINCIPAL TYPE ASSOCIATED WITH NEWTON'S POLYGON

§1. Introduction. Operators of principal and quasi-principal type

1.1. The present chapter is devoted to studying differential operator that majorize locally all lower (in the sense of Newton's polygon) derivatives. The corresponding class of operators is an extension of the class of N quasi-elliptic operators (Chapter 1) to the same degree as the class of dominantly correct operators (Chapter 3) is an extension of the class of stable-correct operators (Chapter 2). Recall that in Chapter 1 we studied a class of differential operators on functions of the variables $(x, y) \in \mathbb{R}^2$, for which in any region $\Omega \subset \mathbb{R}^2$ of a sufficiently small diameter the inequality

$$\sum_{(\alpha,\beta)\in N(P)} \|D_x^\alpha D_y^\beta u\| \leqslant c\|P(x,y;D_x,D_y)u\| \quad \forall u \in \mathcal{D}(\Omega) \tag{1}$$

holded. In the case of constant coefficients, by virtue of the well-known result of Hörmander (see (1.4.20)), inequality (1) is equivalent to the corresponding inequality for the symbol $P(\xi, \eta)$:

$$\sum_{(\alpha,\beta)\in N(P)} |\xi^\alpha \eta^\beta| \leqslant c\widetilde{P}(\xi,\eta), \qquad \widetilde{P} = \left(\sum |P^{(\delta)}|^2\right)^{1/2}. \tag{2}$$

If the polygon $N(P)$ is regular (as in Chapter 1, only in this case we consider estimates for variable coefficients), then inequality (2) is equivalent to the estimate considered in the first half of Chapter 1, namely $\exists c$, $c_0 > 0$ such that

$$\sum_{(\alpha,\beta)\in N(P)} |\xi^\alpha \eta^\beta| \leqslant c|P(\xi,\eta)| \quad \text{for} \quad \xi^2+\eta^2 > c_0^2. \tag{3}$$

Necessary and sufficient conditions for the validity of (3) are stated in terms of quasi-homogeneous parts of the polynomial P corresponding to the sides of the polygon $N(P)$.

In this chapter we shall study differential operators $P(x, y; D_x, D_y)$ for which in an arbitrary bounded region Ω of a sufficiently small diameter $\operatorname{diam}\Omega \leqslant \lambda \leqslant \lambda_0$ the inequality

$$\sum_{(\alpha,\beta)\in\delta(P)} \|D_x^\alpha D_y^\beta u\| \leqslant \varepsilon(\lambda)\|P(x,y;D_x,D_y)u\|$$
$$\forall u \in \mathcal{D}(\Omega) \quad \varepsilon(\lambda)\to 0, \quad \lambda \to 0, \tag{4}$$

holds. Inequality (1) implies (4). This is not an automatic reduction since the lower derivatives must be estimated by means of a small constant.

The constant $\varepsilon(\lambda)$ in inequality (4) being small, the inequality remains true under any perturbations of the coefficients in lower derivatives in P. More precisely, with any symbol $Q(x, y; \xi, \eta) \in S\mathcal{L}_{N(P)}$ a constant $\lambda(Q)$ can be associated, such that for $\lambda \leqslant \lambda(Q)$, $\operatorname{diam}\Omega \leqslant \lambda$, we have

$$\sum_{(\alpha,\beta)\in\delta(P)} \|D_x^\alpha D_y^\beta u\| \leqslant \varepsilon'(\lambda)\|(P+Q)(x, y; D_x, D_y)u\| \quad u \in \mathcal{D}(\Omega). \tag{4'}$$

In the case of constant coefficients this implies the algebraic condition

$$\sum_{(\alpha,\beta)\in\delta(P)} |\xi^\alpha \eta^\beta| \leqslant c(\widetilde{P+Q})(\xi, \eta) \quad \forall Q \in \mathcal{L}_{N(P)} \tag{5}$$

on the symbol. In the first part of this chapter, under some additional conditions on Newton's polygon, we shall find necessary and sufficient conditions on the polynomial P under which inequality (5) is fulfilled. As in the case of inequality (4) (or inequalities in Chapter 3), these conditions will be stated in terms of the principal quasi-homogeneous parts of P corresponding to the sides of Newton's polygon. These parts either do not vanish (the condition of N quasi-ellipticity) or have simple (in the above-mentioned sense) real zeros.

The second part of the chapter is devoted to the proof of inequality (4) for the case of variable coefficients. Here we impose an important additional condition that some of the coefficients of the symbol P should be real. These conditions are such that the symbols $P(x, y; \xi, \eta)$ and $P^*(x, y; \xi, \eta)$ satisfy them simulatneously. It follows that inequality (4) remains valid when P is replaced by P^* or ${}^tP$, whence the local solvability of the operator P is derived.

1.2. The results of this chapter (as well as those in Chapter 6) are a generalization and further development of the well-known result by Hörmander [2], according to which a polynomial $P(\xi_1, \ldots, \xi_n)$ of degree m is called a polynomial of principal type if

$$|\operatorname{grad} P_{(m)}(\xi_1, \ldots, \xi_n)| \neq 0, \quad (\xi_1, \ldots, \xi_n) \in \mathbb{R}^n \setminus \{0\}, \tag{6}$$

where $P_{(m)}$ is the principal homogeneous part of P. Hörmander proved that condition (6) is fulfilled if and only if

$$(1 + |\xi_1| + \cdots + |\xi_n|)^{m-1} \leqslant c_Q(\widetilde{P+Q})(\xi_1, \ldots, \xi_n), \tag{7}$$

where Q is an arbitrary polynomial of degree no higher than $m-1$.

Similarly, a differential operator $P(x; D)$ with variable coefficients is called an operator of principal type if the symbol of its principal homogeneous part satisfies the condition

$$|\operatorname{grad} P_{(m)}(x_1, \ldots, x_n; \xi_1, \ldots, \xi_n)| \neq 0, \quad (\xi_1, \ldots, \xi_n) \in \mathbb{R}^n \setminus \{0\}. \tag{6'}$$

For these operators, under the additional assumption that the symbol $P_{(m)}$ is real, the a priori estimate

$$\|u\|_{(m-1)} \leqslant \varepsilon(\lambda)\|Pu\|, \quad \forall u \in \mathcal{D}(\Omega), \quad \operatorname{diam}\Omega \leqslant \lambda, \quad \varepsilon(\lambda) \to 0, \quad \lambda \to 0, \tag{8}$$

was proved.

1.3. We now pass to the quasi-homogeneous case corresponding to a positive vector $q = (q_1, \dots, q_n)$ whose components are normalized by the condition $\min q_j = 1$. The question is posed as to what are the necessary and sufficient conditions on the polynomial $P(\xi)$ and the symbol $P(x;\xi)$ under which inequalities (7) and (8) hold if in the left-hand sides of the inequalities the sum $|\xi_1|+\dots+|\xi_n|$ and the norm $\| \ \|_{(m-1)}$ are replaced by $|\xi_1|^{1/q_1} + \dots + |\xi_n|^{1/q_n}$ and by the quasi-homogeneous norm

$$\|u\|_{(m-1,q)} = \Big(\int (1 + |\xi_1|^{1/q_1} + \dots + |\xi_n|^{1/q_n})^{2(m-1)} |\widehat{u}(\xi)|^2 \, d\xi\Big)^{1/2},$$

respectively. The corresponding polynomials and operators are called polynomials and operators of q-principal or quasi-principal type. Hörmander's results were generalized to operators of quasi-principal type by Shananin [1, 2] and Lascar [1, 2].

In the quasi-homogeneous case, if we want to state all results in terms of principal quasi-homogeneous parts, we have to impose on the vector q the following additional arithmetical condition.

The numbers $q_1, \dots, q_n$ are integers.

The meaning of this condition will be eludicated below (see Remark 3).

Thus, let a polynomial

$$P(\xi) = \sum a_{\alpha_1 \dots \alpha_n} \xi_1^{\alpha_1} \dots \xi_n^{\alpha_n}, \tag{9}$$

of q-degree m be given, i.e.

$$m = \deg_q P = \max_{(\alpha_1, \dots \alpha_n)} (q_1\alpha_1 + \dots + q_n\alpha_n),$$

and let $P_q = P_{(m;q)}$ be its principal q-homogeneous part. When generalizing condition (7) to the quasi-homogeneous case it should be taken into account that, according to Euler's formula, we have

$$mP_{(m)}(\xi_1, \dots, \xi_n) = \sum \xi_i \partial P_{(m)}(\xi)/\partial \xi_i,$$

and therefore condition (6) is significant only where polynomial $P_{(m)}$ vanishes. Hence, condition (6) can be replaced by

$$\{P_{(m)}(\xi_1, \dots, \xi_n) = 0\} \Rightarrow \{|\operatorname{grad} P_m(\xi_1, \dots, \xi_n)| \neq 0\}.$$

Further, the polynomials $\partial P/\partial\xi_j$ have q-orders $m - q_j$, i.e. if $q_j > 1$, then the derivative $\partial P/\partial\xi_j$ must not affect the estimates for monomials of q-degree $m-1$.

In view of what has been said, a variable ξ_j will be called an *essential variable* of polynomial (9) if $q_j = 1$. The collection of essential variables will be denoted ξ'. Polynomial (9) of q-degree m is called a polynomial of q-principal type if

$$\{P_q(\xi_1,\dots,\xi_n) = 0\} \Rightarrow \{|\operatorname{grad}_{\xi'} P_q(\xi_1,\dots,\xi_n)| \neq 0\}. \tag{10}$$

Similarly, if a symbol $P(x_1,\dots,x_n;\xi_1,\dots,\xi_n)$ is given, then the variables x_j, ξ_j corresponding to $q_j = 1$ are said to be *essential*, and their collections are denoted (respectively) as x' and ξ'. An analog of (6′) is the condition

$$\{P_q(x_1,\dots,x_n;\xi_1,\dots,\xi_n) = 0\} \Rightarrow \{|\operatorname{grad}_{\xi'} P_q(x_1,\dots,x_n;\xi_1,\dots,\xi_n)| = 0\}. \tag{10′}$$

Under the additional assumption that the symbol P_q is real, in the above-mentioned works by Shananin and Lascar analogs of Hörmander's results[1)] were obtained. Our aim is to extend these results from the quasi-homogeneous case to the case of polynomials or operators whose principal part is determined using Newton's polygon (polynomials and operators of N-principal type). To gain a better understanding of these questions, we now present a simple algebraic result relating to the quasi-homogeneous case.

Theorem. *Let $q = (q_1,\dots,q_n) \in \mathbb{Z}^n$ and let $\min q_i = 1$. Then for polynomial (9) the following conditions are equivalent:*

(i) *P is a polynomial of q-principal type (i.e. (10) holds)*

(ii) *for any polynomial $Q(\xi)$, $\deg_q Q < m$, there is a constant $c_Q > 0$ such that*

$$(\rho(\xi)+1)^{m-1} \leqslant c_Q(\widetilde{P+Q})(\xi), \qquad \rho(\xi) \stackrel{\text{def}}{=} \sum |\xi_j|^{1/q_j}. \tag{11}$$

Proof. (i)$\Longrightarrow$(ii). Select on the "sphere" $\rho(\xi) = 1$ a finite covering $\{U_j\}$, the covering being so fine that either $P_q(\xi) \neq 0$, $\xi \in U_j$, or (see (10)) $|\operatorname{grad}_{\xi'} P_q(\xi)| \neq 0$, $\xi \in U_j$. These conditions remain valid in the q-cones

$$KU_j = \{\xi \in \mathbb{R}^n \setminus \{0\}, (\xi_1\rho^{-q_1}(\xi),\dots,\xi_n\rho^{-q_n}(\xi)) \in U_j\}$$

as well, which, obviously cover $\mathbb{R}^n \setminus \{0\}$.

Assume first that $\xi \in KU_j$ and $P_q(\xi) \neq 0$. Since among the polynomials $P^{(\alpha)} + Q^{(\alpha)}$ there is a constant, say a, we have

$$(\widetilde{P+Q})(\xi) > |a| + |P(\xi)+Q(\xi)| > a + \varepsilon|P_q(\xi)| - \varepsilon|P(\xi) - P_q(\xi) + Q(\xi)|. \tag{12}$$

---

1) Using Carleman's estimation technique, Lascar replaced condition (10′) by the weaker condition

$$\{P_q(x;\xi) = 0\} \Rightarrow \{|\operatorname{grad}_{(x',\xi')} P_q(x,\xi)| \neq 0\}.$$

In view of the quasi-homogeneity, there is $\varkappa$ such that

$$|P_q(\xi)| > \varkappa\rho^q(\xi).$$

The polynomial $P - P_q + Q$ is a linear combination of monomials of q-degree less than m. When estimating these monomials the following elementary lemma is of use.

Lemma. *Let $q_1\alpha_1 + \cdots + q_n\alpha_n < m$. Then $\forall\lambda > 0$ there is a constant c_λ such that*

$$|\xi_1^{\alpha_1} \ldots \xi_n^{\alpha_n}| \leqslant \lambda\rho^m(\xi) + c_\lambda. \tag{13}$$

Proof. Since $|\xi_j| \leqslant \rho^{q_j}(\xi)$, we have

$$|\xi_1^{\alpha_1} \ldots \xi_n^{\alpha_n}| \leqslant \rho(\xi)^{\alpha_1 q_1 + \cdots + \alpha_n q_n}.$$

Setting

$$p = m/(\alpha_1 q_1 + \cdots + \alpha_n q_n), \qquad a = \rho(\xi)^{\alpha_1 q_1 + \cdots + \alpha_n q_n}\lambda^{1/p}, \quad b = \lambda^{-1/p}$$

in the elementary inequality

$$ab \leqslant a^p/p + b^{p'}/p', \qquad a, b > 0, \quad 1/p + 1/p' = 1,$$

we obtain (13).

By virtue of the lemma, the third term on the right-hand side of (12) can be estimated from below by means of

$$-\varepsilon(\lambda\rho^m(\xi) + c(\lambda)).$$

Substituting this into (12) we obtain

$$(\widetilde{P + Q})(\xi) > \varepsilon(\varkappa - \lambda)\rho^q(\xi) + a - \varepsilon c(\lambda).$$

Taking $\lambda \leqslant \varkappa/2$ and $\varepsilon < a/2c(\varkappa/2)$ we derive inequality (11) for $\xi \in KU_j$.

In case $\operatorname{grad}_{\xi'} P_q(\xi) \neq 0$, we have

$$\begin{aligned}(\widetilde{P + Q})(\xi) &> |a| + |\operatorname{grad}_{\xi'}(P + Q)| \\ &> a + \varepsilon|\operatorname{grad}_{\xi'} P_q(\xi)| - \varepsilon|\operatorname{grad}_{\xi'}(P(\xi) - P_q(\xi) + Q(\xi))|.\end{aligned}$$

Since

$$|\operatorname{grad}_{\xi'} P_q(\xi)| > \varkappa'\rho^{m-1}(\xi)$$

and $\operatorname{grad}_{\xi'}(P - P_q + Q)$ is a polynomial whose q-degree is less than $m - 1$ (here we use the fact that $q_j \geqslant 1$), it remains to repeat the above estimation.

(ii)$\Longrightarrow$(i). Assume that (10) is not fulfilled, i.e. there is a point $\xi^0 = (\xi_1^0, \dots, \xi_n^0) \in \mathbb{R}^n$ such that

$$P_q(\xi^0) = 0, \qquad (\mathrm{grad}_{\xi'} P_q)(\xi^0) = 0. \tag{10_0}$$

Consider inequality (11) on the curve

$$\xi(t) = (\xi_1^0 t^{q_1}, \dots, \xi_n^0 t^{q_n})$$

for $t \to \infty$. Then we have

$$\mathrm{const}\, t^{m-1} \leqslant \big(1 + \rho(\xi(t))\big)^{m-1} \leqslant c(\widetilde{P+Q})(\xi(t)). \tag{14}$$

We write

$$P(\xi) = P_q(\xi) + P_q'(\xi) + P''(\xi), \qquad Q(\xi) = Q_q(\xi) + Q'(\xi),$$

where P_q' and Q_q are q-homogeneous polynomials of degree $m-1$, and the q-degrees of P'' and Q' do not exceed $m-2$. Then inequality (14) takes the form

$$\mathrm{const}\, t^{m-1} \leqslant c\big(|P_q(\xi^0)| t^m + |P_q'(\xi^0) + Q_q(\xi^0)| t^{m-1} + |\,\mathrm{grad}_{\xi'} P_q(\xi^0)| t^{m-1} + o(t^{m-1})\big). \tag{14'}$$

By virtue of (10_0), the first and third terms on the right-hand side are equal to zero. The polynomial Q can always be chosen so that

$$P_q'(\xi^0) + Q_q(\xi^0) = 0. \tag{15}$$

Indeed, let, for definiteness, $\xi_j^0 \neq 0$ for $j \leqslant s$ and $\xi_j^0 = 0$ for $j > s$. If there are integers $\alpha_1, \dots, \alpha_s$ such that $\alpha_1 q_1 + \dots + \alpha_s q_s = m-1$, then setting $Q(\xi) = c\xi_1^{\alpha_1} \dots \xi_s^{\alpha_s}$, where $c = -P_q'(\xi^0)/({\xi_1^0}^{\alpha_1} \dots {\xi_s^0}^{\alpha_s})$, we attain the fulfilment of condition (15). In case there are no integers $\alpha_1, \dots, \alpha_s$ for which $\alpha_1 q_1 + \dots + \alpha_s q_s = m-1$, we have $P_q'(\xi^0) = Q_q(\xi^0) = 0$, i.e. (15) remains valid.

In view of what had been said, we have $o(t^{m-1})$ on the right-hand side of (14′), and we arrive at a contradiction proving the desired assertion.

Remarks. 1) If $q = (1, \dots, 1)$, then inequality (11) goes into (7). We note that inequality (6) is equivalent to

$$\sum_{|\alpha| \leqslant m-1} |\xi^\alpha| < c(\widetilde{P+Q})(\xi). \tag{16}$$

In the quasi-homogeneous case inequality (11) is stronger than the quasi-homogeneous analog of (16):

$$\sum_{\alpha_1 q_1 + \dots + \alpha_n q_n \leqslant m-1} |\xi_1^{\alpha_1} \dots \xi_n^{\alpha_n}| \leqslant c(\widetilde{P+Q})(\xi). \tag{17}$$

Indeed, according to inequality (11), the function $(\widetilde{P+Q})$ majorizes the expressions $|\xi_j|^{(m-1)/q_j}$. If $q_j > 1$, the number $(m-1)/q_j$ can be non-integral, and then the left-hand side of (17) contains only $|\xi_j|^{k_j}$, $k_j = [(m-1)/q_j]$.

2) We present two examples of quasi-homogeneous polynomials for which condition (10) is violated and, consequently, inequality (11) may not hold. Nevertheless, these polynomials satisfy inequality (17).

Example 1. $n = 2$, $P(\xi) = \xi_1^4 + i\xi_1^2\xi_2$. In this case $q = (1,2)$, and the essential variables is ξ_1. It is obvious that

$$|P(\xi)| + |\partial P/\partial\xi_1| = 0 \quad \text{for} \quad \xi_1 = 0,\ \xi_2 \in \mathbb{R}.$$

On the other hand, it is clear that

$$\sum_{\alpha_1+2\alpha_2 \leqslant 3} |\xi_1|^{\alpha_1}|\xi_2|^{\alpha_2} \leqslant \widetilde{P}(\xi).$$

This inequality (with another constant) also remains true under replacement of P by $P + Q$, where Q is an arbitrary polynomial of $(1,2)$-degree no higher than 3.

Example 2. $n = 3$,

$$P(\xi) = \xi_1^2(\xi_1^4 + \xi_2^2 + \xi_3^2) + i(\xi_2 - \xi_3)^3.$$

This polynomial coincides with its $(1,2,2)$-homogeneous part. The essential variables is ξ_1, and we have

$$|P(\xi)| + |\partial P/\partial\xi_1| = 0 \quad \text{for} \quad \xi_1 = 0,\ \xi_2 = \xi_3 \in \mathbb{R}.$$

On the other hand, the inequalities

$$|P(\xi)| > \xi_1^6 + \xi_1^2(\xi_2^2 + \xi_3^2), \quad |\partial^2 P/\partial\xi_1^2| > 2(\xi_2^2 + \xi_3^2),$$

readily imply (17).

3) In the case of non-integral $q_1, \ldots, q_n$ there can exist monomials ξ^α such that

$$m - 1 < \langle\alpha, q\rangle < m.$$

These monomials can majorize the left-hand side of inequality (11) and guarantee its fulfilment when the q-principal part does not satisfy condition (10). In other words, without the assumption that q_j are integers (10) is not a necessary condition for the validity of (11).

On the other hand, as is seen from the above proof, (10) is a sufficient condition for the validity of inequality (11) for any values of $q_1, \ldots, q_n$.

§2. Polynomials of N-principal type

In this section we shall elaborate a complete description of polynomials in two variables for which inequality (15) holds. In the special case when Newton's polygon is a triangle the established result goes into Theorem 1.3. In other words, the polynomials satisfying inequality (15) are a natural generalization of polynomials of quasi-principal type to the case of arbitrary Newton's polygon. The main result in this section will be proved under some additional arithmetical conditions on Newton's polygon, and we begin the presentation of the material with the description of these conditions.

2.1. Remarks on Newton's polygon. We shall deal with polynomials in two variables

$$P(\xi,\eta)=\sum a_{\alpha\beta}\xi^{\alpha}\eta^{\beta}. \tag{1}$$

As above, $N(P)$ is Newton's polygon, $\Gamma_j^{(0)}=(\alpha_j,\beta_j)$, $j=0,\dots,m+1$, are the vertices of $N(P)$ (where $(\alpha_0,\beta_0)=(0,0)$), and $\Gamma_j^{(1)}$ are the sides joining $\Gamma_j^{(0)}$ and $\Gamma_{j+1}^{(0)}$. All the further statements are essentially simplified if it is assumed that Newton's polygon $N(P)$ *is regular.* In what follows we do not stipulate this condition.

Let $q^{(j)}=(q_1^{(j)},q_2^{(j)})$ be the outer normal vector to $N(P)$. If $j\neq 0$, $m-1$ (i.e. the side Γ_j^1 does not lie on the coordinate axes), then, by virtue of the regularity condition, the vector $q^{(j)}$ has positive components. Since the vector is determined up to within a positive factor, we normalize its components by the condition

$$\min(q_1^{(j)},q_2^{(j)})=1, \qquad \max(q_1^{(j)},q_2^{(j)})=b_j, \tag{2}$$

i.e. either $q^{(j)}=(1,b_j)$ or $q^{(j)}=(b_j,1)$. We also remind the reader that the polygon $N(P)$ can be determined by the system of inequalities

$$\langle(\alpha,\beta),q^{(j)}\rangle\leqslant c_j, \quad j=1,\dots,m, \quad \alpha\geqslant 0,\beta\geqslant 0. \tag{3}$$

If the regularity condition is fulfilled, then all the earlier presented definitions of minor points are equivalent. Therefore $(\alpha,\beta)\in N(P)$ is a minor point if $\exists(\alpha',\beta')\in N(P)$ such that $\alpha<\alpha'$, $\beta\leqslant\beta'$ or $\alpha\leqslant\alpha'$, $\beta<\beta'$. An integral point (α,β) will be called a $\mathbb{Z}$-minor point if there is an integral point $(\alpha'\beta')$ possessing the indicated properties. The convex hull of all $\mathbb{Z}$-minor points will be denoted as $\delta_{\mathbb{Z}}(P)$. Obviously, we have $\delta_{\mathbb{Z}}(P)\subset\delta(P)$. We present an example of a polynomial for which this inclusion relation is strict.

Example. $P(\xi,\eta)=\xi^3+\eta^2+c\xi\eta$. In this case $N(P)$ is a triangle with vertices $(0,0)$, $(0,3)$, and $(2,0)$. The point $(1,1)$ is an interior point of the triangle and, consequently, is minor point. However it is not a $\mathbb{Z}$-minor point.

On the other hand, in the case of polynomials correct in Petrovskiĭ's sense the sets $\delta(P)$ and $\delta_{\mathbb{Z}}(P)$ coincide (see Lemma 2 in Section 3.2.2). We state necessary and sufficient conditions for the coincidence of the polygons $\delta(P)$ and $\delta_{\mathbb{Z}}(P)$.

Lemma. *For polynomial* (1) *with regular Newton's polygon the conditions below are equivalent.*

(i) *The number* b_j, $j=1,\dots,m$, *in* (2) *are integers.*

(ii) *If the polygon* $N(P)$ *is determined by inequalities* (3) *and the vector* $q^{(j)}$ *is normalized by condition* (2), *then*

$$\langle(\alpha,\beta),q^{(j)}\rangle\leqslant c_j-1, \quad j=1,\dots,m, \quad \forall(\alpha,\beta)\in(N(P)\setminus\Gamma_j^{(1)})\subset\mathbb{Z}^2. \tag{4}$$

(iii) $\delta(P)=\delta_{\mathbb{Z}}(P)$.

Proof. (i)$\Longrightarrow$(ii). Recall that

$$c_j = \langle (\alpha_j, \beta_j), q^{(j)} \rangle.$$

Since the vertices (α_j, β_j) have integral coordinates, under condition (i) the numbers c_j in (3) are integers. If a point (α, β) does not lie on the straight line passing along $\Gamma_j^{(1)}$, then $\langle (\alpha, \beta), q^{(j)} \rangle < c_j$. The numbers being integral, we arrive at (4).

(ii)$\Longrightarrow$(iii). Take an arbitrary point $(\alpha^0, \beta^0) \in \delta(P)$. In view of the regularity of $N(P)$, the point cannot lie on the sides $\Gamma_j^{(1)}$, $j = 1, \ldots, m$, and therefore it satisfies inequalities (4). We now show that either for the point $(\alpha, \beta) = (\alpha^0 + 1, \beta^0)$ or for the point $(\alpha, \beta) = (\alpha^0, \beta^0 + 1)$ inequalities (3) hold, i.e. the corresponding point belongs to $N(P)$ and (α^0, β^0) is a $\mathbb{Z}$-minor point.

If all vectors $q^{(j)}$ have the form $q^{(j)} = (1, b_j)$, then inequalities (3) are fulfilled for $(\alpha, \beta) = (\alpha^0 + 1, \beta^0)$, and if $q^{(j)} = (b_j, 1)$ for all j, then (3) holds for $(\alpha, \beta) = (\alpha^0, \beta^0 + 1)$.

We now consider the general case when

$$q^{(j)} = (1, b_j), \quad j = 1, \ldots, h, \qquad q^{(j)} = (b_j, 1), \quad j > h.$$

Denote by N' the intersection of $N(P)$ with the half-plane $\beta \geqslant \beta_h$ and by N'' the

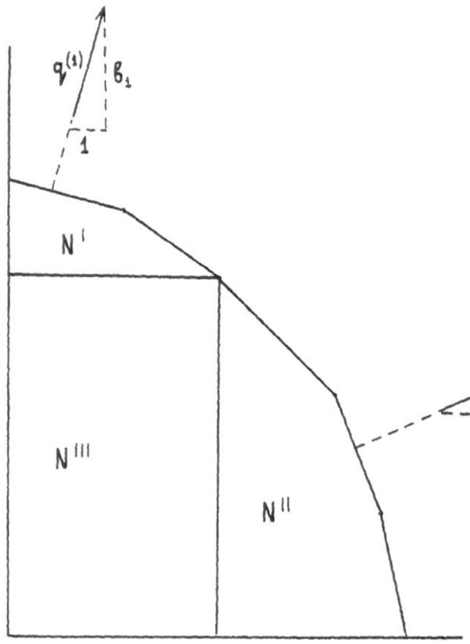

Figure 3

intersection of $N(P)$ with the half-plane $\alpha \geqslant \alpha_h$, and let $N''' = \{(\alpha, \beta), \alpha < \alpha_h, \beta \leqslant \beta_h\}$, (see Figure 3). By what has been said, if $(\alpha^0, \beta^0) \in N'$, then $(\alpha^0 + 1, \beta^0) \in N(P)$, and if $(\alpha^0, \beta^0) \in N''$, then $(\alpha^0, \beta^0 + 1) \in N(P)$. In case $(\alpha^0, \beta^0) \in N'''$, we have either $\alpha^0 < \alpha_h$ or $\beta^0 < \beta_h$, i.e. either $(\alpha^0+1, \beta^0) \in N(P)$ or $(\alpha^0, \beta^0+1) \in N(P)$.

(iii)$\Longrightarrow$(i). Let $q^{(j)} = (1, b_j)$ where $b_j = m/n > 1$, the numbers m and n being natural and coprime. Then the point $(\alpha_{j+1}, \beta_j + [m/n])$ belongs to $\delta(P)$ (as an interior point of $N(P)$) but is not a $\mathbb{Z}$-minor point, which contradicts (iii).

In what follows we impose on the polygon $N(P)$ the arithmetical

Condition (A). *The polygon $N(P)$ satisfies the equivalent conditions of the lemma.*

Denote by $\widehat{\delta}(P)$ the polygon determined by the inequalities

$$\widehat{\delta}(P) = \{(\alpha, \beta) \in \overline{\mathbb{R}^2_+}, \langle(\alpha, \beta), q^{(j)}\rangle \leqslant c_j - 1, j = 1, \dots, m\}.$$

If Condition (A) is fulfilled, then, according to Lemma (ii), the polygon $\widehat{\delta}(P)$ is an extension of $\delta(P)$, i.e.

$$\delta(P) \subset \widehat{\delta}(P) \subset N(P),$$

and (in contrast to $\delta(P)$), the polygon $\widehat{\delta}(P)$ is regular (of course, if $N(P)$ possesses this property).

The example of polynomials whose Newton's polygon is a triangle (Theorem 1.3 and Remarks 1) and 2) in Section 1.3) demonstrates the natural character of the introduction of the polygon $\widehat{\delta}(P)$.

2.2. Statement of the main result. Let $q^{(j)}$ be the outer normal vector to the side $\Gamma_j^{(1)}$ normalized by condition (2) and let $P_{q^{(j)}}(\xi, \eta)$ be the principal $q^{(j)}$-homogeneous part of P. The variable ξ (accordingly, η) is said to be an *essential* variable of the side $\Gamma_j^{(1)}$ if $q_1^{(j)} = 1$ (accordingly, $q_2^{(j)} = 1$). Essential variables will be denoted ξ'.

Definition. Polynomial (1) is called a polynomial of N-principal type if for any side $\Gamma_j^{(1)}$, $j = 1, \dots, m$, with essential variable (or variables) ξ' we have

$$\{P_{q^{(j)}}(\xi, \eta) = 0\} \Rightarrow \{|\operatorname{grad}_{\xi'} P_{q^{(j)}}(\xi, \eta)| \neq 0\}, \quad \xi \neq 0,\ \eta \neq 0. \tag{5}$$

Theorem. *Let $N(P)$ be regular Newton's polygon and let condition (A) be fulfilled. Then the following conditions are equivalent.*

(i) *P is a polynomial of N-principal type, i.e. for $j = 1, \dots, m$ condition (5) holds.*

(ii) *For any polynomial Q, $N(Q) \subset \delta(P)$, there is a constant c_Q such that*

$$\Xi_{\widehat{\delta}(P)}(\xi, \eta) \leqslant c_Q \widetilde{(P+Q)}(\xi) \quad \forall (\xi, \eta) \in \mathbb{R}^2. \tag{6}$$

Proof. (ii)$\Longrightarrow$(i). Let condition (5) be not fulfilled for some j, i.e. let there be $\xi^0 \neq 0$, $\eta^0 \neq 0$ such that

$$P_{q^{(j)}}(\xi^0, \eta^0) = 0, \qquad |\operatorname{grad}_{\xi'} P_{q^{(j)}}(\xi^0, \eta^0)| = 0. \tag{5'}$$

Considering inequality (6) on the curve $\xi(t) = t^{q_1^{(j)}}\xi^0$, $\eta(t) = t^{q_2^{(j)}}\eta^0$ and repeating literally the argument in Theorem 1.3 we show that (5′) contradicts (6).

The implication (i)$\Longrightarrow$(ii) is derived from the following

Proposition. *Let $P(\xi, \eta)$ be a polynomial of N-principal type. Then the inequality*

$$\Xi_{\widehat{\delta}(P)}(\xi, \eta) \leqslant c(|P(\xi, \eta)| + |\operatorname{grad} P(\xi, \eta)| + 1) \tag{7}$$

holds.

Since the definition of a polynomial of N-principal type involves only those terms of the polynomial that correspond to the sides $\Gamma_j^{(1)}$, $j = 1, \ldots, m$, of Newton's polygon, inequality (7) remains valid under replacement of P by $P + Q$, $N(Q) \subset \delta(P)$. Hence, the proposition implies (i)$\Longrightarrow$(ii).

Remarks. 1) If $N(P)$ is a right triangle, then the theorem goes into Theorem 1.3.
2) Inequality (6) implies (1.5). However, as is shown by Example 1 in Section 1.3, conditions (5) are not necessary and sufficient conditions for the fulfilment of (6).
3) We note that the proposition does not use the arithmetical condition (A). Thus, for every polynomial of N-principal type inequality (1.5) holds. In general, the converse assertion is not true.

The remaining part of this section is devoted to the proof of the proposition, which splits into two stages. Inequality (7) is first proved for special regions relating to the vertices or sides $\Gamma_j^{(k)}$, $k = 0, 1$. These regions are selected so that the corresponding principal part $P_{\Gamma_j^{(k)}}$ dominates the other terms. We prove that the specially selected regions cover the entire plane $\mathbb{R}^2$, and therefore inequality (6) results from "pasting together" the inequalities in these special regions. We note that incidentally we obtain a new proof of Theorem 1.3.3.

When proving (7) we can confine ourselves to the case $\xi, \eta > 0$, (the general case reduces to the former by means of the transformations $\xi \to -\xi$ or $\eta \to -\eta$).

2.3. Estimates in the corner regions $G(\Gamma_j^{(0)}, \varepsilon_0)$ relating to the vertices of the polygon $N(P)$. Let $\Gamma_j^{(0)} = (\alpha_j, \beta_j)$, $\Gamma_j^{(0)} \neq (0,0)$, be a vertex of $N(P)$. Denote by $V_{(0)}^j$ the angle between the normals at the vertex, i.e. the region lying between the rays $\{tq^{(j-1)}, t \geqslant 0\}$ and $\{tq^{(j)}, t \geqslant 0\}$ passing along the normals to the sides $\Gamma_{j-1}^{(1)}$ and $\Gamma_j^{(1)}$ at whose intersection the vertex $\Gamma_j^{(0)}$ lies (we note that $q^{(0)} = (-1, 0)$ and $q^{(m+1)} = (0, -1)$). As can easily be seen, $q = (q_1, q_2) \in V_{(0)}^j$ if and only if

$$-q_2^{(j)} q_1 + q_1^{(j)} q_2 \geqslant 0, \qquad q_2^{(j-1)} q_1 - q_1^{(j-1)} q_2 \geqslant 0.$$

Denote by $G(\Gamma_j^{(0)}, \varepsilon_0)$ the set of (ξ, η) such that the vector $(\log \xi, \log \eta)$[1] belongs to the translated angle $V_{(0)}^j$:

$$-q_2^{(j)} \log \xi + q_1^{(j)} \log \eta > \log \frac{1}{\varepsilon_0}, \qquad q_2^{(j-1)} \log \xi - q_1^{(j-1)} \log \eta > \log \frac{1}{\varepsilon_0}. \tag{8}$$

We now show that in the region $G(\Gamma_j^{(0)}, \varepsilon_0)$ the monomial $\xi^{\alpha_j}\eta^{\beta_j}$ corresponding to the vertex $\Gamma_j^{(0)}$ dominates the other monomials.

Lemma. *$\forall \varepsilon > 0$ and for any closed set $K \subset N(P) \setminus \Gamma_j^{(0)}$ there is a constant $\varepsilon_0(K, \varepsilon)$ such that for $\varepsilon_0 = \varepsilon_0(K, \varepsilon)$ the inequality*

$$\xi^\alpha \eta^\beta < \varepsilon \xi^{\alpha_j} \eta^{\beta_j}, \qquad (\xi, \eta) \in G(\Gamma_j^{(0)}, \varepsilon_0), \tag{9}$$

holds.

Since the polygon $\widehat{\delta}(P)$ does not contain the vertices of $N(P)$ distinct from the origin, an immediate consequence of the lemma is the following

Proposition. *If ε_0 is sufficiently small, then for $(\xi, \eta) \in G(\Gamma_j^{(0)}, \varepsilon_0)$ inequality (7) holds:*

$$\Xi_{\widehat{\delta}(P)}(\xi, \eta) \leqslant c|P(\xi, \eta)|. \tag{7'}$$

Proof. We choose the set $K \subset N(P) \setminus \Gamma_j^{(0)}$ so that it contains $\widehat{\delta}(P)$ and all integral points of $N(P)$ distinct from $(\alpha_j, \beta_j) = \Gamma_j^{(0)}$. Then, according to the lemma, for $\varepsilon_0 = \varepsilon_0(K, \varepsilon)$ the left-hand side of (7) does not exceed $\varepsilon \varkappa \xi^{\alpha_j} \eta^{\beta_j}$, where $\varkappa$ is the number of vertices in $\widehat{\delta}(P)$. On the other hand, in view of the lemma, we have

$$\begin{aligned}
\xi^{\alpha_j}\eta^{\beta_j} &= |a_{\alpha_j\beta_j}|^{-1} |P(\xi, \eta) - (P(\xi, \eta) - a_{\alpha_j\beta_j} \xi^{\alpha_j}\eta^{\beta_j})| \\
&\leqslant |a_{\alpha_j\beta_j}|^{-1} \Big(|P(\xi, \eta)| + \sum_{(\alpha,\beta) \in N(P) \setminus \Gamma_j^{(0)}} |a_{\alpha\beta} \xi^\alpha \eta^\beta| \Big) \\
&\leqslant |a_{\alpha_j\beta_j}|^{-1} |P(\xi, \eta)| + \varepsilon \Big(\sum |a_{\alpha\beta} a_{\alpha_j\beta_j}^{-1}| \Big) \xi^{\alpha_j}\eta^{\beta_j}.
\end{aligned}$$

Selecting ε so that the coefficient in $\xi^{\alpha_j}\eta^{\beta_j}$ on the right-hand side does not exceed 1/2 we estimate $\xi^{\alpha_j}\eta^{\beta_j}$ by means of $P(\xi, \eta)$ and thus prove inequality (7′).

The proof of the lemma. Denote by L_j and L_{j-1} the rays passing along the sides $\Gamma_j^{(1)}$ and $\Gamma_{j-1}^{(1)}$ and issuing from the vertex $\Gamma_j^{(0)}$. Then

$$L_j = \{(\alpha, \beta) \in \mathbb{R}^2, \alpha = \alpha_j + t q_2^{(j)}, \beta = \beta_j - t q_1^{(j)}, t \geqslant 0\},$$

[1] We once again remind the reader that under consideration are only nonnegative ξ and η.

and, according to the first inequality (8), for $(\alpha, \beta) \in L_j$ we have

$$\xi^\alpha \eta^\beta = \exp(\alpha \log \xi + \beta \log \eta) = \exp(\alpha_j \log \xi + \beta_j \log \eta)$$
$$\times \exp(t(q_2^{(j)} \log \xi - q_1^{(j)} \log \eta)) \leqslant \varepsilon_0^t \xi^{\alpha_j} \eta^{\beta_j},$$

where $t = t(\alpha, \beta)$. Similarly, for any point $(\alpha, \beta) \in L_{j-1}$ there is $\theta = \theta(\alpha, \beta)$ such that

$$\xi^\alpha \eta^\beta \leqslant \varepsilon_0^\theta \xi^{\alpha_j} \eta^{\beta_j}.$$

It remains to note that the points (α, β), $(\alpha', \beta') \in L_j$, and (γ, δ), $(\gamma', \delta') \in L_{j-1}$, $t(\alpha', \beta') > t(\alpha, \beta) > \varkappa > 0$, $\theta(\gamma', \delta') > \theta(\gamma, \delta) > \varkappa > 0$, can be chosen so that the set K lies in the convex hull of (α, β), (α', β'), (γ, δ), and (γ', δ'). Then for $\varepsilon_0^\varkappa < \varepsilon$ the inequality (7) holds.

2.4. Estimation in half-strips $G(\Gamma_j^{(1)}, \varepsilon_0, \varepsilon_1)$ relating to the sides of the polygon $N(P)$. If $\Gamma_j^{(1)}$ is a side of $N(P)$ not lying on the coordinate axes, i.e. $j = 1 \ldots, m$, and $q^{(j)} = (q_1^{(j)}, q_2^{(j)})$ is the outer normal vector, then we denote by $G(\Gamma_j^{(1)}, \varepsilon_0, \varepsilon_1)$ the half-strip determined (in the plane $(\log \xi, \log \eta)$) by the inequalities

$$\log \varepsilon_0 < -q_2^{(j)} \log \xi + q_1^{(j)} \log \eta < \log \frac{1}{\varepsilon_0}, \qquad q_1^{(j)} \log \xi + q_2^{(j)} \log \eta > \log \frac{1}{\varepsilon_1}. \quad (10)$$

Lemma 1. (i) *For any pair of points (α, β), $(\alpha', \beta') \in \Gamma_j^{(1)}$ there is θ (depending on α, β, α', β') such that*

$$\varepsilon_0^\theta \leqslant \xi^\alpha \eta^\beta \xi^{-\alpha'} \eta^{-\beta'} \leqslant \varepsilon_0^{-\theta} \quad \text{for} \quad (\xi, \eta) \in G(\Gamma_j^{(1)}, \varepsilon_0, \varepsilon_1).$$

(ii) $\forall (\alpha, \beta) \in \Gamma_j^{(1)}$, $\forall \varepsilon > 0$, $\varepsilon_0 > 0$ *and for any closed set $K \subset N(P) \setminus \Gamma_j^{(1)}$ there is $\varepsilon_1(\varepsilon, \varepsilon_0, K)$ such that*

$$\xi^{\alpha'} \eta^{\beta'} < \varepsilon \xi^\alpha \eta^\beta, \qquad (\alpha' \beta') \in K, \qquad (\xi, \eta) \in G(\Gamma_j^{(1)}, \varepsilon_0, \varepsilon_1)$$

for $\varepsilon_1 = \varepsilon_1(\varepsilon, \varepsilon_0, K)$.

Proof. (i) If (α, β), $(\alpha', \beta') \in \Gamma_j^{(1)}$, then $q_1^{(j)}(\alpha - \alpha') + q_2^{(j)}(\beta - \beta') = 0$ since for some θ we have

$$\alpha - \alpha' = -\theta q_2^{(j)}, \qquad \beta - \beta' = \theta q_1^{(j)}.$$

Therefore

$$\xi^\alpha \eta^\beta \xi^{-\alpha'} \eta^{-\beta'} = \exp((\alpha - \alpha') \log \xi + (\beta - \beta') \log \eta) = \exp(\theta(-q_2^{(j)} \log \xi + q_1^{(j)} \log \eta)).$$

With account of the first inequality (10), the right-hand side is estimated from below by means of ε_0^θ and from above by means of $\varepsilon_0^{-\theta}$.

(ii). For any point $(\alpha', \beta') \in N(P) \setminus \Gamma_j^{(1)}$ there is a point $(\gamma, \delta) \in \Gamma_j^{(1)}$ and $t > 0$ such that

$$\alpha' = \gamma - tq_1^{(j)}, \qquad \beta' = \delta - tq_2^{(j)}.$$

Then, in view of (i) and the second inequality (10), we have

$$\xi^{\alpha'}\eta^{\beta'} = \xi^{\gamma}\eta^{\beta}\exp(-t(q_1^{(j)}\log\xi + q_2^{(j)}\log\eta)) \leqslant \varepsilon_1^t\xi^{\gamma}\eta^{\delta} \leqslant \varepsilon_1^t\varepsilon_0^{-|\theta|}\xi^{\alpha}\eta^{\beta}.$$

In this inequality the constants t and θ depend continuously of the point (α', β'), and therefore for $(\alpha', \beta') \in K \subset N(P) \setminus \Gamma_j^{(1)}$ there are unified t and θ for which the indicated inequality is fulfilled. Selecting ε_1 from the condition $\varepsilon_1^t\varepsilon_0^{-|\theta|} < \varepsilon$ we prove the desired assertion.

Lemma 2. *Let $\Gamma_j^{(1)}$ be a side of $N(P)$, let $q^{(j)}$ be the outer normal vector to $\Gamma_j^{(1)}$, and let ξ' be the essential variables of $\Gamma_j^{(1)}$. Let*

$$W(a,b) = \{(\xi,\eta) \in \mathbb{R}^2, a < -q_2^{(j)}\log|\xi| + q_1^{(j)}\log|\eta| < b\}. \tag{11}$$

Then, if condition (5) is fulfilled and the difference $b - a$ is sufficiently small, we have either

$$P_{q^{(j)}}(\xi,\eta) \neq 0 \quad \forall(\xi,\eta) \in W(a,b) \setminus (0,0) \tag{12}$$

or

$$|\operatorname{grad}_{\xi'} P_{q^{(j)}}(\xi,\eta)| \neq 0 \quad \forall(\xi,\eta) \in W(a,b) \setminus (0,0). \tag{12'}$$

Proof. The definition (11) of the region $W(a,b)$ implies that this region can intersect the axes $\{\xi = 0\}$ and $\{\eta = 0\}$ only at the origin. It follows that at all points belonging to $W(a,b) \setminus (0,0)$ one of the expressions $P_{q^{(j)}}(\xi,\eta)$ and $|\operatorname{grad}_{\xi'} P_{q^{(j)}}(\xi,\eta)|$ is nonzero. Further, the q-homogeneity of the polynomial $P_{q^{(j)}}$ implies that the values of these expressions are reconstructed from their values on the arc

$$W^0(a,b) = \{(\xi,\eta) \in W(a,b), |\xi|^{1/q_1^{(j)}} + |\eta|^{1/q_2^{(j)}} = 1\}.$$

Therefore, if $b - a$ is sufficiently small, then, in view of the continuity, at all points in W^0 either $P_{q^{(j)}}$ or $|\operatorname{grad}_{\xi'} P_{q^{(j)}}|$ is nonzero, whence it follows that either (12) or (12') is fulfilled.

Proposition. *Let $\Gamma_j^{(1)}$ be a side of $N(P)$ and let ξ' be the essential variable of this side. Then for any $\varepsilon_0 > 0$ there is ε_1 such that in the region $G(\Gamma_j^{(1)}, \varepsilon_0, \varepsilon_1)$ the inequality*

$$\Xi_{\widehat{\delta}(P)}(\xi,\eta) \leqslant c(|P(\xi,)| + |\operatorname{grad}_{\xi'} P(\xi,\eta)|) \tag{7''}$$

holds.

Proof. 1) We first of all note that, according to definition (11) and the first inequality (10), we have $G(\Gamma_j^{(1)}, \varepsilon_0, \varepsilon_1) \in W(\log\varepsilon_0, \log(1/\varepsilon_0))$. Dividing the closed interval $[\log \varepsilon_0, \log(1/\varepsilon_0)]$ into sufficiently small closed subintervals $[a_\lambda, a_{\lambda+1}]$ and replacing the first inequality in (10) by the set of inequalities

$$a_\lambda \leqslant -q_2^{(j)} \log \xi + q_1^{(j)} \log \eta \leqslant a_{\lambda+1}$$

we split $G(\Gamma_j^{(1)}, \varepsilon_0, \varepsilon_1)$ into a finite number of regions in each of which either (12) or (12′) holds. To simplify the notation, assume that the assertion of Lemma 2 holds throughout the region $G(\Gamma_j^{(1)}, \varepsilon_0, \varepsilon_1)$.

2) Let first $P_{q^{(j)}}(\xi, \eta) \neq 0$ for $(\xi, \eta) \in G(\Gamma_j^{(1)}, \varepsilon_0, \varepsilon_1)$. Then, by virtue of the homogeneity, $\forall(\alpha, \beta) \in \Gamma_j^{(1)}$ we have

$$\xi^\alpha \eta^\beta \leqslant c_0 |P_{q^{(j)}}(\xi, \eta)|, \qquad (\xi, \eta) \in W\Big(\log \varepsilon_0, \log \frac{1}{\varepsilon_0}\Big). \tag{13}$$

Using Lemma 1 (ii) we conclude that for an appropriately chosen ε_1 we have

$$\xi^{\alpha'} \eta^{\beta'} \leqslant \varepsilon c_0 |P_{q^{(j)}}(\xi, \eta)| \quad \text{for} \quad (\xi, \eta) \in G(\Gamma_j^{(1)}, \varepsilon_0, \varepsilon_1),$$
$$(\alpha', \beta') \in N(P) \setminus \Gamma_j^{(1)}. \tag{14}$$

By virtue of these inequalities, for $(\xi, \eta) \in G(\Gamma_j^{(1)}, \varepsilon_0, \varepsilon_1)$ we obtain

$$|P(\xi, \eta) - P_{q^{(j)}}(\xi,)| < \varepsilon c_1 |P_{q^{(j)}}(\xi, \eta)| < |P_{q^{(j)}}(\xi, \eta)|/2$$

provided that ε is sufficiently small. It follows that

$$|P_{q^{(j)}}(\xi, \eta)| < 2|P(\xi, \eta)|.$$

Substituting this inequality into the right-hand sides of (13) and (14) we prove (7″) (without the term $|\operatorname{grad}_{\xi'} P|$ on the right-hand side).

3) The estimate established in 2) can somewhat be strengthened. Denote by $T(\Gamma_j^{(1)})$ the triangle formed by the straight line passing along $\Gamma_j^{(1)}$ and the coordinate axes, i.e.

$$T(\Gamma_j^{(1)}) = \{(\alpha, \beta) \in \mathbb{R}^2, \alpha q_1^{(j)} + \beta q_2^{(j)} \leqslant c_j, \alpha \geqslant 0, \beta \geqslant 0\}.$$

We have in fact proved that

$$\xi^\alpha \eta^\beta \leqslant \text{const}\, |P(\xi, \eta)| \quad \text{for} \quad (\alpha, \beta) \in T(\Gamma_j^{(1)}), \quad (\xi, \eta) \in G(\Gamma_j^{(1)}, \varepsilon_0, \varepsilon_1). \tag{15}$$

4) Now let $|\operatorname{grad}_{\xi'} P_{q^{(j)}}| \neq 0$ in the region in question. Let, for definiteness, $\xi' = \xi$,[1] i.e. $\partial P(\xi, \eta)/\partial \xi \neq 0$, $(\xi, \eta) \in G(\Gamma_j^{(1)}, \varepsilon_0, \varepsilon_1)$. Then for $\partial P/\partial \xi$ an inequality of the type (15) is fulfilled if the triangle $T(\Gamma_j^{(1)})$ is replaced by the smaller triangle

$$T'(\Gamma_j^{(1)}) = \{(\alpha, \beta) \in \mathbb{R}^2, \alpha q_1^{(j)} + \beta q_2^{(j)} \leqslant c_j - 1, \alpha \geqslant 0, \beta \geqslant 0\}.$$

[1] If ξ' consists of the two variables ξ and η, then, narrowing (if necessary) the region under consideration, we can assume that the derivative with respect to one of the essential variables is nonzero.

Thus, we have

$$\Xi_{T'(\Gamma_j^{(1)})} \leqslant c\left|\frac{\partial P(\xi,)}{\partial \xi}\right| \quad \text{for} \quad (\xi,\eta) \in G(\Gamma_j^{(1)}, \varepsilon_0, \varepsilon_1).$$

It now remains to note that $\widehat{\delta}(P) \subset T'(\Gamma_j^{(1)})$.

2.5. The proof of proposition 2.2. We first of all note that since $\widetilde{P}(\xi,\eta) >$ const, inequality (6) is obviously fulfilled in any circle $\xi^2 + \eta^2 \leqslant R^2$. Therefore, by virtue of Proposition 2.3 and 2.4, we have to show that the complement of the set

$$\bigcup_{j=1}^{m} G(\Gamma_j^{(1)}, \varepsilon_0, \varepsilon_1) \cup \bigcup_{j=1}^{m+1} G(\Gamma_j^{(0)}, \varepsilon_0)$$

in the positive quadrant $\overline{\mathbb{R}^2_+}$ is covered by a circle of a sufficiently large radius. We pass to the plane $q = (q_1, q_2)$, $q_1 = \log \xi$, $q_2 = \log \eta$, and denote by $g(\Gamma_j^{(0)}, \varepsilon_0)$ and $g(\Gamma_j^{(0)}, \varepsilon_0, \varepsilon_1)$ the corresponding angles and half-strips:

$$g(\Gamma_j^{(0)}, \varepsilon_0) = \Big\{q \in \mathbb{R}^2, -q_2^{(j)} q_1 + q_1^{(j)} q_2 > \log \frac{1}{\varepsilon_0},$$

$$q_2^{(j-1)} q_1 - q_1^{(j-1)} q_2 > \log \frac{1}{\varepsilon_0}\Big\},$$

$$g(\Gamma_j^{(1)}, \varepsilon_0, \varepsilon_1) = \Big\{q \in \mathbb{R}^2, \log \varepsilon_0 < -q_2^{(j)} q_1 + q_1^{(j)} q_2 > \log \frac{1}{\varepsilon_0},$$

$$q_1^{(j)} q_1 + q_2^{(j)} q_2 > \log \frac{1}{\varepsilon_1}\Big\}.$$

Hence, we have to show that the complement

$$\mathbb{R}^2 \setminus \left(\bigcup_{j=1}^{m} g(\Gamma_j^{(1)}, \varepsilon_0, \varepsilon_1) \cup \bigcup_{j=1}^{m+1} g(\Gamma_j^{(0)}, \varepsilon_0)\right)$$

is contained in the translation of the negative quadrant

$$\overline{\mathbb{R}^2_-} - \{(q_1, q_2) \subset \mathbb{R}^2, q_1 \leqslant 0, q_2 \leqslant 0\}.$$

To prove this fact recall that the regions $g(\Gamma_j^{(0)}, \varepsilon_0)$ result from the translation of the angles $V_{(0)}^j$ between the normals at the vertices $\Gamma_j^{(0)}$, i.e. $g(\Gamma_j^{(0)}, 1) = V_{(0)}^j$. Since $N(P)$ is a convex polygon, we have

$$\mathbb{R}^2 = \bigcup_{j=0}^{m+1} V_{(0)}^j,$$

the angle between the normals at the vertex $\Gamma_0^{(0)}(0,0)$ coinciding with the negative quadrant. Thus,

$$\mathbb{R}^2 = \mathbb{R}^2_- \cup \bigcup_{j=1}^{m+1} g(\Gamma_j^{(0)}, \varepsilon_0), \qquad \varepsilon_0 = 1.$$

Decreasing ε_0 we translate the angles $g(\Gamma_j^{(0)}, \varepsilon_0)$ inside themselves so that "gaps"

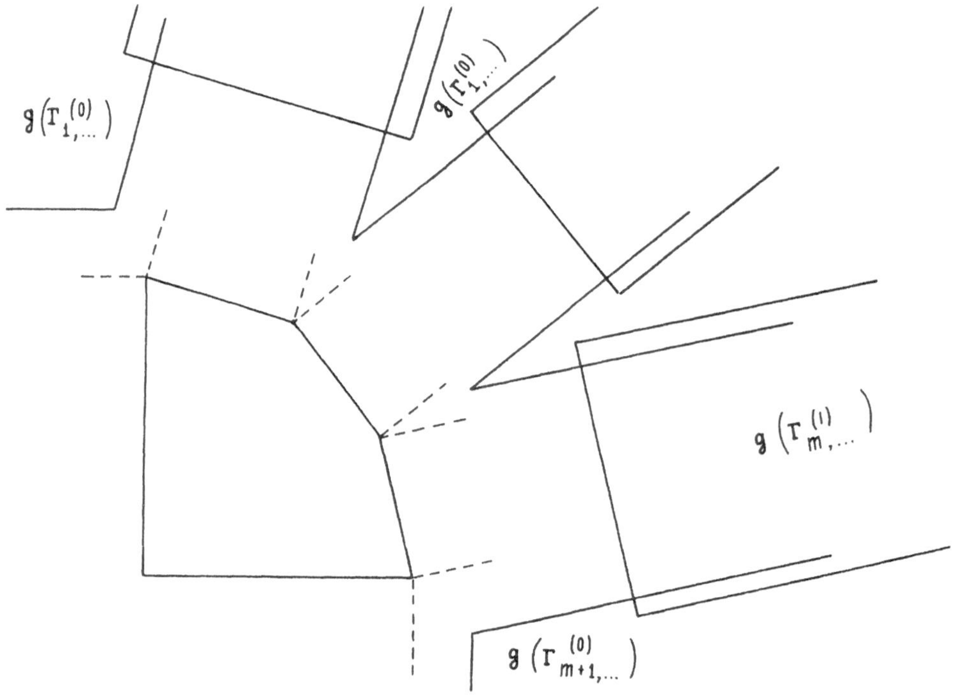

Figure 4

are formed between them (see Figure 4). Outside the neighborhood of the origin these "gaps" are covered by the half-strips $g(\Gamma_j^{(1)}, \varepsilon_0, \varepsilon_1)$. As to the neighborhood of the origin, it can be covered by the translated negative quadrant.

Remark. Comparing Proposition 2.4 with Item 2) in the proof of the proposition in the present section we obtain a new proof of Theorem 1.2.1 for the case of regular Newton's polygon. This method makes it possible to prove the theorem in the general formulation as well.

§3. The main L_2 estimate for operators of N-principal type

3.1. Statement of the results. We consider a differential operator with coefficients belonging to C^∞:

$$P(x,y;D_x,D_y) = \sum a_{\alpha\beta}(x,y)D_x^\alpha D_y^\beta. \tag{1}$$

As above, with the symbol $P(x,y;\xi,\eta)$ the polygons $N(P(x,y))$, $\delta(P(x,y))$, and $\widehat{\delta}(P(x,y))$ are associated, and by $N(P)$, $\delta(P)$, and $\widehat{\delta}(P)$ the convex hulls of the unions of these polygons over all $(x,y)\in\mathbb{R}^2$ [1)] are denoted.

Definition. A differential operator (1) is called an operator of N-principal type if

(i) the polygons $N(P(x,y))$ do not in fact depend on (x,y), i.e. $N(P(x,y)) = N(P)$;
(ii) at every fixed point (x^0,y^0) the polynomial $P(x^0,y^0;\xi,\eta)$ is a polynomial of N-principal type.

The definition implies that the polygon $N(P)$ is regular. Using the considerations in Section 3.2.1 one can weaken condition (i) by replacing the condition of independence of the polygons $N(P(x,y))$ on (x,y) by the condition of independence of the polygons $\delta(P(x,y))$ or $\widehat{\delta}(P(x,y))$. We do not dwell on these questions.

According to Definition 2.1, condition (ii) can be rewritten as

(ii′) the polygon $N(P)$ is regular, and for its any side (not lying on the coordinate axes) we have

$$\{P_{q^{(j)}}(x,y;\xi,\eta)=0\} \Rightarrow \{\mathrm{grad}_{\xi'} P_{q^{(j)}}(x,y;\xi,\eta)\neq 0\},$$
$$\xi' \text{ is the essential variable of } \Gamma_j^{(1)}, \xi\neq 0, \eta\neq 0. \tag{2}$$

To the symbols satisfying the conditions of the above definition all estimates in §2 are extended. More precisely, for any compactum $K\subset\mathbb{R}^2$ there is a constant $c=c(K)$ such that

$$\Xi_{\widehat{\delta}(P)}(\xi,) < c\big(|P(x,y;\xi,\eta)| + |\mathrm{grad}_{(\xi,\eta)} P(x,t;\xi,\eta)| + 1\big)$$
$$\text{for}\quad (\xi,)\in\mathbb{R}^2,\quad (x,y)\in K.$$

The main result in this section is

Theorem 1. *Let* (1) *be an operator of N-principal type and let the following additional condition hold:*

[1)] In what follows we shall prove for operator (1) only results of local character, and it can be assumed that the variables (x,y) range in a fixed region $\Omega\subset\mathbb{R}^2$. For the sake of brevity of the statements, we put $\Omega=\mathbb{R}^2$.

Condition (R). *If an (integral) point $(\alpha, \beta) \in N(P)$ is not integrally minor, then the function $a_{\alpha\beta}(x, y)$ is real.*

Then $\forall\varepsilon > 0$ $\exists\omega(\varepsilon)$ such that in a region Ω of a sufficiently small diameter ($\operatorname{diam}\Omega \leqslant \omega(\varepsilon)$) *the inequality*

$$\|u\|_{\widehat{\delta}(P)} \leqslant \varepsilon\|P(x, y; D_x, D_y)u\| \quad \forall u \in \mathcal{D}(\Omega) \tag{3}$$

holds, where

$$\|u\|^2_{\widehat{\delta}(P)} = \iint \Xi^2_{\widehat{\delta}(P)}(\xi, \eta)|\widehat{u}(\xi, \eta)|^2 \, d\xi \, d\eta.$$

Remark. When $N(P)$ is an isosceles right triangle, operator (1) is an operator of principal type and (3) goes into Hörmander's inequality.

In the case when $N(P)$ is a right triangle with integral leg ratio the estimate (3) was established by Shananin [1] and Lascar [1]. As was already mentioned (see Footnote on page 121), Lascar used a weaker condition than (2).

Corollary 1. *In the conditions of the theorem, for any symbol $Q(x, y; \xi, \eta) \in S\mathcal{L}_{\widehat{\delta}(P)}$ there is $\omega(\varepsilon, Q)$ such that under the condition* diam $< \omega(\varepsilon, Q)$ *the inequality*

$$\|u\|_{\widehat{\delta}(P)} \leqslant \varepsilon\|(P + Q)(x, y; D_x, D_y)u\| \quad \forall u \in \mathcal{D}(\Omega) \tag{4}$$

is fulfilled.

Corollary 2. *If the conditions of Theorem 1 hold, then $\forall\varepsilon > 0$ $\exists\omega^*(\varepsilon)$ such that if* $\operatorname{diam}\Omega < \omega^*(\varepsilon)$, *then we have*

$$\|u\|_{\widehat{\delta}(P)} \leqslant \varepsilon\|\,{}^tP(x, y; D_x, D_y)u\| \quad \forall u \in \mathcal{D}(\Omega) \tag{3'}$$

Proof. If condition (R) holds, then the symbols $P(x, y; \xi, \eta)$ and

$$P^*(x, y; \xi, \eta) = \overline{P}(x, y; \xi, \eta) + \sum \overline{P}^{(\alpha)}_{(\alpha)}(x, y; \xi, \eta)/\alpha!$$

differ by a symbol $Q(x, y; \xi, \eta) \in S\mathcal{L}_{N(P)}$. Therefore, by virtue of (4), we have

$$\|u\|_{\widehat{\delta}(P)} \leqslant \varepsilon\|P^*(x, y; D_x, D_y)u\| \quad \forall u \in \mathcal{D}(\Omega).$$

Using (1.4.6) we obtain (3′).

If the arithmetical condition (A) is imposed on the polygon $N(P)$, then condition (R) means that the coefficients $a_{\alpha\beta}(x, y)$ corresponding to the senior points of $N(P)$ assume real values. This condition can somewhat be be weakened.

Introduce the following notation: if $q \in \mathbb{R}^2$, then

$$\operatorname{char}_q P = \{(x, y; \xi, \eta) \in \mathbb{R}^2 \times \mathbb{R}^2, P_q(x, y; \xi, \eta) = 0\}.$$

Let $M^0 = (x^0, y^0; \xi^0, \eta^0) \in \operatorname{char}_q P$ and let $\xi^0 \neq 0$, $\eta^0 \neq 0$. Let $q = q^{(j)}$ be the outer normal to $\Gamma^{(1)}_j$ and let, for definiteness, ξ be an essential variable. If condition (2) is fulfilled at the point M^0, then, by the implicit function theorem, there exists a smooth function $\lambda(x, y, \eta)$ such that in the neighborhood of M^0 the set $\operatorname{char}_q P$ is determined by the equation $\xi = \lambda(x, y, \eta)$. In other words, $\operatorname{char}_q P$ is a smooth real manifold of codimension 1.

The theorem below is a version of Theorem 1.

Theorem 2. *Let (1) be an operator of N-principal type and let the polygon $N(P)$ satisfy condition (A). Let the following additional condition hold:*

Condition (R′). *Let $\Gamma_j^{(1)}$, $j = 1, \dots, m$, be an arbitrary side of $N(P)$ not lying on the coordinate axes, let $M^0 = (x^0, y^0; \xi^0, \eta^0)$ be an arbitrary point belonging to $\mathrm{char}_{q^{(j)}} P$, where $\xi^0 \neq 0$, $\eta^0 \neq 0$, and let $q^{(j)}$ be the outer normal to $\Gamma_j^{(1)}$. Then there is a neighborhood of the point M^0 where $\mathrm{char}_{q^{(j)}} P$ is a smooth manifold of codimension 1.*

In this case the assertion of Theorem 1 is true.

The generalization of Theorem 1 to the multidimensional case will be proved in Chapter 6. We now present the proof of Theorem 2.

The general plan of the proof of Theorem 2 coincides with that of the proof of Proposition 2.2. A covering of the space $\mathbb{R}^2_{(\xi,\eta)}$ by the regions $G(\Gamma_j^{(0)}, a)$ and $G(\Gamma_j^{(1)}, a, b)$ (see Sections 2.3 and 2.4) is considered. A function $u(x, y) \in \mathcal{D}$ is represented as a sum of functions whose spectra belong to the indicated regions. In these regions the orders of the derivatives involved in the operator are lower relative to those contained in the principal quasi-homogeneous part corresponding to a certain side or vertex. This property substantially simplifies the derivation of estimates. The plan of our further presentation is the following. We first construct (Section 3.2) a partition of unity in $\mathbb{R}^2_{(\xi,\eta)}$ relating to the partition into the regions $G(\Gamma_j^{(0)}, \dots)$ and $G(\Gamma_j^{(1)}, \dots)$. After that (Section 3.3) estimate (3) is microlocolized; Sections 3.4 and 3.6 are devoted to the proof of microlocal estimates.

3.2. Partition of unity. Let a finite covering $\{U_j\}$, $j = 0, \dots, J$, of the plane $\mathbb{R}^2$ of the variables (ξ, η) be given. A system of functions $\{\psi_j(\xi, \eta)\}$ is called a generalized partition of unity subordinate to the covering $\{U_j\}$ if

(i) $\psi_j(\xi, \eta) \in C^\infty$;

(ii) $\psi_j(\xi, \eta) \geqslant 0$ and $\operatorname{supp} \psi_j \subset U_j$;

(iii) there is a constant $K > 0$ such that

$$K^{-1} \leqslant \sum_{j=0}^{J} \psi_j(\xi, \eta) \leqslant K. \tag{5}$$

For $K = 1$ we obtain an ordinary partition of unity:

$$\sum_{j=0}^{J} \psi_j(\xi, \eta) \equiv 1. \tag{5'}$$

We note that from every generalized partition of unity an ordinary partition of unity is obtained by means of the transformation $\psi_j \to \psi_j (\Sigma \psi_k)^{-1}$. However, for our aims it is more convenient not to perform this transformation and to deal with condition (5). When estimating the commutators of the PDO $\psi_j(D_x, D_y)$

corresponding to the functions in (i)–(iii), Hörmander's condition proves extremely useful:

(iv) there is ρ, $0 < \rho < 1$, such that $\forall j$ the inequalities

$$|\psi_j^{(\alpha,\beta)}(\xi,\eta)| < c_{\alpha\beta_j}(1+|\xi|+|\eta|)^{-\rho(\alpha+\beta)} \tag{6}$$

hold, where

$$\psi_j^{(\alpha,\beta)}(\xi,\eta) = \partial^{\alpha+\beta}\psi_j(\xi,\eta)/\partial\xi^\alpha\partial\eta^\beta.$$

Proposition. *Let P be a polynomial with regular Newton's polygon $N(P)$. Consider the covering $\{U_j\}$ by means of the regions in* §2, *i.e.*

$$U_j = G(\Gamma_j^{(0)}, a_j), \qquad j = 1,\dots,m+1, \tag{7}$$

$$U_j = G(\Gamma_j^{(1)}, a_j, b_j), \qquad j = m+2,\dots,2m+1, \tag{7'}$$

$$U_0 = \{(\xi,\eta) \in \mathbb{R}^2, \xi^2+\eta^2 \leqslant R^2\}. \tag{7''}$$

Then there exists a system of functions $\{\psi_j(\xi,\eta)\}$ satisfying conditions (i)–(iv) *and the additional condition*

(v) for any $\alpha \geqslant 0$, $\beta \geqslant 0$, $\alpha+\beta > 0$, $\forall j$ and any integral point $(\gamma,\delta) \in N(P)$ there is a constant $c_{j\alpha\beta\gamma\delta}$ such that

$$|\xi^\gamma\eta^\delta\psi_j^{(\alpha,\beta)}(\xi,\eta)| < c_{j\alpha\beta\gamma\delta}\Xi_{\widehat{\delta}(P)}(\xi,\eta). \tag{8}$$

Proof. Recall that regions (7) are determined by the inequalities

$$-q_2^{(j)}\log|\xi| + q_1^{(j)}\log|\eta| > \log\frac{1}{\varepsilon_0}, \qquad q_2^{(j-1)}\log|\xi| - q_1^{(j-1)}\log|\eta| > \log\frac{1}{\varepsilon_0}.$$

In the case of regions (7′) a third inequality

$$q_1^{(j)}\log|\xi| + q_2^{(j)}\log|\eta| > \log\frac{1}{\varepsilon_1}$$

should be added to the former two. In other words, each of the regions U_j, $j > 0$, is determined by the set of inequalities

$$q_1^{(j,\mu)}\log|\xi| + q_2^{(j,\mu)}\log|\eta| \geqslant \log R_{j\mu}, \tag{9}$$

where $\mu = 1,2$ for $j = 1,\dots,m+1$ and $\mu = 1,2,3$ for $j > m+1$. The numbers $q_k^{(j,\mu)}$ are the components of the outer normals $q^{(j)}$ to the sides of $N(P)$ so that under an appropriate normalization (which is inessential) they can be assumed to be integral or, which is more, even numbers.

Fix a sufficiently small $\varkappa > 0$ and take a function $\theta(t) \in C^\infty$, $\theta(t) \geqslant 0$, such that $\theta(t) = 0$ for $t \leqslant 0$ and $\theta(t) = 1$ for $t \geqslant \varkappa$. Set

$$\psi_j(\xi,\eta) = \theta(\xi^2 + \eta^2 - R^2) \prod_\mu \theta(\exp(q_1^{(j,\mu)} \log|\xi| + q_2^{(j,\mu)} \log|\eta|) - R_{j,\mu}), \quad (10)$$

$$\psi_0(\xi,\eta) = \theta(R_0^2 - \xi^2 - \eta^2). \quad (10')$$

The expressions for these functions immediately imply that

(i') $\psi_j(\xi,\eta) \in C^\infty$ if $\xi \neq 0$, $\eta \neq 0$.

It follows from the definition of the function $\theta(t)$ that for functions (10) and (10′) condition (ii) is fulfilled. Let us verify condition (iii) Since the number of the functions ψ_j is equal to $2m+2$ and each of the functions does not exceed 1, we see that the right-hand inequality (5) with $K = 2m+2$ holds. Further, the constructions in the foregoing section make it possible to select the numbers $R_{j\mu}$ in (9) so that for any point (ξ,η), $\xi^2 + \eta^2 > (R_0 - \varkappa)^2$, there should exist j such that inequality (9) is fulfilled when $\log R_{j\mu}$ is replaced by $\log(R_{j\mu} + \varkappa)$. In this case $\psi_j(\xi,\eta) = 1$, and hence inequality (5) with $K^{-1} = 1$ holds.

The verification of the smoothness of the functions ψ_j on the coordinate axes and the conditions (iv) and (v) requires a rather detailed analysis of the structure of the regions U_j and functions (10); to simplify the presentation of the material we placed the proofs of these facts in the Appendix.

3.3. Microlocalization of estimate (3). An important step in the proof of Theorem 1 or 2 is the following

Proposition. *Let a differential operator* (1) *possess the following property: if* $\{U_j\}$ *is the covering* (7), (7′), (7″) *and* $\{\psi_j\}$ *is its subordinate generalized partition of unity satisfying all conditions of Proposition* 3.2, *then* $\forall \varepsilon > 0$ $\exists \omega(\varepsilon)$ *such that in an arbitrary region* $\Omega \subset \mathbb{R}^2$, $\operatorname{diam}\Omega < \omega(\varepsilon)$, *the inequalities*

$$\|\psi_j(D_x,D_y)u\|_{\widehat{\delta}(P)} \leqslant \varepsilon \|P(x,y;D_x,D_y)(\psi_j(D_x,D_y)u)\| + c(\varepsilon)\|u\|_{(-t)},$$
$$\forall u \in \mathcal{D}(\Omega), \quad t > 1,\ j > 0, \quad (11)$$

are fulfilled. Then estimate (3) *takes place.*

Remark. Since inequality (3) was considered on functions $u(x,y)$ of compact support, the only condition imposed on the coefficients $a_{\alpha\beta}(x,y)$ of the operator P was that they should be smooth. As to inequality (11), here the differential operator is applied to the function $\psi_j(D_x,D_y)u$ which, in general, may not be a function of compact support. Therefore in what follows we assume that the coefficients $a_{\alpha\beta}(x,y)$ are uniformly bounded throughout the plane together with their derivatives of any order, i.e. for any k_1 and k_2 there is a constant $K_{\alpha\beta k_1 k_2}$ such that

$$\max |D_x^{k_1} D_y^{k_2} a_{\alpha\beta}(x,y)| \leqslant K_{\alpha\beta k_1 k_2}.$$

Under this condition on the coefficients the inequality (11) makes sense, and we can apply Hörmander's results on PDO (see Lemma 2 below).

The proof of the proposition. According to (5), the function $a(\xi,\eta) = \sum \psi_j(\xi,\eta)$ is bounded from above and below. Therefore

$$\|u\|_{\widehat{\delta}(P)} = \Big\|a^{-1}(D_x, D_y)\sum_{j=0}^{J}\psi_j(D_x, D_y)u\Big\|_{\widehat{\delta}(P)}$$
$$\leqslant K\sum_{j=0}^{J}\|\psi_j u\|_{\widehat{\delta}(P)} \leqslant \sum_{j=1}^{J}(\varepsilon K\|P(\psi_j u)\| + Kc(\varepsilon)\|u\|_{(-t)}) + K\|\psi_0 u\|_{\widehat{\delta}(P)}.$$

Since $\psi_0(\xi,\eta)$ is a function of compact support, the term $\|\psi_0 u\|_{\widehat{\delta}(P)}$ does not exceed $c(t)\|u\|_{(-t)}$ for any $t \in \mathbb{R}$. Thus, the inequality can be rewritten as

$$\|u\|_{\widehat{\delta}(P)} \leqslant K\varepsilon\sum_{j=1}^{J}\|P(\psi_j u)\| + c'(\varepsilon)\|u\|_{(-t)}$$
$$\leqslant KJ\varepsilon\|Pu\| + c'(\varepsilon)\|u\|_{(-t)} + K\varepsilon\sum_{j=1}^{J}\|(P\psi_j - \psi_j P)u\|. \tag{12}$$

The estimates for the second and third terms on the right-hand side will be stated as separate assertions.

Lemma 1. *If $t > 1/2$, then*

$$\|u\|_{(-t)} \leqslant c(t)\delta\|u\|, \qquad u \in \mathcal{D}(\Omega),\ \delta = \operatorname{diam}\Omega. \tag{13}$$

Lemma 2. *If the function $\psi_j(\xi,\eta)$, $j > 0$, satisfies the conditions of Proposition 3.2, then*

$$\|(P\psi_j - \psi_j P)u\| \leqslant K_1\|u\|_{\widehat{\delta}(P)} + K_2\|u\|_{(-t)}, \tag{14}$$

with some constants K_1 and K_2.

If inequalities (13) and (14) are assumed to be proved, then estimating the right-hand side of (12) by means of these inequalities we find

$$\|u\|_{\widehat{\delta}(P)} \leqslant KJ\varepsilon\|Pu\| + K_1KJ\varepsilon\|u\|_{\widehat{\delta}(P)} + \delta c''(\varepsilon)\|u\|.$$

Fix a sufficiently small ε so that $K_1K_2J\varepsilon < 1/2$. Then take $\omega(\varepsilon)$ such that $\delta c''(\varepsilon) < 1/4$ for $\delta < \omega(\varepsilon)$. As a result, we obtain

$$\|u\|_{\widehat{\delta}(P)} \leqslant 4KJ\varepsilon\|Pu\|.$$

Replacing $4KJ\varepsilon$ by ε we arrive at (3).

The proof of Lemma 1. By the definition of Fourier's transforms of functions belonging to $\mathcal{D}$, we have

$$|\widehat{u}(\xi,\eta)| \leqslant (2\pi)^{-1}\int_{\Omega}|u(x,y)|\,dx\,dy \leqslant (2\pi)^{-1}\sqrt{\operatorname{Area}\Omega}\,\|u\|.$$

Squaring both sides of this inequality, multiplying by $(1+\xi^2+\eta^2)^{-t/2}$, integrating with respect to ξ and η, and noting that $\sqrt{\operatorname{Area}\Omega} \leqslant \operatorname{const}\operatorname{diam}\Omega$ we derive inequality (13).

The proof of Lemma 2. We write the commutator of P and ψ_j in the form

$$\begin{aligned} P\psi_j - \psi_j P &= \sum_{(\alpha,\beta)\in N(P)} (a_{\alpha\beta}(x,y)\psi_j(D_x,D_y) - \psi_j(D_x,D_y)a_{\alpha\beta}(x,y))D_x^\alpha D_y^\beta \\ &= \sum_{\alpha,\beta,j} I_{j\alpha\beta}. \end{aligned}$$

If condition (iv) in the foregoing section holds, then the commutator of $a_{\alpha\beta}(x,y)$ and $\psi_j(D_x,D_y)$ can be written as

$$\psi_j(D_x,D_y)a_{\alpha\beta}(x,y) - a_{\alpha\beta}(x,y)\psi_j(D_x,D_y) = \sum_{0<p+q\leqslant N} (D_x^p D_y^q a_{\alpha\beta})\psi_j^{(p,q)}/p!q! + R_N,$$

where, according to Hörmander [4], for any $t\in\mathbb{R}$ there is N such, that

$$\|R_N D_x^\alpha D_y^\beta u\| \leqslant \operatorname{const}\|u\|_{(-t)} \quad \forall(\alpha,\beta)\in N(P).$$

Further, if condition (v) in the foregoing section is fulfilled, then

$$\begin{aligned} \|(D_x^p D_y^q a_{\alpha\beta})\psi_j^{(p,q)} D_x^\alpha D_y^\beta u\| &\leqslant \operatorname{const}\|\psi_j^{(p,q)} D_x^\alpha D_y^\beta u\| \\ &= \operatorname{const}\left(\iint |\psi_j^{(p,q)}(\xi,\eta)\xi^\alpha\eta^\beta|^2|\widehat{u}(\xi,\eta)|^2\,d\xi\,d\eta\right)^{1/2} \\ &\leqslant \operatorname{const}\left(\iint |\Xi_{\widehat{\delta}(P)}(\xi,\eta)\widehat{u}(\xi,\eta)|^2\,d\xi\,d\eta\right)^{1/2} = \operatorname{const}\|u\|_{\widehat{\delta}(P)}. \end{aligned}$$

3.4. Estimate (3) in regions relating to the vertices of the polygon *N(P)*. Let $\Gamma_j^{(0)}$ be a vertex of Newton's polygon $N(P)$ (distinct from the origin) and let $G(\Gamma_j^{(0)},\varepsilon_0)$ be the corresponding region with a sufficiently small ε_0. We first of all refine the estimate for the symbol in this region.

Lemma 1. *Let $P(x,y;\xi,\eta)$ be the symbol of an operator of N-principal type and let K be a compactum in $\mathbb{R}^2$. Then there is $\varepsilon_0(K)$ such that for $\varepsilon_0 = \varepsilon_0(K)$ there is a constant $c = c(K) > 0$ such that*

$$c^{-1}\Xi_{N(P)}(\xi,\eta) \leqslant |P(x,y;\xi,\eta)| < c\Xi_{N(P)}(\xi,\eta),$$
$$(x,y;\xi,\eta) \in K \times G(\Gamma_j^{(0)},\varepsilon_0). \tag{15}$$

Proof. The right-hand inequality is obvious; the proof of the left-hand inequality is a trivial modification of the proof of Proposition 2.3.

Lemma 2. *If the diameter of the region Ω is sufficiently small, then the inequality*

$$\|\psi_j u\|_{N(P)} \leqslant c_1\|P(x,y;D_x,D_y)(\psi_j u)\| + c_2(t)\|u\|_{(-t)} \tag{16}$$

with an arbitrary $t \in \mathbb{R}$ holds, where ψ_j is the function in Proposition 3.2 relating to the region $G(\Gamma_j^{(0)},\varepsilon_0)$, where $\varepsilon_0 = \varepsilon_0(\Omega)$.

Proof. Inequality (15) allows one to follow the proof of Theorem 1.4.4. Let, for definiteness, the origin belong to Ω, let $P(D_x,D_y) = P(0,0;D_x,D_y)$, and let $P'(x,y;D_x,D_y) = P(x,y;D_x,D_y) - P(D_x,D_y)$. By virtue of (15), for any function v we obtain

$$\|v\|_{N(P)} \leqslant c\|P(D_x,D_y)v\|,$$

whence, putting $v = \psi_j u$, we derive

$$\|\psi_j u\|_{N(P)} \leqslant c\|P(x,y;D_x,D_y)(\psi_j u)\| + c\|P'\psi_j u\|. \tag{17}$$

To estimate the second term on the right-hand we consider a truncating function $\chi(x,y) \in \mathcal{D}$, $\chi(x,y) = 1$ for $x^2+y^2 \leqslant \delta_0^2$, $\chi(x,y) = 0$ for $x^2+y^2 > 2\delta_0^2$, where $\delta_0 > \operatorname{diam}\Omega$. Then

$$c\|P'(\psi_j u)\| \leqslant c\|\chi P'(\psi_j u)\| + c\|(1-\chi)P'(\psi_j u)\|.$$

Since the coefficients of the operator P' vanish at the point $x = y = 0$, they do not exceed const δ_0 for $x^2+y^2 \leqslant 2\delta_0^2$, and therefore

$$\|\chi P'(\psi_j u)\| \leqslant \text{const}\,\delta_0\|\psi_j u\|_{N(P)} < 1/2\|\psi_j u\|_{N(P)}$$

proved that δ_0 is sufficiently small.

We now note that $(1-\chi)u \equiv 0$ for $u \in \mathcal{D}(\Omega)$. If the symbol $\psi_j(\xi,\eta)$ satisfies condition (iv) in Section 3.2, then, according to Hörmander [4], the PDO $\psi_j(D_x,D_y)$ possesses the property of pseudo-locality, whence

$$\|(1-\chi)P'(\psi_j u)\| \leqslant c_t\|u\|_{(-t)} \quad \forall t \in \mathbb{R}.$$

Substituting the resulting estimates into (17) we arrive at (16).

Since $N(P)$ is regular Newton's polygon, we have

$$\Xi_{\widehat{\delta}(P)}(\xi,\eta) \leqslant \varepsilon \Xi_{N(P)}(\xi,\eta) + c(\varepsilon)(\xi^2+\eta^2)^{-t/2},$$

whence

$$\|\psi_j u\|_{\widehat{\delta}(P)} \leqslant \varepsilon \|u\|_{N(P)} + c(\varepsilon)\|u\|_{(-t)}.$$

Substituting the last inequality into (16) and using Lemma 1 in Section 3.3 we obtain inequality (3) in the region in question $G(\Gamma_j^{(0)}, \varepsilon_0)$ with $\operatorname{diam}\Omega < \omega(\varepsilon)$ and $\varepsilon_0 = \varepsilon_0(\varepsilon, \operatorname{diam}\Omega)$.

3.5. The structure of the operators in regions relating to the sides of Newton's polygon. In what follows it will be convenient to deal with narrower regions as compared to $G(\Gamma_j^{(1)}, \varepsilon_0, \varepsilon_1)$. For $a < b$ we denote by $G(\Gamma_j^{(1)}, a, b, c)$ the region in the plane $\mathbb{R}^2_{(\xi,\eta)}$ determined by the inequalities

$$-q_2^{(j)} \log|\xi| + q_1^{(j)} \log|\eta| > \log a, \qquad q_2^{(j)} \log|\xi| + q_1^{(j)} \log|\eta| > \log b,$$
$$q_1^{(j)} \log|\xi| + q_2^{(j)} \log|\eta| > \log c.$$

Using the argument in Lemma 2 of Section 2.4 we shall prove the following

Lemma. *Let the difference $b-a$ be sufficiently small, namely $b-a \leqslant \varkappa(diam\Omega)$. Then in the region $G(\Gamma_j^{(1)}, a, b, c)$ $(j = 1, \ldots, m)$ one of the following two conditions holds:*

(a) $P_{q^{(j)}}(x,y;\xi,\eta) \neq 0$ *for* $(x,y) \in \Omega$, $(\xi,\eta) \in G(\Gamma_j^{(1)}, a, b, c)$;

(b) *there is a variable, say ξ such that*

$$\partial P_{q^{(j)}}(x,y;\xi,\eta)/\partial\xi \neq 0 \quad \text{for} \quad (x,y;\xi,\eta) \in \Omega \times G(\Gamma_j^{(1)} a, b, c).$$

Proposition. *Let conditions* (A) *and* (R′) *of Theorem 2 be fulfilled and let the region Ω and the difference $b-a$ be such that the assertions of the lemma are true for the region $\Omega \times G(\Gamma_j^{(1)}, a, b, c)$, the variable ξ being essential. Let the intersection*

$$\operatorname{char}_{P_{q^{(j)}}} = \{(x,y;\xi,\eta), P_{q^{(j)}}(x,y;\xi,\eta) = 0\} \cap (\Omega \times G(\Gamma_j^{(1)}, a, b, c)),$$

be non-empty. Then the factorization

$$P_{q^{(j)}}(x,y;\xi,\eta) = \Lambda_j(x,y;\xi,\eta) Q_j(x,y;\xi,\eta), \qquad (x,y;\xi,\eta) \in \Omega \times G(\Gamma_j^{(1)}, a, b, c), \tag{18}$$

takes place, where Λ_j *and* Q_j *are* $q^{(j)}$*-homogeneous functions in* (ξ,η) *of orders* 1 *and* c_j-1*, respectively, (where* $q^{(j)}=(1,b_j)$*,* $b_j\geqslant 1$*) depending smoothly on the parameters* $(x,y)\in\Omega$*, and the conditions*

$$Q_j(x,y;\xi,\eta)\neq 0,\qquad (x,y;\xi,\eta)\in\Omega\times G(\Gamma_j^{(1)},a,b,c), \tag{19}$$

$$\Lambda_j(x,y;\xi,\eta)=\xi-\lambda_j(x,y;\eta), \tag{20}$$

hold, where $\lambda_j(x,y,\eta)$ *is a real* b_j*-homogeneous function of order* 1 *of the variable* η*.*

Proof. According to Section 1.1.3, the symbol $P_{q^{(j)}}$ is represented in the form

$$P_{q^{(j)}}(x,y;\xi,\eta)=a_{\alpha_j\beta_j}(x,y)\xi^{\alpha_j}\eta^{\beta_j+1}P^{[j]}(x,y;\xi,\eta)\overset{\text{def}}{=}A_jP^{[j]}, \tag{21}$$

where the symbol $P^{[j]}$ is a polynomial in ξ and η solved with respect to the highest powers of ξ and η and having the coefficient 1 in the highest power of ξ. Since Newton's polygon does not depend on (x,y), we have

$$a_{\alpha_j\beta_j}(x,y)\neq 0\quad\text{for}\quad (x,y)\in\Omega.$$

Further, as was already mentioned, the region $G(\Gamma_j^{(1)},a,b)$ does not intersect the coordinate axes, and therefore $\xi^{\alpha_j}\eta^{b_j+1}\neq 0$, whence

$$A_j(x,y;\xi,\eta)\neq 0\quad\text{for}\quad (x,y;\xi,\eta)\in\Omega\times G(\Gamma_j^{(1)},a,b,c).$$

By the hypothesis, the symbol $P_{q^{(j)}}$ and, consequently, the symbol $P^{[j]}(x,y;\xi,\eta)$ have a real zero $(x_0,y_0;\xi_0,\eta_0)$, i.e. the point ξ_0 is a real root of the polynomial $\pi_j(\xi)=P^{[j]}(x_0,y_0;\xi,\eta_0)$. According to condition (b) of the lemma, this root is simple. Therefore in a neighborhood of (x_0,y_0,η_0) there exists a smooth branch $\lambda_j(x,y,\eta)$ of this root. Hence, we have obtained the factorization

$$P^{[j]}(x,y;\xi,\eta)=(\xi-\lambda_j(x,y,\eta))h_j(x,y;\xi,\eta)\overset{\text{def}}{=}\Lambda_jh_j, \tag{18'}$$

where the functions Λ_j and h_j are $q^{(j)}$-homogeneous and smooth in the neighborhood of the point (x_0,y_0,ξ_0,η_0), and we have

$$h_j(x,y;\xi,\eta)\neq 0,$$

in the neighborhood. In view of the $q^{(j)}$-homogeneity, the functions Λ_j and h_j can be defined for all $(\xi,\eta)\in G(\Gamma_j^{(1)},a,b,c)$ provided that $b-a$ is sufficiently small. Further, if the diameter of Ω is sufficiently small, we can assume that factorization (18′) holds for all $(x,y)\in\Omega$. Setting $Q_j=A_jh_j$ we arrive at factorization (18).

3.6. Estimate (3) in regions relating to the sides of the polygon $N(P)$.

1) We are going to prove inequalities (11) in region (7′) relating to the side $\Gamma_j^{(1)}$. Dividing the closed interval $[\log\varepsilon_0, \log 1/\varepsilon_0]$ into small subintervals $[a_\lambda, a_{\lambda+1}]$ we partition the region in question into the subregions $G(\Gamma_j^{(1)}, a, b, c)$ in each of which either condition (a) or condition (b) of Proposition 3.5 is fulfilled. As can easily be seen, in the microlocalization of estimate (3) the regions (7′) can be replaced by the indicated smaller regions. Indeed, each of the regions $G(\Gamma_j^{(1)}, a, b, c)$ is determined by inequalities (9) where it is only necessary to replace the constants $R_{i\mu}$ on the right-hand side by the constants $R_{i\mu\lambda}$ depending on the additional parameter λ. Replacing $R_{i\mu}$ in (10) by $R_{j\mu\lambda}$ we obtain functions $\psi_{j\lambda}$ whose supports lie in $G(\Gamma_j^{(1)}, a_\lambda, a_{\lambda+1}, \varepsilon_1)$. To simplify the notation, assume that one of the conditions of Lemma 3.5 holds throughout the region (7′). If condition (a) holds, then, by virtue of the results in Section 2.4, inequality (15) is fulfilled. Since this case has already been considered in Section 3.4, in what follows we shall assume that condition (b) holds, i.e. the $q^{(j)}$-homogeneous part of $P_{q^{(j)}}$ has the form (18) ($q^{(j)} = (1, b_j)$).

2) By the definition of the region $G(\Gamma_j^{(1)}, a, b, c)$, we have

$$a \leqslant |\eta||\xi|^{-b_j} \leqslant b, \qquad |\xi||\eta|^{b_j} \geqslant c.$$

It follows that in this region we have

$$c^{-1} \leqslant \frac{\Xi_{\delta(P)}(\xi,\eta)}{(1+|\xi|+|\eta|^{1/b_j})^{c_j-1}} \leqslant c, \qquad c_j = d_P(q^{(j)}). \tag{22}$$

We associate with the vector $q^{(j)} = (1, b_j)$ the norm

$$\|u\|^2_{(q^{(j)},s)} = \int (1+|\xi|+|\eta|^{1/b_j})^{2s} |\hat{u}(\xi,\eta)|^2\, d\xi\, d\eta.$$

Then (22) implies

Lemma 1. *Let U be a region of the type of $G(\Gamma_j^{(1)}, a, b, c)$ and let* $\operatorname{supp}\psi(\xi,\eta) \subset U$. *Then there is a constant $K = K(U)$ such that*

$$K^{-1}\|\psi u\|_{(q^{(j)},c_j-1)} \leqslant \|\psi u\|_{\delta(P)} \leqslant K\|\psi u\|_{(q^{(j)},c_j-1)} \tag{23}$$

3) In view of Lemma 1, instead of (11) it suffices to prove the inequality

$$\|\psi u\|_{(q^{(j)},c_j-1)} \leqslant \varepsilon\|P(\psi u)\| + c(\varepsilon)\|u\|_{(-t)}. \tag{24}$$

If condition (A) is fulfilled, then

$$\|(P - P_{q^{(j)}})(\psi u)\| \leqslant \text{const}\, \|\psi u\|_{(q^{(j)},c_j-1)}.$$

Thus, in (24) the operator P can be replaced by its $q^{(j)}$-homogeneous principal part, and hence the inequality

$$\|\psi u\|_{(q^{(j)},c_j-1)} \leqslant \varepsilon \|P_{q^{(j)}}(\psi u)\| + c(\varepsilon)\|u\|_{(-t)}, \tag{24'}$$

has to be proved.

4) We associate with the symbols Λ_j and Q_j in (18) the PDO $\Lambda_j(x, y; D_x, D_y)$ and $Q_j(x, y; D_x, D_y)$. Using the standard technique of PDO with quasi-homogeneous symbols it is proved that if $R_j = P - \Lambda_j Q_j$, then

$$\|R_j v\| \leqslant \text{const}\, \|v\|_{(q^{(j)},c_j-1)}.$$

It follows that (24′) is equivalent to the inequality

$$\|\psi u\|_{(q^{(j)},c_j-1)} \leqslant \varepsilon \|\Lambda_j Q_j(\psi u)\| + c(\varepsilon)\|u\|_{(-t)}. \tag{25}$$

5) The main point in the proof of (25) is the estimation of the operator Λ_j with real symbol (20).

Lemma 2. *Let the symbol $\Lambda_j(x, y; \xi, \eta)$ possess the properties indicated in Proposition* 3.5. *Then the inequality*

$$\|w\| \leqslant 2\|x\Lambda_j(x, y; D_x, D_y)w\| + c\|xw\|, \tag{26}$$

is fulfilled.

Proof. We have

$$\begin{aligned} -\operatorname{Im}(xw, \Lambda_j w) &= -\operatorname{Im}\left(xw, \overline{\frac{1}{i}\frac{\partial w}{\partial x}}\right) + \operatorname{Im}(xw, \overline{\lambda_j(x, y, D_y)w}) \\ &= -\frac{1}{2}\iint x\frac{\partial |w|^2}{\partial x}\,dx\,dy + \iint x\left(\frac{\lambda_j^* - \lambda_j}{2i}\right) w\overline{w}\,dx\,dy. \end{aligned}$$

Since the symbol $\lambda_j(x, y; \eta)$ is real, the difference $\lambda_j^* - \lambda_j$ is a bounded operator in L_2. On integrating by parts in the first term on the right-hand side we obtain

$$\frac{1}{2}\|w\|^2 \leqslant -\operatorname{Im}(w, x\Lambda_j w) + c\|xw\|\|w\| \leqslant \|x\Lambda_j w\|\|w\| + c\|xw\|\|w\|.$$

Dividing both sides of this inequality by $\|w\|$ we obtain (26).

6) We now complete the derivation of inequality (11). To this end we substitute $w = Q_j(x, y; D_x, D_y)(\psi u)$ into (26). As in the proof of Lemma 2 in Section 3.4, taking a truncating function $\chi(x, y)$ equal to 1 in a (small) neighborhood of the region Ω we find

$$\begin{aligned} \|Q_j(\psi u)\| &\leqslant 2\|x\Lambda_j Q_j(\psi u)\| + c\|xQ_j(\psi u)\| \\ &\leqslant 2\|x\chi\Lambda_j Q_j(\psi u)\| + 2\|(1-\chi)x\Lambda_j Q_j(\psi u)\| \\ &\quad + c\|x\chi Q_j(\psi u)\| + c\|x(1-\chi)Q_j(\psi u)\|. \end{aligned} \tag{27}$$

If the diameter of Ω is sufficiently small, then $2x\chi < \varepsilon$, whence

$$2\|x\chi\Lambda_j Q_j(\psi u)\| \leqslant \varepsilon\|\Lambda_j Q_j(\psi u)\|,$$
$$c\|x\chi Q_j(\psi u)\| \leqslant c\varepsilon\|Q_j(\psi u)\|.$$

As in Section 3.4, in view of the pseudo-locality, the second and fourth terms on the right-hand side of (27) are estimated by means of $\|u\|_{(-t)}$ with any t. Consequently,

$$\|Q_j(\psi u)\| \leqslant \varepsilon\|\Lambda_j Q_j(\psi u)\| + c(\varepsilon)\|u\|_{(-t)} + c\varepsilon\|Q_j(\psi u)\|.$$

Taking ε satisfying the condition $c\varepsilon < 1/2$ and replacing 2ε by ε we obtain

$$\|Q_j(\psi u)\| \leqslant \varepsilon\|\Lambda_j Q_j(\psi u)\| + c'(\varepsilon)\|u\|_{(-t)}. \tag{28}$$

The PDO Q_j is $q^{(j)}$ quasi-elliptic. Applying the technique of such PDO (which does not practically differ from the technique of elliptic PDO) it is easy to derive the inequality

$$\|\psi u\|_{(q^{(j)},c_j-1)} \leqslant \text{const}\,\|Q_j(\psi u)\| + c''\|u\|_{(-t)}.$$

Substituting this into (28) we arrive at (25). Theorem 2 is proved completely.

Appendix

Below we present a complete proof of Proposition 3.2, i.e. we shall prove the smoothness of functions (3.10) in the neighborhood of the coordinate axes and verify properties (iv) and (v) playing a decisive role (see the proof of Proposition 3.3) in the microlocalization of the estimates in question. We shall assume that $\xi^2 + \eta^2$ is sufficiently large, and therefore the first factor in (3.10) will be omitted.

1) We begin with the verification of (3.8) in the special case when $\gamma = \alpha$ and $\delta = \beta$, i.e. we shall prove the inequality

$$|\xi^\alpha \eta^\beta \psi_j^{(\alpha,\beta)}(\xi,\eta)| < c_{j\alpha\beta}. \tag{1}$$

For the sake of simplicity, consider the case $\alpha = 1$, $\beta = 0$ since the other derivatives are estimated in an analogous manner. Differentiating in (3.10) we obtain

$$\xi\frac{\partial\psi_j}{\partial\xi} = \sum_\mu q_1^{j\mu}\xi^{q_1^{j\mu}}\eta^{q_2^{j\mu}}\theta'(\xi^{q_1^{j\mu}}\eta^{q_2^{j\mu}} - R_{j\mu})\prod_{\nu\neq\mu}\theta(\xi^{q_1^{j\nu}}\eta^{q_2^{j\nu}} - R_{j\nu}).$$

By the definition of the function $\theta(t)$, its derivative $\theta'(t)$ is nonzero for $0 \leqslant t \leqslant \varkappa$, i.e. the right-hand side of the above relation is nonzero when

$$R_{j\mu} \leqslant \xi^{q_1^{j\mu}}\eta^{q_2^{j\mu}} \leqslant R_{j\mu} + \varkappa,$$

whence trivially follows (1) for $\alpha = 1$, $\beta = 0$.

The proof of the proposition will be presented in the following order. We first consider functions (3.10) with supports in the regions corresponding the vertices $\Gamma_j^{(0)}$ not lying on the coordinate axes. Then we consider the case of vertices lying on coordinate axes and the case of regions corresponding to the sides.

2) Thus, let $\Gamma_j^{(0)} = (\alpha_j, \beta_j)$; $\alpha_j, \beta_j > 0$, be a vertex of $N(P)$. In this case for large $\xi^2 + \eta^2$ the function (3.10) has the form

$$\psi_j(\xi, \eta) = \theta(|\xi|^{-q_2^{(j)}} |\eta|^{q_1^{(j)}} - \varepsilon_0^{-1})\theta(|\xi|^{q_2^{(j-1)}} |\eta|^{-q_1^{(j-1)}} - \varepsilon_0^{-1}), \tag{2}$$

and is nonzero when

$$\varepsilon_0^{-1} \leqslant |\xi|^{-q_2^{(j)}} |\eta|^{q_1^{(j)}}, \qquad \varepsilon_0^{-1} \leqslant |\xi|^{q_1^{(j-1)}} |\eta|^{-q_2^{(j-1)}}.$$

Putting $\varkappa_j = q_2^{(j)}/q_1^{(j)}$ and $p_j = \varepsilon_0^{-1/q^{(j)}}$ we rewrite the inequalities in the form

$$p_j |\xi|^{\varkappa_j} \leqslant |\eta|, \qquad p_{j-1}|\eta| \leqslant |\xi|^{\varkappa_{j-1}}, \tag{3}$$

whence

$$|\xi|^{\varkappa_{j-1} - \varkappa_j} \geqslant p_j p_{j-1}.$$

By virtue of the convexity of the polygon $N(P)$, we have $\varkappa_1 > \cdots > \varkappa_m$, whence, with account of inequalities (3), on the support of (2) we find

$$|\xi| > (p_j p_{j-1})^{1/(\varkappa_{j-1} - \varkappa_j)}, \qquad |\eta| > p_j (p_j p_{j-1})^{\varkappa_j/(\varkappa_{j-1} - \varkappa_j)}. \tag{4}$$

Thus, the support of ψ_j does not intersect the coordinate lines, and therefore $\psi_j \in C^\infty$.

Inequalities (3) imply that there are $\rho > 0$, $\omega > 0$ such that

$$\log|\xi|, \log|\eta| > \rho \log(1 + |\xi| + |\eta|) \quad \text{for} \quad (\xi, \eta) \in \operatorname{supp} \psi_j. \tag{5}$$

In view of (5), it trivially follows from (1) that condition (iv) holds.

We now show that the function (2) satisfies (v). In view of Lemma 2.3, we can confine ourselves to the case $\gamma = \alpha_j$, $\delta = \beta_j$.

Let, for definiteness, $\alpha > 0$ in condition (v). Then, with account of (1) and (4), we obtain

$$|\psi_j^{(\alpha,\beta)}(\xi, \eta)| \leqslant \text{const}\, |\xi|^{-\alpha} |\eta|^{-\beta} \leqslant \text{const}\, |\xi|^{-1}.$$

Multiplying this inequality by $|\xi|^{\alpha_j} |\eta|^{\beta_j}$ we obtain (v).

3) Now let the vertex $\Gamma_j^{(0)}$ lie on a coordinate axis, say $j = 1$. Then $q^{(j-1)} = (-1, 0)$ in (2), and the function (2) is nonzero when

$$\varepsilon_0^{-1} \leqslant |\xi|^{-q_2^{(1)}} |\eta|^{q_1^{(1)}}, \qquad \varepsilon_0^{-1} \leqslant |\eta|. \tag{6}$$

Hence, (2) is a smooth function in a neighborhood of the axis $\eta = 0$. Further, since $|\eta| > \varepsilon_0^{-1}$, we have $|\xi|^{-q_2^{(1)}}|\eta|^{q_1^{(1)}} \to \infty$ as $|\xi| \to 0$, and therefore the second function on the right-hand side of (2) is identically equal to 1 for small $|\xi|$, that is the smoothness of (2) is completely proved.

We proceed to condition (iv). It was noted in the foregoing section that if inequalities (5) hold on the support of the function (2), then condition (iv) is a consequence of (5). Conditions (5) can in fact be replaced by more delicate conditions:

$$\log|\xi| > \rho\log(1 + |\xi| + |\eta|) - \log\omega, \qquad (\xi, \eta) \in \partial\psi_j/\partial\xi, \tag{7}$$

$$\log|\eta| > \rho\log(1 + |\xi| + |\eta|) - \log\omega, \qquad (\xi, \eta) \subset \partial\psi_j/\partial\eta. \tag{7'}$$

Inequality (7′) holds by virtue of the first inequality (6). Further, we have

$$\frac{\partial\psi_j}{\partial\xi} = -q_2^{(1)}|\xi|^{-q_2^{(1)}-1}|\eta|^{q_1^{(1)}}\theta'(|\xi|^{-q_2^{(1)}}|\eta|^{q_1^{(1)}} - \varepsilon_0^{-1})\theta(|\eta| - \varepsilon_0^{-1}), \tag{8}$$

i.e. the function in question is nonzero when

$$|\xi|^{-q_2^{(1)}}|\eta|^{q_1^{(1)}} \leqslant \varkappa + \varepsilon_0^{-1}, \tag{9}$$

i.e. condition (7) is fulfilled.

When verifying condition (v) we can confine ourselves (Lemma 2.3) to the case $\gamma = 0$, $\delta = \beta_1$, i.e. to proving the inequality

$$|\eta|^{\beta_1}|\psi^{(\alpha,\beta)}(\xi, \eta)| < \text{const}\, \Xi_{\widehat{\delta}(P)}(\xi, \eta).$$

If $\beta > 0$, then this inequality readily follows from (1). Thus, it remains to consider the case $\beta = 0$. To simplify the notation, assume that $\alpha = 1$. By virtue of (8) and (9), we have

$$\left|\eta^{\beta_1}\frac{\partial\psi_j}{\partial\xi}\right| \leqslant \text{const}\,|\eta|^{\beta_1}|\xi|^{-1} \leqslant \text{const}\,|\eta|^{\beta_1 - q_1^{(1)}/q_2^{(1)}}.$$

We note that $\Gamma_1^{(1)}$ is determined by the equation

$$(q_1^{(1)}/q_2^{(1)})\alpha + \beta = \beta_1,$$

i.e. the point $(1, \beta_1 - q_1^{(1)}/q_2^{(1)})$ belongs to $N(P)$. Therefore the point $(0, \beta_1 - q_1^{(1)}/q_2^{(1)})$ belongs to $\widehat{\delta}(P)$, and thus (7) is proved for the region $G(\Gamma_1^{(0)}, \varepsilon_0, \varepsilon_1)$.

4) We now proceed to the case of the function (3.10) corresponding the region $G(\Gamma_j^{(1)}, \varepsilon_0, \varepsilon_1)$. Setting $q^{(j)} = (a, b)$, where a and b are natural numbers, we write

$$\psi_j(\xi, \eta) = \theta(|\xi|^{-b}|\eta|^a - \varepsilon_0)\theta(|\xi|^b|\eta|^{-a} - \varepsilon_0)\theta(|\xi|^a||^b - \varepsilon_1).$$

This function is nonzero when

$$\log \varepsilon_0 \leqslant -b \log|\xi| + a \log|\eta| \leqslant \log \frac{1}{\varepsilon_0}, \qquad a \log|\xi| + b \log|\eta| > \log \frac{1}{\varepsilon_1}. \tag{10}$$

These inequalities imply that

$$\log|\xi| > -(b \log \varepsilon_0 + a \log \varepsilon_1)/(a^2 + b^2), \tag{11}$$

$$\log|\eta| > (a \log \varepsilon_0 + b \log \varepsilon_1)/(a^2 + b^2), \tag{11'}$$

i.e. the support of ψ_j does not intersect the coordinate axes, and therefore ψ_j is a smooth function. Inequality (10) implies (5) and, consequently, condition (iv) for the function ψ_j.

By virtue of Lemma 1 in Section 2.4, when verifying condition (v) we may confine ourselves to the points $(\gamma, \delta) \in \Gamma_j^{(1)}$. We have to estimate the derivative of $\psi_j(\alpha, \beta)$. Let, for definiteness, $\alpha > 0$. On the side $\Gamma_j^{(1)}$ there always exists a point (γ, δ) such that $\gamma \geqslant 1$. By virtue of the inequalities (1), (11), and (11′), we have

$$|\psi_j^{(\alpha,\beta)}(\xi, \eta)| < \text{const}\, |\xi|^{-1},$$

whence

$$|\xi^\gamma \eta^\delta \psi_j^{(\alpha,\beta)}(\xi, \eta)| \leqslant \text{const}\, |\xi^{\gamma-1} \eta^\delta| \leqslant \text{const}\, \Xi_{\widehat{\delta}(P)}(\xi, \eta).$$

Proposition 3.2 is proved completely.

§4. Local solvability of differential operators of *N*-principal type

In Section 1.4.6 the definition of local solvability was stated for a differential operator, and the local solvability was established for N quasi-elliptic differential operators with variable coefficients. We now extend this result to those operators of N-principal type for which inequality (3.3) holds.

4.1. We associate with an operator $P(x, y; D_x, D_y)$ of N-principal type the family of norms

$$[u]_s^2 = \iint \Xi_{\widehat{\delta}(P)}^{2s}(\xi, \eta) |\widehat{u}(\xi, \eta)|^2 \, d\xi \, d\eta, \tag{1}$$

and denote by $\mathfrak{G}^{(s)}$ the space of distributions with finite norm (1). Some simple facts concerning these spaces are presented in the Appendix. It is shown there that they relate to the class of spaces H^μ introduced in §1.4. In this case for any s there are $\varkappa_s$, $\varkappa'_s$ such that $H^{(\varkappa_s)} \subset \mathfrak{G}^{(s)} \subset H^{(\varkappa'_s)}$. Hence, the results on local solvability in $\mathfrak{G}^{(s)}$ make it possible to obtain cruder results on local solvability in the spaces $H^{(s)}$.

As in Section 1.4.6, we denote by $\mathfrak{G}_{\overline{\Omega}}^{(s)}$ the set of those distributions belonging to $\mathfrak{G}^{(s)}$ whose supports belong to $\overline{\Omega}$ and by $\mathfrak{G}^{(s)}(\Omega)$ the quotient space $\mathfrak{G}^{(s)} \setminus \mathfrak{G}_{\mathbb{R}^2 \setminus \Omega}^{(s)}$ endowed by the natural quotient norm.

The main result of this chapter is the following

Theorem. *Let a differential operator $P(x, y; D_x, D_y)$ of N-principal type satisfy the additional conditions guaranteeing the fulfilment of inequality (3.3). Then $\forall s \in \mathbb{R}$ there is $\omega(s)$ such that in a region Ω, $\operatorname{diam}\Omega < \omega(s)$, the equation*

$$P(x, y; D_x, D_y)u = f, \tag{2}$$

possesses a solution $u \in \mathfrak{G}^{(s+1)}(\Omega) \quad \forall f \in \mathfrak{G}^{(s)}(\Omega)$.

The major part of this section is devoted to the proof of the a priori estimate (cf. Section 1.4.6)

$$[v]_{-s} \leqslant c_{s,\Omega}[P(x, y; D_x, D_y)v]_{-s-1} \quad \forall v \in \mathcal{D}(\Omega). \tag{3}$$

The theorem is derived from (3) using some general concepts of functional analysis.

4.2. We begin with extending (3.3) to the scales of norms $\| \quad \|_\mu$.

Proposition. *Let inequality (3.3) hold. Let the function $\mu(\xi, \eta)$ satisfy (1.4.39) and the additional conditions (cf. Proposition 3.2)*

$$|\partial^{\alpha+\beta}\mu(\xi,\eta)/\partial\xi^\alpha\partial\eta^\beta| < c_{\alpha\beta}\mu(\xi,\eta)(1 + |\xi| + |\eta|)^{-\rho(\alpha+\beta)}, \quad 0 < \rho \leqslant 1, \tag{4}$$

$$|\xi^\gamma\eta^\delta\partial^{\alpha+\beta}\mu(\xi,\eta)/\partial\xi^\alpha\partial\eta^\beta| < c_{\alpha\beta\gamma\delta}\Xi_{\widehat{\delta}(P)}(\xi,\eta)\mu(\xi,\eta), \ \forall(\gamma,\delta) \in N(P). \tag{5}$$

Then there are $t(\mu)$ and $\omega(\mu)$ such that in the region Ω, $\operatorname{diam}\Omega < \omega(\mu)$, the inequality

$$\|\mu(D)u\|_{\widehat{\delta}(P)} \leqslant c_{\mu\Omega}(\|\mu(D)Pu\| + \|u\|_{(-t)}), \tag{6}$$

is fulfilled for any $t < t(\mu)$.

Proof. Take ω_0 so small that in a region of diameter no great than ω_0 the inequality (3.3) holds, and let $\operatorname{diam}\Omega < \omega < \omega_0$. Select a truncating function $\psi(x, y)$ equal to 1 on Ω with support lying in the ball $x^2 + y^2 < 2\omega_0^2$. Applying inequality (3.3) to the function $\psi(x, y)\mu(D_x, D_y)u$ we obtain

$$\begin{aligned}\|\mu(D_x, D_y)u\|_{\widehat{\delta}(P)} &\leqslant \|\psi\mu u\|_{\widehat{\delta}(P)} + \|(1-\psi)\mu u\|_{\widehat{\delta}(P)} \\ &\leqslant \varepsilon(\Omega)\|P(\psi\mu u)\| + \|(1-\psi)\mu u\|_{\widehat{\delta}(P)}.\end{aligned} \tag{7}$$

By Leibniz' formula, we have

$$P(\psi\mu u) = \psi P(\mu u) + \sum \frac{1}{\alpha!\beta!} D_x^\alpha D_y^\beta \psi P^{(\alpha,\beta)}(\mu u),$$

whence

$$\|P(\psi\mu u)\| \leqslant \|P(\mu u)\| + c\|\mu(D)u\|_{\widehat{\delta}(P)}.$$

Substituting this into (7) and using the smallness of the constant $\varepsilon(\Omega)$ we obtain

$$\begin{aligned}\|\mu u\|_{\widehat{\delta}(P)} &\leqslant \varepsilon_1(\Omega)\|P(\mu u)\| + \|(1-\psi)\mu u\|_{\widehat{\delta}(P)} \\ &\leqslant \varepsilon_1\|Pu\|_\mu + \varepsilon_1\|(P\mu - \mu P)u\| + \|(1-\psi)\mu u\|_{\widehat{\delta}(P)}. \qquad (7')\end{aligned}$$

We now prove that

$$\|(P\mu - \mu P)u\| \leqslant c_1\|\mu u\|_{\widehat{\delta}(P)} + c_2\|u\|_{(-t)}, \qquad (8)$$

$$\|(1-\psi)\mu u\|_{\widehat{\delta}(P)} \leqslant c_3\|u\|_{(-t)}. \qquad (9)$$

Substituting (8) and (9) into (7′) and taking into account the smallness of ε_1 we arrive at (6).

The proof of (8) is obtained by means of a trivial modification of the proof of Lemma 2 in Section 3.3.

The proof of (9) is based in the property of pseudolocality of the PDO $\mu(D_x, D_g)$ whose symbol satisfies condition (4).

Take a function $\varphi(x,y) \in \mathcal{D}$ equal to 1 on the support of the function $\psi(x,y)$ and, consequently, in the region Ω. Then

$$\begin{aligned}(1-\psi)\mu(D_x,D_y)u &= (1-\psi)\mu(D_x,D_y)(\varphi u) \\ &= \sum_{\alpha+\beta=0}^{N}(1-\psi)D_x^\alpha D_y^\beta\varphi\mu^{(\alpha,\beta)}(D_x,D_y)u/\alpha!\beta! + (1-\psi)R_N u = (1-\psi)R_N u.\end{aligned}$$

Taking sufficiently large N one can attain the fulfilment of (9) for any $t \in \mathbb{R}$.

4.3. Proposition. *For any $s \in \mathbb{R}$ there are $\omega(s)$ and $t(s)$ such that in a region Ω, $\operatorname{diam}\Omega < \omega(s)$, for any $t < t(s)$ the inequality*

$$[u]_{s+1} \leqslant c_{s\Omega}([P(x,y;D_x,D_y)u]_s + \|u\|_{(-t)}) \quad \forall u \in \mathcal{D}(\Omega) \qquad (10)$$

holds.

As is shown in the Appendix, $\forall s \in \mathbb{R}$ there is a smooth function $M_s(\xi,\eta)$ that satisfies conditions (1.4.39), (4), and (5) with $\mu = M_s$, and for some $c_s > 0$ we have

$$c_s^{-1} \leqslant \Xi^s_{\widehat{\delta}(P)}(\xi,\eta)M_s^{-1}(\xi,\eta) \leqslant c_s \quad \forall(\xi,\eta) \in \mathbb{R}^2. \qquad (11)$$

By virtue of these inequalities, the inequality for the norms

$$c_s^{-1}\|M_s(D_x,D_y)u\| \leqslant [u]_s \leqslant c_s^{-1}\|M_s(D_x,D_y)\|$$

holds. Therefore $\|M_s(D_x,D_y)u\|$ can be taken as the norm (1), and inequality (10) is obtained if we set $\mu(\xi,\eta) = M_s(\xi,\eta)$ in (6).

4.4. We now show that in the case of a region Ω of a small diameter the second term on the right-hand side of (10) can be dropped.

Proposition. *Let inequality* (10) *be fulfilled. Then* $\forall s \in \mathbb{R}$ *there is* $\omega(s)$ *such that in the region* Ω, $\operatorname{diam}\Omega < \omega(s)$, *the inequality*

$$[u]_{s+1} \leqslant c_s[P(x,y;D_x,D_y)u]_s \quad \forall u \in \mathcal{D}(\Omega) \tag{12}$$

holds.

Proof. Since $[u]_{s+1} \geqslant \text{const}\,\|u\|$ for $s \geqslant -1$, in this case $\|u\|_{(-t)}$ can be estimated by means of $(1/2)c(s,\Omega)[u]_{s+1}$ if the diameter of Ω is sufficiently small (see Lemma 1 in Section 3.3).

To prove (12) for $s < -1$ we use the argument in the paper by Nirenberg and Treves [1].

Assume that in an arbitrary small region Ω the inequality (12) does not hold. Then there is a sequence $u_j \in \mathcal{D}$ possessing the following properties:

$$\|u_j\|_{(-t)} = 1, \quad [Pu_j]_s \to 0, \quad \operatorname{supp} u_j \to (0,0). \tag{13}$$

By virtue of (10), the norms $[u_j]_{s+1}$ are uniformly bounded. If t is sufficiently large, then the set of functions

$$\{u, [u]_{s+1} \leqslant \text{const}, \operatorname{supp} u \in \Omega, \operatorname{diam}\Omega < \infty\},$$

is compact in $H^{(-t)}$. Therefore the sequence $\{u_j\}$ contains a subsequence (we again denote it as $\{u_j\}$) possessing the following properties:

$$u_j \to u \quad \text{in} \quad H^{(-t)}, \quad P(x,y;D_x,D_y)u = 0, \quad \operatorname{supp} u = (0,0).$$

Hence, if (12) does not hold, then for a sufficiently large t there is a distribution u in the space $H^{(-t)}$ with support at the point $(0,0)$ such that $P(x,y;D_x,D_y)u = 0$. We shall prove that $u \equiv 0$. Since, by virtue of the first condition (13), we have $\|u\|_{(-t)} = 1$, the resulting contradiction proves inequality (12).

If the support of a distribution $u \in H^{(-t)}$ is at the origin, there is a differential operator $Q(D_x,D_y)$ such that $u = Q(D_x,D_y)\delta(x,y)$, where $\delta(x,y)$ is Dirac's delta function. Expanding the coefficients of $P(x,y;D_x,D_y)$ into Taylor's series at $(0,0)$ we write the operator in question in the form

$$P(x,y;D_x,D_y) = P(D_x,D_y) + \sum_{\gamma+\delta>0} x^\gamma y^\delta P_{\gamma\delta}(D_x,D_y).$$

Since the degrees of the differential operators $P_{\gamma\delta}$ do not exceed the degree of the operator $P(x,y;D_x,D_y)$, there is an integer $N > 0$ such that $x^\gamma y^\delta P_{\gamma\delta}(D_x,D_y)\delta(x,y)$ is equal to zero for $\gamma+\delta > N$. Thus, the condition $Pu = 0$ means that

$$(PQ)(D_x,D_y)\delta + \sum_{\gamma+\delta=1}^{N} x^\gamma y^\delta (P_{\gamma\delta}Q)(D_x,D_y)\delta = 0.$$

Passing to the Fourier transform we find

$$(PQ)(\xi,\eta)+\sum_{\gamma+\delta=1}^{N}\left(i\frac{\partial}{\partial\xi}\right)^{\gamma}\left(i\frac{\partial}{\partial\eta}\right)^{\delta}(P_{\gamma\delta}Q)(\xi,\eta)\equiv 0. \tag{14}$$

Take a positive vector $q=(q_1,q_2)$, for example, $q=(1,1)$. Then the q-degree of the sum in (14) is less than that of PQ. Therefore (14) can be rewritten as

$$P_q(\xi,\eta)Q_q(\xi,\eta)+H(\xi,\eta)\equiv 0, \qquad \deg_q H<\deg_q P+\deg_q Q.$$

It follows that $P_q(\xi,\eta)Q_q(\xi,\eta)\equiv 0$, $H(\xi,\eta)\equiv 0$. Since $P_q(\xi,\eta)\not\equiv 0$, we have $Q_q(\xi,\eta)\equiv 0$, and, consequently, $Q(\xi,\eta)=0$, whence $u=0$.

4.5. The proof of Theorem 4.1.

1) We first of all note that inequality (12) is a consequence of (3.3), i.e. it applies to those operators for which (3.3) is fulfilled. Since the operator $P(x,y;D_x,D_y)$ and the transposed operator ${}^tP(x,y;D_x,D_y)$ simultaneously satisfy the conditions of Theorem 1 or 2 in Section 3.1, inequality (3.3) holds for the former operator, and therefore the inequality

$$[v]_{-s}\leqslant \mathrm{const}[\,{}^tP(x,y;D_x,D_y)u]_{-s-1}, \qquad v\in\mathcal{D}(\Omega), \tag{3'}$$

is fulfilled.

2) We consider the unbounded operator

$$\mathfrak{G}_{\overline{\Omega}}^{(-s)}\to\mathfrak{G}_{\overline{\Omega}}^{(-s-1)} \quad (v\mapsto {}^tP(x,y;D_x,D_y)v), \tag{15}$$

with domain $H_{\Omega}^{(-s)}$. This operator is a restriction of a continuous operator, and, consequently (cf. the proof of Theorem 2.2.3) is a closable operator. Indeed, if $s\geqslant 0$, then, according to Lemma 3 in the Appendix, $\mathfrak{G}^{(-s)}\subset H^{N(P),-s}$, $\mathfrak{G}^{(-s-1)}\subset H^{N(P),-s-1}$, and (15) is a restriction of the continuous operator

$$H_{\overline{\Omega}}^{N(P),-s}\to H_{\overline{\Omega}}^{N(P),-s-1} \quad (v\mapsto {}^tPv).$$

In case $s\leqslant 0$, we have $\mathfrak{G}^{(-s)}\subset H^{N(P),-s/\theta}$, $\mathfrak{G}^{(-s-1)}\subset H^{N(P),(-s-1)/\theta}$ (for $s\leqslant -1$), and it is necessary to consider a continuous operator from $H_{\overline{\Omega}}^{N(P),-s/\theta}$ into $H^{N(P),-s/\theta-1/\theta}$.

3) Let us assume that the closure of operator (15) has been performed. Since in the case of a region Ω with a smooth boundary the space $\mathfrak{G}_{\overline{\Omega}}^{(-s)}$ is the closure of $\mathcal{D}(\Omega)$, inequality (3′) is extended throughout the domain of operator (15). This implies the injectivity of operator (15). Repeating the argument in the proof of Theorem 2.2.3 we conclude from inequality (13) that the image of operator (15) is closed.

4) The domain of operator (15) is dense in $\mathfrak{G}^{(-s)}_{\overline{\Omega}}$, and therefore this operator has an adjoint operator which is also a closed operator with closed image. The injectivity of (15) implies the surjectivity of the adjoint operator. The duality of $\mathfrak{G}^{(-s)}_{\overline{\Omega}}$ and $\mathfrak{G}^{(s)}(\Omega)$ (cf. Section 1.4.6) and the definition of the operator ${}^tP$ (see (1.4.54)) imply that the operator

$$\mathfrak{G}^{(s)}(\Omega) \to \mathfrak{G}^{(s-1)}(\Omega) \quad (u \mapsto P(x, y; D_x, D_y)u),$$

with domain containing $H^{N(P),s}(\Omega)$ is surjective, which proves Theorem 4.1.

Appendix

Lemma 1. *There is $\varkappa > 0$ such that*

$$(1 + |\xi| + |\eta|)^{\varkappa} \Xi_{\hat{\delta}(P)}(\xi, \eta) \leqslant \mathrm{const}\, \Xi_{N(P)}(\xi, \eta) \quad \forall (\xi, \eta) \in \mathbb{R}^2. \tag{1}$$

The proof is based in the fact that the polygon $N(P)$ is regular and the polygon $\hat{\delta}(P)$ is the convex hull of a finite number of minor (not necessarily integral) points of $N(P)$.

Lemma 2. *There is $\theta > 1$ such that*

$$\Xi_{N(P)}(\xi, \eta) \leqslant \mathrm{const}\, \Xi^{\theta}_{\hat{\delta}(P)}(\xi, \eta). \tag{2}$$

Proof. Let (γ_j, δ_j) be the vertices of $\hat{\delta}(P)$. Instead of (2) it suffices to prove the inequality

$$\Xi_{N(P)}(\xi, \eta) \leqslant \mathrm{const} \sum |\xi|^{\gamma_j \theta} |\eta|^{\delta_j \theta}. \tag{2'}$$

This inequality follows from the purely geometrical fact that there is $\theta > 1$ such that

$$N(P) \subset \theta \hat{\delta}(P).$$

Indeed, the polygon $N(P)$ is determined by the system of inequalities[1)]

$$\langle q^{(j)}, (\alpha, \beta) \rangle \leqslant c_j, \quad j = 1, \ldots, m, \quad \alpha \geqslant 0,\ \beta \geqslant 0,$$

and the polygon $\hat{\delta}(P)$ is determined by the system

$$\langle q^{(j)}, (\alpha, \beta) \rangle \leqslant c_j - 1.$$

Putting $\theta = \max c_j/(c_j - 1)$ we prove the desired assertion.

As a consequence of Lemmas 1 and 2 we have

[1)] Under an appropriate normalization of the outer normals (see Section 2.1).

Lemma 3. *Let, as in §1.4, $H^{N(P),s}$ denote the space of distributions with finite norm*

$$\iint \Xi^{2s}_{N(P)}(\xi,\eta)|\widehat{u}(\xi,\eta)|^2\,d\xi\,d\eta.$$

Then the following embeddings take place:

$$\begin{aligned} &H^{N(P),s} \subset \mathfrak{G}^{(s)} \subset H^{N(P),s/\theta} && \text{for} \quad s \geqslant 0,\\ &H^{N(P),s/\theta} \subset \mathfrak{G}^{(s)} \subset H^{N(P),s} && \text{for} \quad s < 0. \end{aligned}$$

Proposition. *$\forall s \in \mathbb{R}$ there is a smooth function $M_s(\xi,\eta)$ satisfying conditions* (1.4.39), (4.4), *and* (4.5), *for which inequality* (4.11) *holds.*

We first of all reduce the proof of the proposition to the case of a single fixed value of s, say $s = 1$.

Lemma 4. *Let a smooth function $M(\xi,\eta)$ satisfy the conditions*

$$c^{-1} \leqslant \Xi_{\widehat{\delta}(P)}(\xi,\eta)M^{-1}(\xi,\eta) \leqslant c, \tag{3}$$

$$|\partial^{\alpha+\beta}M(\xi,\eta)/\partial\xi^\alpha\partial\eta^\beta| < c_{\alpha\beta}(1+|\xi|+|\eta|)^{-\rho(\alpha+\beta)}M(\xi,\eta), \quad 0<\rho<1, \tag{4}$$

$$|\xi^\gamma\eta^\delta\partial^{\alpha+\beta}M(\xi,\eta)/\partial\xi^\alpha\partial\eta^\beta| < c_{\alpha\beta\gamma\delta}M^2(\xi,\eta),$$
$$\forall(\gamma,\delta)\in N(P), \quad \forall(\xi,\eta)\in\mathbb{R}^2. \tag{5}$$

Then the function $M_s(\xi,\eta) = M^s(\xi,\eta)$ satisfies the conditions of the proposition.

Proof. Inequality (4.11) for $M_s = M^s$ is a trivial consequence of (3). Let us show that (4) implies (4.4) for $\mu = M_s$. For the sake of brevity, we shall write $\mu^{(\alpha,\beta)} = \partial^{\alpha+\beta}\mu/\partial\xi^\alpha\partial\eta^\beta$. By the chain rule, $(M^s)^{(\alpha,\beta)}$ is a linear combination of expressions of the form of

$$M^s(\xi,\eta)\prod M^{(\alpha_k,\beta_k)}(\xi,\eta)M^{-1}(\xi,\eta), \qquad \sum(\alpha_k+\beta_k) = \alpha+\beta,$$

whence it follows that (4.4) is fulfilled for $\mu = M^s$. Similarly, $\xi^\gamma\eta^\delta(M^s)^{(\alpha,\beta)}$ is a linear combination of expressions

$$(\xi^\gamma\eta^\delta M^{(\alpha_1,\beta_1)})M^{s-1}\prod_{k>1} M^{(\alpha_k,\beta_k)}M^{-1}. \tag{6}$$

By virtue of (4), the factor is bounded, and, by virtue of (5), it does not exceed const M^2, i.e. all expressions (6) do not exceed const $M^{s+1} \leqslant$ const $\Xi_{\widehat{\delta}(P)}M^s$ and hence (4.5) with $\mu = M^s$ holds.

The proof of the proposition. We first assume that $\widehat{\delta}(P)$ has integral vertices (γ_j, δ_j) and set

$$M^2(\xi,\eta) = \sum \xi^{2\gamma_j}\eta^{2\delta_j}.$$

In view of the lemma, it suffices to verify conditions (4.4) and (4.5) for the polynomial M^2. Since only a finite number of derivatives $(M^2)^{(\alpha,\beta)}$ are nonzero, condition (4.4) follows directly from the regularity of the polygon $N(M^2) = 2\widehat{\delta}(P)$. To verify (4.5) we note that $\xi^\gamma\eta^\delta(M^2)^{(\alpha,\beta)}$ is a linear combination of monomials of the form of

$$\xi^\gamma\eta^\delta\xi^{2\gamma_k-\alpha}\eta^{2\delta_k-\beta}, \qquad \alpha+\beta \geqslant 1, \quad (\gamma,\delta)\in N(P), \tag{7}$$

where the $q^{(j)}$-degree of the monomial $\xi^\gamma\eta^\delta$ does not exceed c_j and the $q^{(j)}$-degree of $\xi^{2\gamma_k-\alpha}\eta^{2\delta_k-\beta}$ does not exceed $2\langle q^{(j)},(\gamma_k,\delta_k)\rangle - \langle q^{(j)},(\alpha,\beta)\rangle \leqslant 2(c_j-1)-1 \leqslant 2c_j-3$. Hence, the $q^{(j)}$-degree of (7) does not exceed $3(c_j-1)$. This shows that (7) can be estimated by means of const M^3.

In the case when some of the numbers γ_j, δ_j are not integers we take a positive function $\widetilde{|t|} \in C^\infty$ such that $\widetilde{|t|} = |t|$ for $|t| \geqslant 2$ and $\widetilde{|t|} = 1$ for $|t| < 1$. Setting

$$M^2(\xi,\eta) = \sum \widetilde{|\xi|}^{2\gamma_j}\widetilde{|\eta|}^{2\delta_j} + 1,$$

we obtain a function satisfying all the desired conditions.

CHAPTER V

TWO-SIDED ESTIMATES IN SEVERAL VARIABLES RELATING TO NEWTON'S POLYHEDRA

Introduction

Beginning with this chapter we develop the method of Newton's polyhedron in the theory of differential equations. The passage from Newton's polygon to Newton's polyhedron is quite natural. For polynomials we consider the convex hull of the set of monomial multiexponents completed in a certain way. In the present chapter we are interested in conditions under which the polynomials admit of an adequate estimate by means of the sum of the moduli of the constituent monomials. Of course, this relates to adequate estimates for differential operators whose symbols possess the indicated property.

In the case of polygons the main technique was the factorization of symbols, up to within lower terms, as a product of quasi-homogeneous symbols. This technique is not generalized to the multidimensional case. Nevertheless, it turns out that, as before, the essential properties of polynomials can be determined by considering simultaneously all quasi-homogeneous parts of the polynomial (corresponding to the faces of Newton's polyhedron). As was shown by Mikhaĭlov [1, 2], in these terms conditions for the existence of an adequate estimate are stated. In the presentation of Mikhaĭlov's theorem we follow the paper by Gindikin [1]. Some other conditions for the existence of adequate estimates are also retained, namely the stability condition for the property that the polynomial does not vanish and the absence of zeros in a complex neighborhood of the real space (see Volevich and Gindikin [2]). In the analytical aspect no new considerations are required as compared to the two-dimensional case.

§1. Estimates for Polynomials in $\mathbb{R}^n$ relating to Newton's polyhedra

In this section we derive conditions under which a polynomial admits of a two-sided estimate throughout the space $\mathbb{R}^n$ by means of the sum of the moduli of the constituent monomials. This is the basic estimate for deriving other estimates in various types of regions. Among them there are regions that go into those considered in the foregoing chapters in the case $n = 2$.

1.1. Newton's polyhedron of a polynomial. Let $P(\xi)$ be a polynomial in $\xi = (\xi_1, \dots, \xi_n) \in \mathbb{R}^n$:

$$P(\xi) = \sum_{\alpha \in \nu(P)} a_\alpha \xi^\alpha, \tag{1}$$

where $\xi^\alpha = \xi^{\alpha_1} \dots \xi_n^{\alpha_n}$ and $\alpha = (\alpha_1, \dots, \alpha_n) \in \mathbb{R}^n_\alpha$. Here $\nu(P)$ is a finite set in $\overline{\mathbb{R}^n_+}$ (the closure of the positive coordinate n-hedron in $\mathbb{R}^n_\alpha$). It is convenient to admit the value $a_\alpha = 0$ for some $\alpha \in \nu(P)$ and it is only important that $\nu(P)$ is finite.

Thus, we shall sometimes regard $\nu(P)$ as being broader than the set of multiindices corresponding to the nonzero coefficients.

The statement of the problem. To find out for what polynomials there is $c > 0$ such that the following estimates hold:

$$c|\xi^\alpha| \leqslant |P(\xi)| \quad \forall\alpha \in \nu(P), \quad \forall\xi \in \mathbb{R}^n. \tag{2}$$

Naturally, the answer to the question is given in terms of Newton's polyhedron, a multidimensional generalization of Newton's polygon.

By *Newton's polyhedron* $N(P)$ of a polynomial P will be means the convex hull of the set $\nu(P)$.

We note that it is not assumed that $\dim N(P) = n$. Let $V(N)$ be the set of vertices of the polyhedron N; $V(P) \stackrel{\text{def}}{=} V(N(P))$

Consider the space $\mathbb{R}^n_q$ dual to the space of exponents, and let the duality be determined by the form

$$\langle q, \alpha\rangle = q_1\alpha_1 + \cdots + q_n\alpha_n. \tag{3}$$

For a polyhedron $N \subset \mathbb{R}^n_\alpha$ we put

$$d(N, q) = \max_{\alpha\in N}\langle q, \alpha\rangle, \tag{4}$$

i.e.

$$\langle q, \alpha\rangle = d(N, q), \tag{5}$$

is a supporting hyperplane to N.

In the case of Newton's polyhedron $N(P)$ we shall call the expression

$$d_P(q) \stackrel{\text{def}}{=} d(N(P), q), \qquad d_P(\lambda q) = \lambda d_P(q), \quad \lambda \in \mathbb{R}_+,$$

the q-degree of the polynomial P. Let N_q be the face of N lying in the supporting hyperplane (5). By the q-principal part of P will be meant the polynomial

$$P_q(\xi) = \sum_{\alpha\in N_q(P)} a_\alpha\xi^\alpha = \sum_{\langle q,\alpha\rangle = d_P(q)} a_\alpha\xi^\alpha, \qquad P_{\lambda q} = P_q. \tag{6}$$

It is clear that P_q is a quasi-homogeneous (q-homogeneous) polynomial:

$$P_q(\lambda^q\xi) = P_q(\lambda^{q_1}\xi_1, \ldots, \lambda^{q_n}\xi_n) = \lambda^{d_P(q)}P_q(\xi), \qquad \lambda \in \mathbb{R}_+. \tag{7}$$

We obviously have the following generalizations of Lemmas 1 and 2 in Section 1.1.2.

Lemma 1. *We have*

$$P(t^q\xi) = t^{d_P(q)}P_q(\xi) + o(t^{d_P(q)}), \qquad t \to \infty, \tag{8}$$

$$(PQ)_q = P_q Q_q, \qquad \deg_{PQ}(q) = \deg_P(q) + \deg_Q(q), \tag{9}$$

$$(P+Q)_q = \begin{cases} P_q + Q_q & \text{for} \quad d_P(q) = d_Q(q), \\ P_q & \text{for} \quad d_P(q) > d_Q(q). \end{cases} \tag{10}$$

Lemma 2. *If two polynomials P and Q satisfy the conditions*

$$P_q \equiv Q_q, \qquad \dim N(P) = n,$$

for all q, then $P - Q$ is expanded into monomials ξ^α, where α is an interior point of $N(P) = N(Q)$.

For some reasons that will be eludicated in what follows it is convenient to set

$$P_{(0)}(\xi) = P_{(0,\dots,0)}(\xi) \stackrel{\text{def}}{=} P(\xi). \tag{11}$$

1.2. Basic estimate. Generalizing definition (1.2.2) we set

$$\Xi_N(\xi) = \sum_{\alpha\in V(N)} |\xi^\alpha|, \qquad \Xi_P(\xi) = \Xi_{N(P)}(\xi). \tag{12}$$

Recall that $V(N)$ is the set of vertices of the polyhedron N. The problem stated in the foregoing section (estimate (2)) is equivalent to the question of existence of $c > 0$ such that

$$c\Xi_P(\xi) \leqslant |P(\xi)|. \tag{2'}$$

The equivalence follows from (cf. Lemma 1.2.1)

Lemma 1. *For $\alpha \in N$ we have*

$$|\xi^\alpha| \leqslant \Xi_N(\xi). \tag{13}$$

The proof is carried out as in the case of $n = 2$ (Lemma 1.2.1).

For $c > 0$ we denote by $\Omega(c)$ the set determined by the conditions

$$c^{-1} \leqslant |\xi_j| \leqslant c \quad \forall j \in \{1, \dots, n\}. \tag{14}$$

Hence, $\Omega(c)$ is a union of nonintersecting parallelepipeds lying in coordinate n-hedrons.

Lemma 2. (i) *Let* $\dim N = n$ *and let* α *be an interior point of* N. *Then for any* $c_1 > 0$ *there is* c_2 *such that outside the set* $\Omega(c_2)$ *the inequality*

$$c_1|\xi^\alpha| \leqslant \Xi_N(\xi) \tag{15}$$

holds.

(ii) *If*

$$c|\xi^{\alpha^0}| \leqslant \Xi_N(\xi), \qquad \xi \in \mathbb{R}^n, \tag{13'}$$

for some $c > 0$, *then* $\alpha^0 \in N$. *And if for any* $c > 0$ *there is* c_1 *such that inequality* (15) *is fulfilled outside* $\Omega(c_1)$, *then* α *is an interior point of* N.

Proof. (i) For a given $j = 1, \dots, m$ we denote by $\beta_{(j)}$ and $\delta_{(j)}$ the points of intersection of the straight line passing through $\alpha \in N$ and parallel to the jth coordinate axis with the boundary of the polyhedron N. By virtue of Lemma 1, there is $a > 0$ such that

$$a\sum_{j=1}^{n}(|\xi^{\beta_{(j)}}| + |\xi^{\delta_{(j)}}|) \leqslant \Xi_N(\xi) \quad \text{for} \quad \xi \in \mathbb{R}^n.$$

Hence,

$$a\sum_{j=1}^{n}(|\xi^{\beta_{(j)}}| + |\xi^{\delta_{(j)}}|) = a\sum_{j=1}^{n}(|\xi_j|^{\lambda_j} + |\xi_j|^{-\mu_j})|\xi^\alpha| \leqslant \Xi_N(\xi),$$
$$\xi \in \mathbb{R}^n, \quad \lambda_j, \mu_j > 0,\ 1 \leqslant j \leqslant n.$$

Select a number c_2 so that the inequalities

$$|\xi_j|^{\lambda_j} > \frac{c_1}{2na} \quad \text{for} \quad |\xi_j| > c_2, \qquad j = 1, \dots, n,$$
$$|\xi_j|^{-\mu_j} > \frac{c_1}{2na} \quad \text{for} \quad |\xi_j| > \frac{1}{c_2}, \qquad j = 1, \dots, n,$$

hold simultaneously. Then in the complement to $\Omega(c_2)$ the inequality (15) holds.

(ii) The first part of the assertion is proved by repeating literally the argument in Lemma 1.2.2. To prove the second part of the assertion we note that if the point α belongs to the boundary of N, then the supporting hyperplane $\langle q, \alpha\rangle = \langle q, \alpha^0\rangle$ passes through it. Therefore along the curve $\{\xi_j(t) = t^{q_j}\xi_j^0, j = 1, \dots, n\}$ inequality (13′) is violated for $t \to +\infty$ or $t \to 0$ for any c.

Lemma 3. *Let* $P(\xi)$ *and* $Q(\xi)$ *be arbitrary polynomials. Then there is* $c = c(P, Q)$ *such that*

$$c\Xi_P(\xi)\Xi_Q(\xi) \leqslant \Xi_{PQ}(\xi) \leqslant \Xi_P(\xi)\Xi_Q(\xi) \quad \forall\xi \in \mathbb{R}^n.$$

The proof is a literal repetition of that in the case $n = 2$ (see Lemmas 1 and 2 in Section 1.2.3).

Theorem. *A polynomial $P(\xi)$ admits of estimate (2′) for all $\xi \in \mathbb{R}^n$ if and only if the principal quasi-homogeneous parts $P_q(\xi)$ are nonzero for all $q \in \mathbb{R}^n$ outside the coordinate planes, i.e.*

$$P_q(\xi) \neq 0 \quad \text{for} \quad \xi^{(1)} = \xi_1 \dots \xi_n \neq 0. \tag{16}$$

Remark. By the accepted convention, conditions (16) include the condition

$$P(\xi) \neq 0 \quad \text{for} \quad \xi^{(1)} \neq 0, \tag{16′}$$

corresponding to $q = (0, \dots, 0)$.

Proof. Necessity. Let $\xi^{(1)} \neq 0$. Then $\Xi_P(\xi) > 0$, and hence, by virtue of (2′), we have $|P(\xi)| > 0$. Further, for $\xi^{(1)} \neq 0$ we have

$$\Xi_P(\rho^q \xi) \geqslant c' \rho^{d_P(q)}, \qquad c' = c'(\xi) > 0.$$

Indeed, to prove this inequalities it suffices to retain in sum (12) only those terms corresponding to $\alpha \in N_q(P)$ for which $|(\rho^q \xi)^\alpha| = \rho^{d_P(q)} |\xi^\alpha|$. It now follows from (8) that

$$P_q(\xi) \neq 0.$$

Sufficiency. To begin with, we make two preliminary simplifications. First, instead of the case of the entire space $\mathbb{R}^n$ we can consider the case of the positive coordinate n-hedron. The matter is that the case of $\mathbb{R}^n$ obviously reduces to considering all coordinate n-hedrons and the case of an arbitrary coordinate n-hedron is reduced by means of a trivial transformation to the case of the positive n-hedron.

Second, one can confine oneself to the case when the polynomial $P(\xi)$ has real coefficients (cf. Mikhaĭlov [1]). Indeed, if $P(\xi)$ is a polynomial with complex coefficients for which $P_q(\xi) \neq 0$ for $\xi^{(1)} \neq 0$, then the polynomial $Q(\xi) = |P(\xi)|^2$ with real coefficients possesses the same property since, according to (9), we have

$$Q_q(\xi) = P_q \overline{P_q(\xi)} = |P_q(\xi)|^2.$$

If the theorem is proved for polynomials with real coefficients, then for $c > 0$ we have

$$c\Xi_Q(\xi) \leqslant Q(\xi) = |P(\xi)|^2.$$

By virtue of Lemma 3, we conclude that for some $c_1 > 0$ we have $c_1(\Xi_P(\xi))^2 \leqslant \Xi_Q(\xi)$, whence

$$c_2 \Xi_P(\xi) \leqslant |P(\xi)|,$$

for some $c_2 > 0$.

Thus, we consider a polynomial $P(\xi)$ with real coefficients on $\overline{\mathbb{R}^n_+}$ for which (16) is fulfilled. Therefore $P(\xi)$ retains sign, and it can be assumed without loss of generality that

$$P(\xi) > 0 \quad \text{for} \quad \xi_j > 0, \quad j = 1, \dots, n.$$

Then, by virtue of (8), we have

$$P_q(\xi) > 0 \quad \text{for all} \quad q \in \mathbb{R}^n \quad \text{and} \quad \xi_j > 0, \quad j = 1, \dots, n.$$

We shall prove the validity of the desired estimate by double induction, namely on the dimension n and on the number of (integral) points in $N(P)$.

For $n = 1$ we have the following obvious assertion: if a polynomial

$$P(\xi) = \sum_{\alpha=k}^{m} a_\alpha \xi^\alpha, \qquad \xi \in \mathbb{R}^1,$$

is nonzero for $\xi \neq 0$, then it satisfies an inequality of the form of

$$|P(\xi)| \geqslant c(\xi^m + \xi^k) \quad \text{for} \quad \xi \in \mathbb{R}_+.$$

Assume that the desired assertion is true for polynomials in $n-1$ variables. We first of all prove it for quasi-homogeneous polynomials in n variables. So, let

$$N(P) = N_q(P), \qquad P(\xi) = P_q(\xi), \qquad q \neq (0).$$

Let, for definiteness, $q_1 \neq 0$. We set

$$\eta_j = \xi_j \xi_1^{-q_j/q_1}, \quad j \geqslant 2, \qquad Q(\eta_2, \dots, \eta_n) = P_q(1, \eta_2, \dots, \eta_n).$$

Then

$$P(\xi) = P_q(\xi) = \xi_1^{d_P(q)/q_1} P_q(1, \eta_2, \dots, \eta_n) = \xi^{d_P(q)/q_1} Q(\eta),$$
$$Q(\eta) = \xi_1^{-d_P(q)/q_1} P_q(\xi).$$

Newton's polyhedron $N(Q)$ is obtained from the polyhedron $N(P_q)$ by projecting the latter on the subspace $\{\alpha_1 = 0\}$. Here the faces of $N(Q)$ are obtained by projecting the faces $N_{\tilde{r}}(P_q)$ lying on the boundary of $N(P_q)$ (i.e. the faces of dimension no higher $n-2$), and we have

$$Q_r(\eta) = P_{\tilde{r}}(1, \eta_2, \dots, \eta_n), \qquad r_j = \tilde{r}_j - q_j \tilde{r}_1 / q_1, \quad j \geqslant 2.$$

It is obvious that

$$Q_r(\eta) \neq 0 \quad \text{for} \quad \eta^{(1)} \neq 0;$$

including $r = (0)$ and with account of

$$\Xi_Q(\eta) = \Xi_{P_q}(1, \eta_2, \dots, \eta_n),$$

we obtain (2′) for P_q.

We now begin the induction on the number of points. A subset $\mathcal{D}$ of integral points belonging to $N(P)$ is said to be *regular* if

(i) $\mathcal{D}$ contains all the vertices of $N(P)$, i.e. $\mathcal{D} \supset V(P)$;

(ii) if $\mathcal{D}$ contains an interior point of a face N_q, then $\mathcal{D}$ contains all integral points of that face.

Further, for the sake of brevity, we shall say that $\mathcal{D}$ contains N_q if $\mathcal{D}$ contains all integral points belonging to this face. The above definition implies that $\mathcal{D}$ is the union of a set of faces of $N(P)$ (including all the vertices) and a number of interior points of $N(P)$. We put

$$P_{\mathcal{D}}(\xi) = \sum_{\alpha \in \mathcal{D}} a_\alpha \xi^\alpha,$$

where a_α are the coefficients of the polynomial P. It follows from (i) that

$$V(P_{\mathcal{D}}) = V(P),$$

for all regular sets $\mathcal{D}$.

By means of induction on the number of points in $\mathcal{D}$ we now prove that for some $c > 0$ and c_1 we have

$$c\Xi_P(\xi) \leqslant P_{\mathcal{D}}(\xi) \quad \forall \xi \in \mathbb{R}^n \setminus \Omega(c_1). \tag{17}$$

We remind the reader that we have $P_q(\xi) \geqslant 0$ for all q. In particular, the coefficients a_α corresponding to the vertices $\alpha \in V(P)$ are positive. Hence, inequality (17) holds for $\mathcal{D} = V(P)$. Lemma 2 (i) allows one to add to $V(P)$ an arbitrary set of interior points (provided that c_1 in (17) is sufficiently large). Further, since (17) has already been proved for quasi-homogeneous polynomials in n variables, the estimate we are interested in remains valid when $\mathcal{D} = \mathcal{D}(q)$, where $\mathcal{D}(q)$ is the set obtained by adding to the points belonging to the face $N_q(P)$ the other vertices of $N(P)$ not belonging to that face. This estimate is taken as the basis of the induction hypothesis.

Now let $\mathcal{D}$ be a regular set, and let us assume that estimate (17) has already been proved for all regular sets with a smaller number of points. We shall prove the estimate for $\mathcal{D}$. It is natural to assume that $\mathcal{D}$ contains some boundary points distinct from the vertices. Then, according to (ii), the set $\mathcal{D}$ contains a face $N_r(P)$, $r \in \mathbb{R}^n$, of maximum dimension. We associate with this face two regular subsets of $\mathcal{D}$, namely the subset $\mathcal{D}(r)$ obtained as the union of N_r and the other vertices of $N(P)$ not belonging to N_r and the subset $\mathcal{D}\{r\}$ obtained by removing from $\mathcal{D}$ the interior points of N_r. By the induction hypothesis, for the sets $\mathcal{D}(r)$ and $\mathcal{D}\{r\}$

inequalities of the type (17) hold. Multiplying these inequalities we obtain the estimate for the polynomial $Q(\xi) = P_{\mathcal{D}(r)}(\xi)P_{\mathcal{D}\{r\}}(\xi)$:

$$c(\Xi_P(\xi))^2 \leqslant Q(\xi) = P_{\mathcal{D}(r)}(\xi)P_{\mathcal{D}\{r\}}(\xi), \qquad \xi \in \mathbb{R}^n \setminus \Omega(c_1). \tag{17'}$$

We now perform the main rearrangement of the introduced polynomials. Denote by $\widetilde{\mathcal{D}}(r)$ the regular set obtained by adding all faces lying on the boundary of the face $N_r(P)$ to the set of vertices. We put

$$\widetilde{Q}(\xi) = P_{\mathcal{D}}(\xi)P_{\widetilde{\mathcal{D}}(r)}(\xi).$$

Let us show that the polynomials $Q(\xi)$ and $\widetilde{Q}(\xi)$ differ by a linear combination of monomials ξ^α corresponding to the interior points of the duplicated polyhedron $2N(P)$. Then, by Lemma 2 (i), $\forall\varepsilon > 0$ there is $c_1 = c_1(\varepsilon)$ such that

$$|Q(\xi) - \widetilde{Q}(\xi)| < \varepsilon\Xi_{P^2}(\xi).$$

According to Lemma 3, $\Xi_{P^2}(\xi) \leqslant (\Xi_P(\xi))^2$. Taking $\varepsilon = c/2$ and substituting into (17′) we see that for $\xi \in \mathbb{R}^n \setminus \Omega(c_1)$ we have

$$\frac{c}{2}(\Xi_P(\xi))^2 \leqslant \widetilde{Q}(\xi) = P_{\mathcal{D}}(\xi)P_{\widetilde{\mathcal{D}}(r)}(\xi) \leqslant c_2 P_{\mathcal{D}}\Xi_P(\xi).$$

Cancelling by $\Xi_P(\xi)$ we arrive at (17).

Thus, with account of Lemma 2 in Section 1.1, we have to show that

$$Q_q(\xi) = \widetilde{Q}_q(\xi) \quad \forall q \in \mathbb{R}^n.$$

If vector q is not parallel to r (i.e. $q \neq \lambda r$), then from the definitions of the sets $\mathcal{D}(r)$, $\mathcal{D}\{r\}$, and $\widetilde{\mathcal{D}}\{r\}$ it follows that

$$(P_{\mathcal{D}})_q = (P_{\mathcal{D}\{r\}})_q, \qquad (P_{\widetilde{\mathcal{D}}(r)})_q = (P_{\mathcal{D}(r)})_q.$$

Multiplying these relations we obtain $Q_q = \widetilde{Q_q}$. In case $q = \lambda r$ we have

$$(P_{\mathcal{D}})_r = (P_{\mathcal{D}(r)})_r, \qquad (P_{\widetilde{\mathcal{D}}(r)})_r = (P_{\mathcal{D}\{r\}})_r$$

Hence, inequalities of the type (17) have been proved for all regular sets $\mathcal{D}$ and, in particular, for the boundary of the polyhedron $N(P)$. Therefore, by virtue of Lemma 2 (ii), inequality (17) holds for $\mathcal{D} = N(P)$ as well. Consequently, we have proved inequality (2′) outside the compact set $\overline{\Omega(c_1)}$. However, since we have $P(\xi) \neq 0$ for $\xi^{(1)} \neq 0$, we see that (2′) with some constant $c > 0$ is also fulfilled everywhere on $\overline{\Omega(c_1)}$ (we have not yet used this condition). The proof is completed.

Remark. As is seen from the proof, if the condition $P(\xi) \neq 0$ for $\xi^{(1)} \neq 0$ is dropped, then inequality (2′) is fulfilled for some c_1 outside the set $\Omega(c_1)$.

1.3. Some other conditions for the existence of an adequate estimate (2′). In this subsection we shall derive an analog of Theorem 1.2.4, namely we shall prove that the existence of estimate (2′) is equivalent to the property that the polynomial $P(\xi)$ remains nonzero for any small perturbations of the coefficients or that it has no zeros in a complex neighborhood of a special form of the real space $\mathbb{R}^n$.

Theorem. *For a polynomial* (1) *the following conditions are equivalent:*

(I) *inequality* (2′) *holds for all* $\xi \in \mathbb{R}^n$;

(II) *there is* $\varepsilon > 0$ *such that all the polynomials*

$$P_\delta(\xi) = \sum_{\alpha \in N(P)} (a_\alpha + \delta_\alpha)\xi^\alpha, \qquad |\delta_\alpha| < \varepsilon, \tag{18}$$

are nonzero for $\xi^{(1)} \neq 0$;

(III) *there is* $c > 0$ *such that for all* $\xi \in \mathbb{R}^n$ *the polynomial* $P(\xi)$ *has no zeros in the polycylinder*

$$|\zeta_j - \xi_j| < c|\xi_j|, \qquad 1 \leqslant j \leqslant n,\ \zeta \in \mathbb{C}^n. \tag{19}$$

Proof. (I)$\Longrightarrow$(II). By virtue of the lemma in Section 1.2, inequality (2′) is equivalent to the existence of $c' > 0$ such that

$$c' \sum_{\alpha \in N(P)} |\xi^\alpha| \leqslant |P(\xi)| \quad \forall \xi \in \mathbb{R}^n.$$

Therefore it suffices to take $\varepsilon = c'/2$ for the condition $P_\delta(\xi) \neq 0$ to be fulfilled for $|\delta_\alpha| < \varepsilon$.

(II)$\Longrightarrow$(III). Denote polycylinder (19) as $\mathcal{D}_\xi(c)$. If $\zeta \in \mathcal{D}_\xi(c)$, then ζ can be represented in the form

$$\zeta_j = (1 + d_j)\xi_j, \quad |d_j| < c, \quad 1 \leqslant j \leqslant n, \quad d = (d_1, \dots, d_n).$$

We have $P(\zeta) = \Sigma a_\alpha \zeta^\alpha = \Sigma(a_\alpha + \delta_\alpha)\xi^\alpha$, where $\delta_\alpha = a_\alpha[(1+d)^\alpha - 1]$. We take $c > 0$ satisfying the condition $(1 + c)^m - 1 < \varepsilon$ where m is the degree of the polynomial P and ε is the constant involved in (II). For this choice of c we have $|\delta_\alpha| < \varepsilon$, and, by virtue of (II), $P(\zeta) \neq 0$ for $\zeta \in \mathcal{D}_\xi(c)$.

(III)$\Longrightarrow$(I). In view of Theorem 1.2, it suffices to show that (III) implies that $P_q(\xi) \neq 0$ for all $q \in \mathbb{R}^n$, $\xi^{(1)} \neq 0$. Let $P_q(\xi) = 0$ for some q, ξ, and $\xi^{(1)} \neq 0$. We fix these values of q and ξ.

It follows from (III) that for some $\varepsilon > 0$ we have

$$P(\rho^q \zeta) \neq 0 \quad \text{for} \quad |\zeta_j - \xi_j| < \varepsilon, \quad 1 \leqslant j \leqslant n,\ \rho > 0. \tag{20}$$

It suffices to take $\varepsilon = c \min |\xi_j|$, where c is the constant in (III).

We select an integral vector $\pi = (\pi_1, \dots, \pi_n)$ such that the face $N_\pi(P)$ is a vertex belonging to the face $N_q(P)$. This can be done since the vectors $\pi \in \mathbb{R}^n$ for which a certain vertex is the face $N_\pi(P)$ form a polyhedral cone of maximum dimension. Therefore $P_\pi(\zeta)$ is a monomial with a nonzero coefficient, and we have $P_\pi(\zeta) \neq 0$ for $\zeta \in \mathbb{C}^n$, $\zeta^{(1)} \neq 0$. Consider the expression $P(z^\pi \xi)$ which is a linear combination of integral powers of z with exponents lying between $-d_P(-\pi)$ and $d_P(\pi)$, the coefficient in the highest power of z being equal to $P_\pi(\xi)$ and nonzero ($\xi^{(1)} \neq 0$). If we set $r = \max\{d_P(-\pi), 0\}$, then $z^r P(z^\pi \xi)$ is a polynomial in z of degree $s = r + d_P(\pi)$. We put

$$Q_\rho(z) = \rho^{-d_P(q)} z^r P(\rho^q z^\pi \xi).$$

This is a polynomial in z whose coefficients are continuous functions of the parameter ρ for $\rho \to +\infty$ (see(8)). The coefficient in the highest power z^s is equal to

$$\rho^{-d_P(q)} P_\pi(\rho^q \xi) = P_\pi(\xi).$$

Here we make use of the fact that the vertex N_π belongs to the face N_q. So, we have found that the leading coefficient in the polynomial $Q_\rho(z)$ does not depend on ρ, and hence its roots depend on ρ continuously in the neighborhood of $\rho = +\infty$. For $\rho = +\infty$ we have

$$Q_\infty(z) = z^r P_q(z^\pi \xi),$$

i.e. $z = 1$ is a root of $Q_\infty(z)$. Therefore for any $\varepsilon' > 0$ there is M such that for $\rho > M$ the polynomial $Q_\rho(z)$ has a root $z(\rho)$ for which

$$|z(\rho) - 1| < \varepsilon'.$$

We have

$$P(\rho^q z(\rho)^\pi \xi) = 0.$$

Selecting $\varepsilon' > 0$ such that $|z^\pi \xi_j - \xi_j| < \varepsilon$, $1 \leqslant j \leqslant n$, we arrive at a contradiction to (20) for $|z - 1| < \varepsilon'$. The proof is completed.

1.4. Estimates with account of summarized degrees with respect to groups of variables. When applying Newton's polygons to polynomials in several variables in Chapters 2 and 3 we have already taken into account the summarized degree with respect to all spatial variables. We now consider this technique in a more general situation.

Let the space $\mathbb{R}^n$ be represented as a direct product of subspaces $\mathbb{R}^{l_j}$:

$$\mathbb{R}^n = \mathbb{R}^{l_1} \times \cdots \times \mathbb{R}^{l_k}, \qquad \sum l_j = n.$$

We set

$$\xi = (\xi_{(1)}, \dots, \xi_{(k)}), \quad \xi_{(j)} \in \mathbb{R}^{l_j}, \quad |\xi|^{[l]} = (|\xi_{(1)}|, \dots, |\xi_{(k)}|) \in \overline{\mathbb{R}^k_+},$$

where $|\xi_{(j)}|$ is the Euclidean norm of the vector $\xi_{(j)} \in \mathbb{R}^{l_j}$. Accordingly, we split the set of monomial exponents $\alpha \in \overline{\mathbb{Z}^n_+}$:

$$\alpha = (\alpha_{(1)}, \dots, \alpha_{(k)}), \quad \alpha_{(j)} \in \overline{\mathbb{Z}^{l_j}_+}, \quad j = 1, \dots, k,$$
$$\alpha^{[l]} = (|\alpha_{(1)}|, \dots, |\alpha_{(k)}|) \in \overline{\mathbb{R}^k_+},$$

where $|\alpha_{(j)}|$ denotes the sum of the components of the vector $\alpha_{(j)}$.

Proceeding from polynomial (1) we construct the set

$$\nu^{[l]}(P) = \{\beta \in \overline{\mathbb{R}^k_+}, \beta = \alpha^{[l]}, \alpha \in \nu(P)\},$$

and a reduced polyhedron $N^{[l]}(P)$, the convex hull of $\nu^{[l]}(P)$. Let $V^{[l]}(P)$ be the set of vertices of the polyhedron $N^{[l]}(P)$. For $q \in \mathbb{R}^k$ we put

$$d^{[l]}_P(q) = \max_{\beta \in \nu^{[l]}(P)} \langle q, \beta \rangle = \max_{\alpha \in \nu(P)} \langle q, \alpha^{[l]} \rangle.$$

Let $N^{[l]}_q(P)$ be the face of $N^{[l]}(P)$ lying in the supporting plane $\langle q, \beta \rangle = d_P(q)$ and let

$$P^{[l]}_q(\xi) = \sum_{\alpha^{[l]} \in N^{[l]}_q(P)} a_\alpha \xi^\alpha.$$

Finally, let

$$\Xi^{[l]}_P(\xi) = \sum_{\beta \in V^{[l]}(P)} (|\xi|^{[l]})^\beta.$$

Theorem. *For polynomial* (1) *the following conditions are equivalent.*

(I) *There is $c > 0$ such that*

$$c\Xi^{[l]}_P(\xi) \leqslant |P(\xi)| \quad \forall \xi \in \mathbb{R}^n. \tag{21}$$

(II) *For all $q \in \mathbb{R}^k$ we have*

$$P^{[l]}_q(\xi) \neq 0 \quad \text{for} \quad |\xi_{(1)}| \dots |\xi_{(k)}| \neq 0. \tag{22}$$

In particular,

$$|P^{[l]}_{(0,\dots,0)}| \stackrel{\text{def}}{=} |P(\xi)| \neq 0 \quad \text{for} \quad |\xi_{(1)}| \dots |\xi_{(k)}| \neq 0. \tag{22'}$$

(III) *There is $\varepsilon > 0$ such that all the polynomials*

$$P^{[l]}_\delta(\xi) = \sum_{\alpha^{[l]} \in N^{[l]}(P)} (a_\alpha + \delta_\alpha)\xi^\alpha, \qquad |\delta_\alpha| \leqslant \varepsilon, \tag{24}$$

are nonzero for $|\xi_{(1)}| \dots |\xi_{(k)}| \neq 0$.

(IV) *There is $c > 0$ such that for each $\xi \in \mathbb{R}$ the polynomial $P(\zeta)$ is nonzero in the complex polysphere (a direct product of complex spheres)*

$$|\zeta_{[j]} - \xi_{[j]}| < c|\xi_{[j]}|, \qquad 1 \leqslant j \leqslant k, \ \zeta \in \mathbb{C}^n.$$

This theorem is derived in a simple way from the corresponding assertions in Sections 1.2 and 1.3. We eludicate the proof of the implication (II)$\Longrightarrow$(I).

In each group of the variables $\xi_{(j)}$ we pass to polar coordinates:

$$\xi_{(j)} = |\xi_{(j)}|\omega_{(j)}, \qquad |\omega_{(j)}| = 1,\ 1 \leqslant j \leqslant k.$$

Then

$$P(\xi) = P(|\xi_{(1)}|\omega_1, \dots, |\xi_{(k)}\omega_k|) \overset{\text{def}}{=} P_\omega(\xi^{[l]}).$$

To the polynomial P_ω in k variables with coefficients depending continuously on ω we apply Theorem 1.2, which results in

$$c(\omega)\Xi_{P_\omega}(\xi) \leqslant P(\xi) \quad \text{for} \quad \xi \in \mathbb{R}^n.$$

It remains to note that $N(P_\omega) = N^{[l]}(P)$, the constant $c(\omega)$ can be regarded as a continuous function of ω, and ω is a point belonging to a compact set.

Remark. We note that the set $\{|\xi_{(1)}|\dots|\xi_{(k)}| = 0\}$ is contained in the set $\{|\xi_1\dots\xi_n| = 0\}$ and the conditions of the theorem are stronger as compared to those in Sections 1.2 and 1.3 guaranteeing the existence of estimate (2). Let us discuss the corresponding geometrical pattern. Let $\widetilde{N}(P)$ be the set of those $\alpha \in \overline{\mathbb{R}^n_+}$ for which $\alpha^{[l]} \in N^{[l]}(P)$. It is clear that

$$N(P) \subset \widetilde{N}(P),$$

and, generally, this is a strict inclusion relation. The conditions of the theorem correspond to the existence of the estimate

$$c\Xi_{\widetilde{N}(P)} \leqslant P(\xi) \quad \text{for} \quad \xi \in \mathbb{R}^n, \tag{21'}$$

which is equivalent to (21). According to Lemma 1.2, to this end it is necessary that

$$N(P) = \widetilde{N}(P).$$

This is a rather strong condition on Newton's polyhedron $N(P)$.

§2. Two-sided estimates in some regions in $\mathbb{R}^n$ relating to Newton's polyhedron. Special classes of polynomials and differential operators in several variables

The results of the foregoing section cannot be applied directly to differential operators. Such applications require estimates holding not throughout the space but for part (or all) of the variables tending to infinity in some way. Moreover, in the case of differential operators Newton's polyhedron is usually completed by adding some "minor" points that are selected proceeding from the character of the desired analytical estimates.

2.1. Estimates on condition that part of the variables tend to infinity. We divide the variables in $\mathbb{R}^n$ into two groups $\xi \in \mathbb{R}^k$ and $\eta \in \mathbb{R}^l$, $k+l=n$, and let

$$P(\xi,\eta) = \sum_{(\alpha,\beta)\in\nu(P)} a_{\alpha\beta}\xi^\alpha\eta^\beta, \qquad \alpha\in\mathbb{R}^k,\ \beta\in\mathbb{R}^l. \tag{1}$$

We shall prove the following generalization of Theorem 1.2.

Theorem. *For a polynomial* (1) *the conditions below are equivalent:*

(i) *there are $c_1 > 0$ and c_2 such that*

$$c_1\Xi_P(\xi,\eta) \leqslant |P(\xi,\eta)| \quad \text{for} \quad |\eta| = |\eta_1| + \cdots + |\eta_l| > c_2; \tag{2}$$

(ii) *if at least one of the components of a vector $r \in \mathbb{R}^l$ is positive, then*

$$P_{(q,r)}(\xi,\eta) \neq 0 \quad \text{for} \quad (\xi,\eta)^{(1)} = \xi_1\ldots\xi_k\eta_1\ldots\eta_l \neq 0. \tag{3}$$

For $l=0$ the theorem goes into Theorem 1.2.

In the situation under consideration there naturally appears the notion of minor points.

Definition. A point $(\alpha,\beta) \in \mathbb{Z}^n_+$ is said to be *minor* for a polyhedron N if for some $\lambda > 0$ and all $j \leqslant l$ we have

$$(\alpha, \beta + \lambda e_j) \in N,$$

where e_j is a vector belonging to $\mathbb{R}^l$, whose jth component is 1 and the other components are zeros.

The points belonging to N that are not minor will be called senior points. Of course, the notion of minor points depends on the way the variables are divided into groups.

Denote by N^σ the polyhedron obtained from N by adding all the minor points, by δ^σ the set of minor points of N, by $\pi^\sigma N$ the set of senior points, and by V^σ the set of senior vertices of N. When $N = N(P)$ is Newton's polyhedron of polynomial (1), we shall add (P) to these symbols.

Lemma. (i) *If (α,β) is a minor point of N, then for every $\varepsilon > 0$ there is $a > 0$ such that*

$$\varepsilon\Xi_N(\xi,\eta) \geqslant |\xi^\alpha||\eta^\beta| \quad \text{for} \quad |\eta| = |\eta_1| + \cdots + |\eta_l| > a \quad \forall\xi\in\mathbb{R}^k; \tag{4}$$

(ii) *$\pi^\sigma N$ is the union of all faces $N_{(q,r)}$ such that at least one of the components r_j is positive.*

The assertions of the lemma are verified directly. To prove (i) we take ε satisfying the condition $\varepsilon(a/l)^\lambda \geqslant 1$, where λ is the number involved in the definition of a minor point (α, β). Then $\varepsilon|\xi^\alpha||\eta^{\beta+\lambda e_j}| \geqslant |\xi^\alpha\eta^\beta|$ for $|\eta_j| > a/l$ for all j, whence follows (4). As to (ii), in this case if $(\alpha, \beta) \in N_{(q,r)}$ where $r_j > 0$, then $(\alpha, \beta+\lambda e_j) \notin N$ for all $\lambda > 0$. In case $(\alpha, \beta) \in N_{(q,r)}$, where all $r_j \leqslant 0$ and (α, β) does not belong to the faces with $r_j > 0$, the expression $(\alpha, \beta + \lambda e_j)$ satisfies all inequalities determining N for all j and a sufficiently small λ, namely if $(\alpha, \beta) \in N_{(q,r)}$, then all $r_j \leqslant 0$, and the inequality holds for all $\lambda > 0$; and if $(\alpha, \beta) \notin N_{(q,r)}$, then for (α, β) the inequality is strict and is retained for small λ.

We now proceed to the proof of the theorem. The implication (i)$\Longrightarrow$(ii) is proved as in Theorem 1.2. Consider the implication (ii)$\Longrightarrow$(i). Let $\widetilde{N}$ be the convex hull of the set of vertices $V^\sigma(P)$. Then the union of the senior faces $\pi^\sigma N(P)$ is a regular set for $\widetilde{N}$ (the definition is contained in the proof of Theorem 1.2). Therefore for some $a_1, a_2 > 0$ we have

$$a_1 \Xi_{\widetilde{N}}(\xi, \eta) \leqslant |P_{\pi^\sigma N}(\xi, \eta)|$$

outside the compact set $\Omega(a_2)$ (with respect to the variables ξ and η; see (1.14)). Using assertion (i) of the lemma we can replace this inequality by the inequality

$$a_1 \Xi_P(\xi, \eta) \leqslant |P(\xi, \eta)|, \qquad \xi \notin \Omega(a_2) \subset \mathbb{R}^n_\xi,\ |\eta| > a_2, \tag{4'}$$

for some other $a_1, a_2 > 0$.

For any fixed $\xi^0 \in \Omega(a_2)$, considering $P_{(0,r)}(\xi^0, \eta)$ (i.e. $q = (0, \dots, 0)$) and applying Theorem 1.2 and part (i) of the lemma we conclude that for some b_1 and b_2 we have

$$b_1 \Xi_P(\xi^0, \eta) \leqslant P(\xi^0, \eta) \quad \text{for} \quad |\eta| > b_2, \tag{5}$$

and from the proof it is clearly seen that the constants b_1 and b_2 can be regarded as being independent of the point ξ^0 belonging to the compact set $\Omega(a_2)$. The combination of (4′) and (5) yields (2). We note that in the theorem the expression Ξ_P can be replaced by Ξ_P^σ where the summation extends only over senior vertices $V^\sigma(P)$.

Remarks. Following the same idea it is possible to construct many versions of these estimates. Below are several examples (for more detail see Gindikin [1]).

1) If the existential quantifier with respect to c_2 in the statement of the theorem is replaced by the universal quantifier ($\forall c_2\ \exists c_1$), then in condition (3) the existence of a positive component of the vector r should be replaced by the existence of a nonnegative component.

2) The cases indicated in 1) can be unified by dividing the set of indices into three groups Σ_1, Σ_2, Σ_3. For any $a > 0$ there exist $c_1, c_2 > 0$ such that

$$c_1 \Xi_P(\xi) \leqslant P(\xi)$$

for all $j \in \Sigma_1 : |\xi_j| > a$, $j \in \Sigma_2 : |\xi_j| > c_2$, $j \in \Sigma_3$, if and only if

$$P_q(\xi) \neq 0 \quad \text{for} \quad \xi^{(1)} \neq 0, \quad q_i \geqslant 0,\ i \in \Sigma_2, \quad \text{and} \quad q_i > 0,\ i \in \Sigma_3.$$

2.2. Estimates in the case of complete polyhedra. In Chapter 1 we meant by Newton's polygon the convex hull of not only monomial exponents but also their projections on the coordinate axes. Let us consider a multidimensional generalization of this construction.

We shall write $\gamma \preccurlyeq \alpha$; $\alpha, \gamma \in \mathbb{R}^n$, if $\gamma_j \leqslant \alpha_j$ for all $j \leqslant n$. By the completion of a polyhedron $N \subset \overline{\mathbb{R}^n_+}$ will be meant the polyhedron $\widetilde{N}$ obtained from N by adding those $\gamma \in \overline{\mathbb{R}^n_+}$ for which $\gamma \preccurlyeq \alpha$ for some $\alpha \in N$. A polyhedron N is said to be *complete* if $N = \widetilde{N}$.

Lemma. 1) *A polyhedron N is complete if and only if it is determined in $\mathbb{R}^n_+$ by a finite system of inequalities of the form of*

$$\langle q, \alpha\rangle \leqslant d(q), \qquad q_i \geqslant 0, \quad j \leqslant n. \tag{6}$$

2) *For a complete polyhedron N all the points not belonging to the faces N_q, where $q_j \geqslant 0$, $j \leqslant n$, are minor (in the sense of the definition in Section 2.1 for $k = 0$).*

3) *For a complete polyhedron N a face N_q, where $q_j < 0$, lies in the coordinate plane $\{\alpha_j = 0\}$ and coincides with the intersection of the face $N_{q(j)}$ with this plane, where $q(j)$ is obtained from q by replacing $q_j < 0$ by $q_j = 0$. In particular, if $N = N(P)$, then*

$$P_q(\xi) = P_{q(j)}(\xi)\big|_{\xi_j=0}. \tag{7}$$

If $q_j < 0$ in the vector q for $j \in J$ and if $q(J)$ is obtained from q by replacing these q_j by zeros, then

$$P_q(\xi) = P_{q(J)}(\xi)\big|_{\xi_j=0}, \qquad j \in J. \tag{7'}$$

In particular, if $q_j < 0$ for $j \in J$ and $q_j = 0$ for $j \notin J$, then

$$P_q(\xi) = P(\xi)\big|_{\xi_j=0}, \qquad j \in J. \tag{8}$$

Recall that $P_{(0,\dots,0)}(\xi) = P(\xi)$.

Proof. It is clear that, along with α, all $\gamma \preccurlyeq \alpha$ satisfy system (6), i.e. system (6) determines a complete polyhedron. Now let N be a complete polyhedron. Consider N_q where there is a component $q_j < 0$. If N contains a point α with $\alpha_j > 0$, then the point $\widetilde{\alpha}$ obtained from α by decreasing α_j does not belong to N but we have $\widetilde{\alpha} \preccurlyeq \alpha$. Consequently, N_q lies in the plane $\{\alpha_j = 0\}$. By virtue of the completeness of N, either $N_{q(j)}$ lies in $\{\alpha_j = 0\}$ or $N_{q(j)}$ has a non-empty intersection with $\{\alpha_j = 0\}$ (the projections of the points of $N_{q(j)}$). In both cases this intersection coincides with N_q. We simultaneously conclude that the inequality $\langle q, \alpha\rangle < d(q)$ is inessential in the determination of N. It can easily be seen that this argument remains valid when q has no positive components. Relations (7), (7′), and (8) are a direct expression for these geometrical properties for the case $N = N(P)$. Finally, if a point $\alpha \in N$ belongs to none of those faces N_q for which all $q_j \geqslant 0$, then all inequalities in (6) are strict, and the point obviously remains in the complete polyhedron under a small positive perturbation of any coordinate α_j, i.e. is minor.

It is often convenient to think of a complete polyhedron as the union of the parallelepipeds $\{\gamma \preccurlyeq \alpha, \gamma \in \overline{\mathbb{R}^n_+}\}$ over the senior points $\alpha \in N$.

Theorem. *For a polynomial* (1) *the following conditions are equivalent.*

(i) *There are* $c_1, c_2 > 0$ *such that*

$$c_1 \Xi_{\widetilde{N}(P)}(\xi, \eta) \leqslant |P(\xi, \eta)| \quad \text{for} \quad |\eta| > c_2,\ \xi \in \mathbb{R}^k, \tag{9}$$

where $\widetilde{N}(P)$ *is the completion of Newton's polyhedron* $N(P)$.

(ii) *If* $q_j \geqslant 0$, $r_i \geqslant 0$, $j \leqslant k$, $i \leqslant l$, *and the vector* r *has a positive component, then*

$$P_{(q,r)}(\xi, \eta) \neq 0 \quad \text{for} \quad \xi^{(q)}\eta^{(r)} \neq 0. \tag{10}$$

We note that the condition $\xi^{(q)}\eta^{(r)} \neq 0$ means that $\xi_j \neq 0$ if $q_j \neq 0$ and $\eta_i \neq 0$ if $r_i \neq 0$. Without loss of generality, we can assume that $N(P) = \widetilde{N}(P)$ regarding some $a_{\alpha\beta}$ in (1) as being equal to zero. Then this theorem is merely a specialization of Theorem 2.1 for the case of complete Newton's polyhedron $N(P)$. To prove the theorem use should be made of the description (7), (7'), (8) of the principal parts $P_{(q,r)}(\xi, \eta)$ when among q_j, r_i there are negative components. The description allows one to confine oneself in the case of complete Newton's polyhedra to (q, r) with nonnegative components under the requirement that $P_{(q,r)}$ should be nonzero on some coordinate planes.

We note that for $l = 0$, along with the conditions $P_{(q)}(\xi) \neq 0$ for $\xi^{(q)} \neq 0$, for the faces $N_{(q)}(P)$ the condition $P(\xi) \neq 0$ also takes part for all $\xi \in \mathbb{R}^n$.

Among the complete polyhedra we separate out a class of regular polyhedra. This notion generalizes the notion of regular Newton's polygons considered in Section 1.1.

Definition. A complete polyhedron is said to be *regular* if it contains no faces parallel to the coordinate planes and not belonging to them.

In view of the lemma, this condition is equivalent to the property that N is determined in $\mathbb{R}^n_+$ by a finite system of inequalities of the form of

$$\langle q, \alpha \rangle \leqslant d(q), \qquad q_j > 0,\ j \leqslant n. \tag{11}$$

Remarks. 1) Regular polyhedra can also be separated out among the complete polyhedra by means of the following condition: if $\alpha \in N$ and $\gamma = \alpha - e_j \in \mathbb{Z}^n_+$, then γ is a minor point of N.

2) For a regular polyhedron the face N_q, where $q_j = 0$, $j \in J$, and $q_j > 0$, $j \notin J$, lies in the coordinate plane $\{\alpha_j = 0, j \in J\}$, and it is the intersection of a

face $N_{q\{J\}}$ with this plane, where all the components of $q\{J\}$ are positive and the components with indices $j \in J$ coincide with q_j. Therefore

$$P_q(\xi) = P_{q\{J\}}(\xi)\big|_{\xi_j=0,\ j\in J}. \tag{12}$$

Here $N_q = N_{\widetilde{q}}$ where $\widetilde{q}_j = q_j$ for $j \notin J$ and $\widetilde{q}_j = \varepsilon > 0$, $j \in J$, the number $\varepsilon > 0$ being sufficiently small.

This means that whereas in the case of arbitrary complete polyhedra one can confine oneself to considering the principal parts P_q with $q_j \geqslant 0$, $j \leqslant n$, in the case of regular polyhedra it is possible to confine oneself to the consideration of P_q with $q_j > 0$ for all j. In particular, if in the conditions of the theorem it is additionally required that the polyhedron $N(P)$ should be regular, then (ii) can be replaced by the condition

(ii′) if $q_i > 0$, $0 \leqslant i \leqslant k$, and $r_j > 0$, $1 \leqslant j \leqslant l$, then

$$\begin{aligned} &P_{(q,r)}(\xi,\eta) \neq 0 \quad \text{for} \quad (\xi,\eta)^{(1)} \neq 0, \\ &P_{(0,r)}(\xi,\eta) \neq 0 \quad \text{for} \quad \eta^{(1)} \neq 0, \quad \xi \in \mathbb{R}^k. \end{aligned} \tag{13}$$

We remind the reader that last the group of conditions guarantees that $P_{(q,r)}(\xi,\eta) \neq 0$ for $(\xi,\eta)^{(1)} \neq 0$ if $q_j \leqslant 0$, $j \leqslant k$.

Using Remark 2) we can combine some of the conditions, namely if $N_{(\widetilde{q},\widetilde{r})}$ is the intersection of the face $N_{(q,r)}$ with a coordinate plane, then the condition on $P_{(\widetilde{q},\widetilde{r})}$ can be added to the condition on $P_{(q,r)}$.

2.3. *N* quasi-elliptic polynomials and operators.

Definition. A polynomial $P(\xi)$, $\xi \in \mathbb{R}^n$, is said to be *N quasi-elliptic* if its Newton's polyhedron $N(P)$ is regular and the equivalent conditions of Theorem 2.2 for $k = 0$ hold, i.e. if $q_j > 0$, $1 \leqslant j \leqslant n$, then

$$P_q(\xi) \neq 0 \quad \text{for} \quad \xi^{(1)} \neq 0. \tag{13′}$$

Under these conditions for some c_1 and c_2 we have

$$c_1 \Xi_P(\xi) \leqslant |P(\xi)| \quad \text{for} \quad |\xi| > c_2. \tag{3′}$$

To N quasi-elliptic differential operators (i.e. operators with N quasi-elliptic symbols) the result of §1.4, which were obtained for the case of two variables, are automatically extended. In particular, N quasi-elliptic differential operators are hypoelliptic, and an analog of Theorem 1.4.2 holds for them. Similarly, if the symbol $P(x,\xi)$ of an operator $P(x,D)$ is N quasi-elliptic for each x and Newton's polyhedron $N(P(x))$ does not depend on x, then an analog of the results in Section 1.4.4 takes place. In this case we have hypoelliptic differential operators of constant strength.

2.4. N-parabolic polynomials and operators. Consider a polynomial

$$P(\xi,\tau) = \sum_{(\alpha,\beta)\in\nu(P)} a_{\alpha\beta}\xi_1^{\alpha_1}\dots\xi_n^{\alpha_n}\tau^\beta, \qquad \xi\in\mathbb{R}^n,\ \tau\in\mathbb{C}^1. \tag{14}$$

By its Newton's polyhedron will be meant the convex hull of the point $(\alpha,\beta)\in\nu(P)$ and those minor points (α,β) for which $(\alpha_1,\dots,\alpha_n\widetilde{\beta})\in\nu(P)$ for some $\widetilde{\beta}>\beta$. This corresponds to the notion of minor points in the sense of Section 2.1 if τ is included in the second group of variables. In this section the completed polyhedron will simply be called Newton's polyhedron and denoted as $N(P)$.

Definition. A polynomial (14) is said to be N stable-correct if

(i) $P(\xi,\eta)$ is solved with respect to the highest power of τ;

(ii) the polyhedron $N(P)$ is complete and regular with respect to α, i.e. is determined by a finite system of inequalities of the form of

$$\langle q,\alpha\rangle + r\beta \leqslant d_P(q,r), \qquad q_j>0,\ j\leqslant n,\ r\geqslant 0; \tag{15}$$

(iii) there are $c_1>0$ and c_2 such that

$$c_1\Xi_P(\xi,\tau)\leqslant |P(\xi,\tau)| \quad \text{for} \quad \xi\in\mathbb{R}^n,\ \operatorname{Im}\tau\leqslant c_2. \tag{16}$$

If instead of (ii) the stronger condition holds:

(ii′) the polyhedron $N(P)$ is regular;

then the polynomial is said to be N-parabolic.

Theorem. *For a polynomial satisfying the conditions* (i) *and* (ii) *the condition* (iii) *is fulfilled if and only if*

$$P_{(q,r)}(\xi,\tau)\neq 0 \quad \textit{for} \quad q_j>0,\ j\leqslant n,\ r>0,\ \tau\xi_1\dots\xi_n\neq 0,\ \operatorname{Im}\tau\leqslant 0. \tag{17}$$

This theorem is derived directly from Theorem 2.2 with account of (13′) and the following circumstances. Instead of $\tau\in\mathbb{C}^1$ we consider the pair of real variables $\eta_1=\operatorname{Re}\tau$ and $\eta_2=\operatorname{Im}\tau$, and with respect to these variables we take into account the summarized degree (see Section 1.4 and the remark in the end of the present section). This means that we consider the principal parts $P_{(q,r_1,r_2)}$ where either $r_1=r_2=r$ or $r_1=0$ or $r_2=0$. In the last two cases we have faces lying in coordinate planes, and, with account of (12), $P_{(q,r,0)}$ is obtained from $P_{(q,r,r)}$ for $\eta_2=\operatorname{Im}\tau=0$ and $P_{(q,r,0)}$ is obtained for $\eta_1=\operatorname{Re}\tau=0$. Therefore we can confine ourselves to $P_{q,r,r}$ on condition that the condition $P_{(q,r,r)}\neq 0$ is considered not only for $\operatorname{Im}\tau\cdot\operatorname{Re}\tau\neq 0$ but also for $\operatorname{Im}\tau\leqslant 0$, $|\tau|\neq 0$. Finally, by virtue of condition (i), the second set of conditions (13′) is fulfilled automatically since these conditions relate only to the face corresponding to the highest power of τ.

Remark. If the above argument is continued, we conclude that for a polynomial $P(\xi, \tau)$ satisfying condition (ii) and not obeying (i) the condition (iii) holds if the following condition is added to (17): $Q(\xi) \neq 0$ for $\xi \in \mathbb{R}^n$, where $Q(\xi)$ is the coefficient in the highest power of τ in the expansion of P in powers of τ.

N-stable polynomials are exponentially correct, and N-parabolic polynomials are, in addition, hypoelliptic. If the polyhedron $N(P(x,t))$ of a symbol $P(x,t;\xi,\tau)$ does not depend on the point (x,t), then the differential operator $P(x,t;D_x,D_t)$ corresponding to this symbol is of constant strength. Therefore the results on Cauchy's problem for N stable-correct and N quasi-parabolic operators are direct consequences of Theorem 2.5.5.

CHAPTER VI

OPERATORS OF PRINCIPAL TYPE ASSOCIATED WITH NEWTON'S POLYHEDRON

Introduction

In this chapter the results of Chapter 4 are extended to the case of polynomials in $n > 2$ variables, and accordingly, to differential operators in n variables. The general plan of the presentation of the material is the same as in Chapter 4. However, the passage to the case of n variables involves some additional difficulties, mainly of geometrical character. The thing is that a polygon in the plane has faces of only two types, namely faces of zero dimension (vertices) and faces of maximum dimension (sides). For $n > 2$ there appear faces of intermediate dimension between the zero and maximum dimensions. The notion of essential variables of a side in Section 4.2.2 is readily generalized to the notion of essential variables of a face of maximum dimension whereas the definition of essential variables of faces of lower dimension is rather intricate. Similar difficulties arise when constructing a covering by regions corresponding to faces of different dimensions, which serves as an analog of the covering by the regions $G(\Gamma_j^{(0)}, \varepsilon_0)$ and $G(\Gamma_j^{(1)}, \varepsilon_0, \varepsilon_1)$ in §4.2.

Briefly, the plan of the presentation of the material in this chapter is the following. In §1 we define polynomials of N-principal type and state a multidimensional generalization of Theorem 4.2.2.

In §2 special regions are defined and analogs of estimates (4.2.7) are proved for them. §3 presents the main geometrical construction of a covering of $\mathbb{R}^n$ by regions associated with the faces of Newton's polyhedron. This construction allows us to complete the proof of the basic theorem in §1. In §4 differential operators with variable coefficients are considered whose symbols $P(x;\xi)$ are polynomials of N-principal type with respect to ξ. For these operators a multidimensional analog of estimate (4.3.3) is proved. The resulting estimate makes it possible to reproduce almost literally the argument in §4.4 and to prove a local solvability theorem for differential operators of N-principal type with variable coefficients. In the Appendix to the chapter we present the construction of a partition of unity subordinate to the covering in §3.

§1. Polynomials of *N*-principal type

In this section all necessary notions involved in the definition of polynomials of N-principal type are introduced and a multidimensional analog of Theorem 4.2.2 is stated.

1.1. Newton's polyhedron and polyhedra of minor terms. Consider a polynomial

$$P(\xi) = \sum_{\alpha \in \gamma(P)} a_\alpha \xi^\alpha. \qquad (1)$$

If the coefficient a_α corresponding to $\alpha = 0$ is equal to zero, then we shall add the origin to the set $\gamma(P)$. Let $\widetilde{\gamma}(P) = \gamma(P) \cup \{0\}$. In this chapter by Newton's polyhedron $N(P)$ of polynomial (1) will be meant the convex hull of the finite set $\widetilde{\gamma}(P)$. Moreover, we shall deal only with those polynomials for which

$$\textit{the polyhedron } N(P) \textit{ is regular.} \tag{2}$$

This means that the polyhedron $N(P)$ is complete (i.e. along with every point $\alpha \in N(P)$ it contains all its projections on the faces of the various dimensions lying in the coordinate planes) and is determined by the systems of inequalities

$$\langle q, \alpha\rangle \leqslant d_P(q), \qquad q \in \mathbb{R}^n_+,\ \alpha \in \overline{\mathbb{R}^n_+}. \tag{3}$$

This description implies that $N(P)$ has vertices at $\{0\}$ and on each of the coordinate axes and that the $(n-1)$-dimensional faces of $N(P)$ lie in the coordinate hyperplanes $\{\alpha_j = 0, j = 1, \dots, n\}$, and in the hyperplanes

$$\langle q^{(j)}, \alpha\rangle \leqslant d_P(q), \qquad j = 1, \dots, J, \tag{3$'$}$$

where all components of the vectors $q^{(j)}$ are positive.

As in Chapter 5, a point $\alpha \in N(P)$ is said to be *minor* if there is j, $1 \leqslant j \leqslant n$, such that $\alpha + e_j \in N(P)$, where $e_j = (0, \dots, 1, 0, \dots, 0)$[1]. The points of $N(P)$ that are not minor are said to be *senior.*

Denote by $\delta(P)$ the convex hull of the set of minor points of $N(P)$. As was noted in Chapter 4 (for the case $n = 2$), as a rule, the polyhedron $\delta(P)$ is not regular (although it is of course complete). In what follows we shall deal not with the polyhedron $\delta(P)$ but with its extension $\widehat{\delta}(P)$. We set

$$\widehat{d}_P(q) = d_P(q) - \min_{1 \leqslant i \leqslant n} q_i, \qquad q = (q_1, \dots, q_n) \in \mathbb{R}^n. \tag{4}$$

Let $\widehat{\delta}(P)$ denote the convex solid determined by the inequalities

$$\langle \alpha, q\rangle \leqslant \widehat{d}_P(q), \qquad q \in \mathbb{R}^n_+,\ \alpha \in \overline{\mathbb{R}^n_+}. \tag{5}$$

Lemma. *If the polyhedron $N(P)$ is regular, then the solid $\widehat{\delta}(P)$ possesses the following properties:*

(i) *$\widehat{\delta}(P)$ is a convex polyhedron;*

(ii) *$\widehat{\delta}(P)$ is a regular polyhedron;*

(iii) *the inclusion*

$$\delta(P) \subset \widehat{\delta}(P) \subset N(P), \tag{6}$$

takes place.

The proof of the lemma will be given in the next section and will be preceded by some general remarks on polyhedra.

[1] In the terminology of Chapter 4, the point α should be called a $\mathbb{Z}$-minor point. Since no minor points in the sense of Chapters 1 and 4 are involved in our further presentation, the term $\mathbb{Z}$-minor points will not be used.

1.2. The properties of the polyhedron $\widehat{\delta}(P)$. We remind the reader that if N is a convex polyhedron, then

$$\langle \alpha, q \rangle \leqslant d \tag{7}$$

is called a supporting half-space to N if (7) is fulfilled for all $\alpha \in N$, and for at least one point $\alpha \in N$ the equality is attained. By the convexity of N, to every supporting half-space there corresponds a single face $\Gamma^{(k)} \subset N$ ($k = \dim \Gamma^{(k)}$) of maximum dimension belonging to it. On the other hand, in general, to every face there correspond many supporting half-spaces containing it. The vector q in (7) is called the direction vector of the half-space. The set of direction vectors of all supporting half-spaces containing a face $\Gamma^{(k)} \subset N$ is called the normal cone of the face $\Gamma^{(k)}$ and is denoted $V_{(k)}$. We note that $V_{(k)}$ is a closed convex polyhedral angle (cone), and we have

$$\dim V_{(k)} = n - k.$$

It should be noted that the interior points of $V_{(k)}$ are direction vectors of the half-spaces for which the face of N of maximum dimension contained in them is the face $\Gamma^{(k)}$. The vectors belonging to the boundary $\partial V_{(k)}$ are direction vectors of supporting half-spaces containing the faces $\Gamma^{(k')}$, where $k' = \dim \Gamma^{(k')} > k$, the face lying on the boundary of the face $\Gamma^{(k')}$ being $\Gamma^{(k)}$.

Thus, with a convex polyhedron N finite set of closed convex polyhedral angles $V^j_{(k)}$, $k = 0, \ldots, n-1$, $j = 1, \ldots, J$ is associated, the angles $V^j_{(0)}$ (the normal cones of the vertices of N) covering the whole space:

$$\mathbb{R}^n = \bigcup_{j=1}^{J} V^j_{(0)}.$$

The boundary $\partial V^j_{(k)}$ of each of the angles in this set is a union of a finite number of angles $V^{(l)}_{(k+1)}$ of lower dimensions. For $k = n-1$ we obtain one-dimensional rays $\{\lambda q^{(j)}\}$, where $q^{(j)}$ are the direction vectors of the supporting half-spaces containing the $(n-1)$-dimensional faces of N. The function

$$d(N, q) = \max_{\alpha \in N} \langle \alpha, q \rangle$$

is called the supporting function to the polyhedron N. It is a positively homogeneous function of degree 1, i.e. $d(N, \lambda q) = \lambda q(N, q)$, $\lambda > 0$.

If $q \in \mathbb{R}^n$ and $\Gamma^{(k)}$ is the face of maximum dimension lying in the supporting half-space for which q is the direction vector, then obviously

$$d(N, q) = \max_{\alpha \in \Gamma^{(k)}} \langle \alpha, q \rangle.$$

And it follows that if $V_{(k)}$ is one of the normal cones of the polyhedron N, then

$$d(N, q' + q'') = d(N, q') + d(N, q''), \qquad q', q'' \in V_{(k)} \setminus \partial V_{(k)}. \tag{8}$$

Remark. We presented above a dual description of a polyhedron. More precisely, given the above-mentioned finite set of convex polyhedral angles $V^j_{(k)}$, $0 \leqslant k \leqslant n-1$, $j = 1, \dots, J_0$, and a homogeneous function $d(N, q)$ satisfying conditions (8), the convex solid

$$\langle \alpha, q \rangle \leqslant d(N, q) \quad \forall q \in \mathbb{R}^n$$

is a convex polyhedron and the angles $V^j_{(k)}$ are the normal cones of its faces of the various dimensions.

The proof of Lemma 1.1. (i) Let us show that all inequalities (5) are consequences of a finite number of inequalities among them. Since the polyhedral angles $V^j_{(k)}$ of the faces $\Gamma^{(k)}_j \subset N(P)$ cover the whole space $\mathbb{R}^n$, it suffices to separate out a finite number of inequalities among

$$\langle \alpha, q \rangle \leqslant \widehat{d}(q), \quad q \in V^j_{(k)}, \quad 0 \leqslant k \leqslant n-1, \quad j = 1, \dots, J, \tag{5'}$$

so that the remaining inequalities are their consequences.

We introduce the following notation. Let S be a subset of $\mathbb{R}^n$ and let $I = (i_1, \dots, i_n)$, $s \leqslant n$, be a set of indices assuming the values $1, \dots, n$. Then

$$S_I = \{q = (q_1, \dots, q_n) \in S, q_{i_1} = \dots = q_{i_s} = \min_{1 \leqslant i \leqslant n} q_i\}.$$

We divide the system of inequalities (5′) into a finite set of subsystems corresponding to the various sets I of indices:

$$\langle \alpha, q \rangle \leqslant \widehat{d}(q), \qquad q \in (V^j_{(k)})_I. \tag{5''}$$

It now remains to note that on the set $(V^j_{(k)} \setminus \partial V^j_{(k)})_I$ the function $\widehat{d}_P(q)$ is positively homogeneous and additive.

(ii) As has been proved, one can select a finite number of vectors $q^{(j)} \in \mathbb{R}^n_+$, $j = 1, \dots, J$, such that the polyhedron $\widehat{\delta}(P)$ is determined by the inequalities

$$\langle \alpha, q \rangle \leqslant \widehat{d}_P(q^{(j)}), \qquad j = 1, \dots, J, \quad \alpha \in \overline{\mathbb{R}^n_+}.$$

According to Section 5.2.3, such polyhedra are said to be *regular*.

(iii) The right-hand inclusion (6) is obvious. Since the polyhedron $\delta(P)$ is the convex hull of a finite set of integral points, it suffices to show that all integral minor points $\alpha \in N(P)$ belong to $\widehat{\delta}(P)$. According to the definition of minor points, for every such point α there is j such that $\beta = \alpha + e_j \in N(P)$. Therefore

$$\langle \alpha, q \rangle = \langle \beta, q \rangle - q_j \leqslant d_P(q) - \min_{1 \leqslant i \leqslant n} q_i = \widehat{d}_P(q).$$

For our further aims we need the following

Lemma. *If a hyperplane $\langle \alpha, q\rangle = d_P(q)$, $q \in \mathbb{R}^n_+$, contains an integral point $\alpha^0 \in N(P)$, then $\langle \alpha, q\rangle = \widehat{d}_P(q)$ is a supporting hyperplane to both $\widehat{\delta}(P)$ and $\delta(P)$.*

Proof. According to Lemma 1.1 (i), $\delta(P)$ lies in the half-space $\langle \alpha, q\rangle \leqslant \widehat{d}_P(q)$. If $\min q_j = q_{j_0}$, then the point $\alpha^0 - e_{j_0}$ belongs simultaneously to $\delta(P)$, $\widehat{\delta}(P)$, and the hyperplane $\langle \alpha, q\rangle = \widehat{d}_P(q)$.

1.3. Essential variables corresponding to the faces of the polyhedron $N(P)$. When defining polynomials of N-principal type in two variables in §4.2 we associate with each face of $N(P)$ of nonzero dimension certain variables that were called the essential variables of that face. We now extend this definition to the multidimensional case.

A vector $q \in \mathbb{R}^n_+$ is said to be *essential* if $\langle \alpha, q\rangle \leqslant \widehat{d}(q)$ is a supporting half-space to $\widehat{\delta}(P)$; the set of essential vectors will be denoted as $\widehat{\mathbb{R}}^n_+$. If $V_{(k)}$ is the normal cone of a face $\Gamma^{(k)} \subset N(P)$, then we set

$$\widehat{V}_{(k)} = V_{(k)} \cap \widehat{\mathbb{R}}^n_+.$$

Lemma. (i) *If a face $\Gamma^{(k)} \subset N(P)$ does not lie in the coordinate hyperplanes, then $\widehat{V}_{(k)} = V_{(k)}$.*

(ii) *Let a face $\Gamma^{(k)}$ lie in the coordinate hyperplane $\{\alpha_i = 0, i \in I\}$, where I is a subset of the set $\{1, \dots, n\}$, and not lie in the coordinate hyperplanes of lower dimensions. Then*

$$V_{(k)} \setminus \widehat{V}_{(k)} = \{q \in V_{(k)}, q_i = \min q_j \Rightarrow i \in I\}.$$

Proof. (i) follows automatically from Lemma 1.2. The same lemma implies that if $q \in V_{(k)} \cap \mathbb{R}^n_+$, and there is an index $j_0 \notin I$ such that $\min q_j = q_{j_0}$, then $q \in \widehat{V}_{(k)}$. Therefore the vectors $q \in V_{(k)} \setminus \widehat{V}_{(k)}$ belong to the right-hand set in (8).

We note that in the situation (ii) the cone $V_{(k)}$ has faces of codimension 1 in the hyperplanes $\{q_i = 0, i \in I\}$. It follows (see (8)) that $\dim(V_{(k)} \setminus \widehat{V}_{(k)}) = \dim V_{(k)}$. As to $\widehat{V}_{(k)}$, the dimension of this closed set can be less than that of $V_{(k)}$. In particular, if the smallest components of the vectors $q \in V_{(k)}$ have indices $i \in I$, then $\widehat{V}_{(k)}$ is an empty set.

The face which $\widehat{V}_{(k)} = \varnothing$ will be called an *inessential* face of $N(P)$, and the other principal faces will be called *essential* faces.

Let $J = (j_1, \dots, j_s)$, $s \leqslant n$, be a set of indices assuming the values $1, \dots, n$.

Definition. $\xi_J = (\xi_{j_1}, \dots, \xi_{j_s})$ is called the set of essential variables of a face $\Gamma^{(k)}$ if

a) $\widehat{V}_{(k),J} = \{q \in \widehat{V}_{(k)}, \min q_i = q_j \Rightarrow j \in J\}$ is a relatively open set in $\Gamma^{(k)}$;

b) the set J is maximal in the sense that there is no broader set $J' \supset J$ for which $\widehat{V}_{(k),J'}$ satisfies condition a).

Let Γ^{n-1} be a face of $N(P)$ of maximum dimension and let q^0 be the outer normal to Γ^{n-1}. Then (according to the lemma) q^0 is an essential vector, and the essential variables are those ξ_j for which $q_j^0 = \min q^0$.

In the case of faces $\Gamma^{(k)}$ of dimension $k < n-1$ there can appear several sets of essential variables.

For the sake of visuality, we present another, less formal definition of essential variables.

Let $\Gamma^{(k)}$ be a face of $N(P)$ and let $\pi(\Gamma^{(k)})$ be the k-dimensional hyperplane passing through it:

$$\pi(\Gamma^{(k)}) = \{\alpha \in \mathbb{R}^n_\alpha, \langle \alpha, q\rangle = d_P(q) \quad \forall q \in V_{(k)}\}.$$

Assume that for some j the intersection $\pi(\Gamma^{(k)} - e_j) \cap \widehat{\delta}(P)$ is non-empty and belongs to the boundary of $\widehat{\delta}(P)$. Then there is a face $\Delta^{(k')} \subset \widehat{\delta}(P)$ lying in the plane $\pi(\Gamma^{(k)} - e_j)$. The faces $\Delta^{(k')}$ and $\Gamma^{(k)}$ are called an associated pair of faces.

Let $J = (j_1, \ldots, j_s)$, $j_1 = j$, be the maximal set of indices such that the planes $\pi(\Gamma^{(k)} - e_{j_\varkappa})$, $\varkappa = 1, \ldots, s$, coincide with $\pi(\Gamma^{(k)} - e_j)$; the set ξ_J is called the set of essential variables of the associated pair $\Gamma^{(k)}$ and $\Delta(k')$.

A face $\Gamma^{(k)}$ is said to be inessential if there are no faces $\Delta^{(k')} \subset \widehat{\delta}(P)$ associated with it ($\Gamma^{(k)}$ must lie in a coordinate plane).

We note that every face of $N(P)$ of maximum dimension has exactly one face of $\widehat{\delta}(P)$ associated with it. However, as is shown by simple examples (even in the case of the plane), the polyhedron $\widehat{\delta}(P)$ may have faces of maximum dimension that have no faces of $N(P)$ of maximum dimension associated with them.

We demonstrate the above definitions by an example of a quasi-homogeneous polynomial P for which $N(P)$ is determined by the inequality

$$\langle \alpha, q^0\rangle \leqslant m, \qquad \alpha \in \overline{\mathbb{R}^n_+},\ q^0 \in \mathbb{R}^n_+, \tag{9}$$

where it is assumed (cf. Theorem 4.1.3) that $\min q_j^0 = 1$ and that the numbers $q_1^0, \ldots, q_n^0$ are integers. We shall show that the polyhedron $\widehat{\delta}(P)$ has the form

$$\widehat{\delta}(P) = \{\alpha \in \mathbb{R}^n_+, \langle \alpha, q^0\rangle \leqslant m - 1\}. \tag{10}$$

To give an accurate proof of this fact we calculate the supporting function $d_P(q)$. As can easily be seen,

$$d_P(q) = \max_{\alpha \in N(P)} \langle \alpha, q\rangle = m \max_{1 \leqslant j \leqslant n} q_j / q_j^0. \tag{11}$$

Indeed, set $\alpha^{(j)} = (m/q_j^0) e_j$, $j = 1, \ldots, n$. Then each point $\alpha \in N(P)$ is represented as

$$\alpha = \sum_{j=1}^n \varkappa_j \alpha^{(j)}, \qquad \varkappa_j = \frac{\alpha_j q_j^0}{m} \geqslant 0, \qquad \sum \varkappa_j \leqslant 1,$$

i.e. $\alpha^{(j)}$ is a vertex of $N(P)$. Further, $\forall \alpha \in N(P)$ we have

$$\langle \alpha, q\rangle = \sum \varkappa_j \langle \alpha^{(j)}, q\rangle \leqslant m \max q_j/q_j^0.$$

If the maximum on the right-hand side is attained for $j = k$, then for $\alpha = \alpha^{(k)}$ this inequality goes into an equality, which proves (11).

Since $\widehat{d}_P(q^0) = d_P(q^0) - \min q_j^0 = m - 1$, the polyhedron $\widehat{\delta}(P)$ is contained in the polyhedron Δ determined by the right-hand side of (10). Let us check the opposite inclusion, i.e. $\Delta \subset \widehat{\delta}(P)$. To this end it suffices to show that the points $\widehat{\alpha}^{(j)} = ((m-1)/q_j^0)e_j$, the vertices of Δ, belong to $\widehat{\delta}(P)$, i.e. to verify the inequalities

$$\langle \widehat{\alpha}^{(j)}, q\rangle \leqslant \widehat{d}_P(q),$$

or, assuming that $\min q_j = 1$, the inequality

$$(m-1)q_j/q_j^0 \leqslant m \max q_j/q_j^0.$$

The latter inequality is equivalent to

$$(m-1)\Big(\max_i \frac{q_i}{q_i^0} - \frac{q_j}{q_j^0}\Big) + \Big(\max_i \frac{q_i}{q_i^0} - 1\Big) \geqslant 0.$$

The first term on the left-hand side is obviously nonnegative. If q and q^0 are normalized by the conditions $\min q_i = \min q_i^0 = 1$, then the second term is also nonnegative.

We now describe the essential variables of the faces of $N(P)$. Polyhedron (9) has exactly one face of maximum dimension:

$$\Gamma^{(n-1)} = \{\alpha \in \overline{\mathbb{R}_+^n}, \langle \alpha, q^0\rangle = m\}.$$

If $J(q^0)$ is a set of indices $j \in \{1, \dots, n\}$ for which $q_j^0 = \min q_j^0 \overset{\text{def}}{=} 1$, then the set of variables $\xi_{J(q^0)}$ is the set of essential variables of that face.

We now consider the face $\Gamma_1^{(n-2)} = \Gamma^{(n-1)} \cap \{q_1 = 0\}$. Here the following two cases are possible:

(a) $J(q^0) = \{1\}$, i.e. $1 = q_1^0 < q_j^0$, $j \geqslant 2$;

(b) the set $J_1(q^0) = J(q^0) \setminus \{1\}$ is non-empty.

In case (a), according to the lemma, $\Gamma_1^{(n-2)}$ is an inessential face. In case (b) the essential variables of $\Gamma_1^{(n-2)}$ are $\xi_{J_1(q^0)}$. The faces that are intersections of $\Gamma^{(n-1)}$ with the faces of higher codimension lying in coordinate planes are considered in like manner. Polyhedron (9) has no other faces not lying in the coordinate planes.

Since, by the hypothesis, polyhedron (9) has integral vertices, the numbers m/q_j^0 are integers. Therefore, as can easily be seen,

$$\delta(P) = \Big\{\alpha \in \mathbb{R}_+^n, \langle \alpha, q^0\rangle \leqslant m - 1, \alpha_j \leqslant \frac{m}{q_j^0} - 1, j = 1, \dots, n\Big\}.$$

If $q_j^0 > 1$ then $(m-1)/q_j^0 > m/q_j^0 - 1$ so that $\delta(P)$ is obtained from $\widehat{\delta}(P)$ by "cutting off", using planes parallel to coordinate planes, those vertices $((m-1)/q_j^0)e_j$ for which $q_j^0 > 1$.

We note that in the general case as well $\delta(P)$ is obtained from $\widehat{\delta}(P)$ by cutting off, using planes parallel to coordinate planes, neighborhoods of some of the vertices of $\widehat{\delta}(P)$ lying in the coordinate planes.

1.4. An additional arithmetical condition. We present a multidimensional generalization of condition (A) in Section 4.2.1. It relates to a face $\Gamma^{(k)} \subset N(P)$ with normal cone $V_{(k)}$.

Condition (A). (i) *Let $\Gamma^{(k)}$ be an essential face; this means that the set $\widehat{V}_{(k)}$ is non-empty and decomposes into non-empty components $\widehat{V}_{(k),J}$. Then in each of the components there is a vector q, $q \notin \partial V_{(k)}$*[1]*, for which*

$$\langle \beta, q\rangle \leqslant \widehat{d}_P(q) \quad \forall \beta \in (N(P)\setminus \Gamma^{(k)}) \cap \mathbb{Z}^n. \tag{12}$$

(ii) *Let $\Gamma^{(k)}$ be an inessential face. Then there is a vector $q \in V_{(k)} \setminus \partial V_{(k)}$ such that*

$$\langle \beta, q\rangle \leqslant \max_{\alpha\in\widehat{\delta}(P)} \langle \alpha, q\rangle \quad \forall \beta \in (N(P)\setminus \Gamma^{(k)}) \cap \mathbb{Z}^n. \tag{13}$$

This condition is employed when proving the necessity in the main theorem stated below (cf. Theorem 4.2.2). It is rather difficult to verify condition (i). We note that an obvious sufficient condition for the validity of (i) is the condition

(i′) *if $\Gamma^{(k)} \subset N(P)$ is an essential face, then in every component of $\widehat{V}_{(k),J}$ there is an integral vector q normalized by the condition* $\min q_i = 1$.

Remarks. 1) Condition (i′) means, in particular, that the outer normals $q^{(j)}$ to the faces of maximum dimension are integral vectors if they are normalized by the condition $\min_k q_k^{(j)} = 1$.

2) According to Lemma 1.3, for $n = 2$ all the faces of the polygon $N(P)$ not lying on the coordinate axes are essential, i.e. condition (A) reduces to (i). It is easy to see that in this case condition (i) coincides with condition (i′) and condition (ii) in Lemma 2.1.

1.5. Polynomials of *N*-principal type.

Definition. A polynomial $P(\xi)$ is called a polynomial of *N-principal* type if for each senior face $\Gamma^{(k)}$ of the polynomial the following condition hold.

(i) If $\Gamma^{(k)}$ is an essential face, then for any set ξ_J of essential variables of this face we have

$$\{P_{\Gamma^{(k)}}(\xi) = 0\} \Rightarrow \{\operatorname{grad}_{\xi_J} P_{\Gamma^{(k)}}(\xi) \neq 0\}, \qquad \xi^{(1)} = \xi_1 \ldots \xi_n \neq 0; \tag{14}$$

[1]That is, in the supporting half-space $\langle \alpha, q\rangle \leqslant d_P(q)$ there are no faces $\Gamma^{(k)} \subset N(P)$ of dimension $k' > k$ containing $\Gamma^{(k)}$.

(ii) If the face $\Gamma^{(k)}$ is inessential, then

$$P_{\Gamma^{(k)}}(\xi) \neq 0 \quad \text{for} \quad \xi^{(1)} \neq 0. \tag{15}$$

Theorem. *Let $N(P)$ be a regular polyhedron and let condition* (A) *hold. Then the conditions below are equivalent.*

(i) *The polynomial P is a polynomial of N-principal type, i.e. the conditions of the above definition are fulfilled.*

(ii) *For any polynomial $Q(\xi)$, $N(Q) \subset \delta(P)$, there is a constant $c = c(Q)$ such that*

$$\Xi_{\widehat{\delta}(P)}(\xi) \leqslant c(\widetilde{P+Q})(\xi) \quad \forall \xi \in \mathbb{R}^n. \tag{16}$$

The necessity of conditions (14), (15) for the fullfilment of inequality (16), i.e. the implication (ii)$\Longrightarrow$(i), is proved following the scheme that has already been used many times.

Let for some $\eta \in \mathbb{R}^n$, $\eta_1, \ldots, \eta_n \neq 0$, the relations

$$P_{\Gamma^{(k)}}(\eta) = 0, \qquad \operatorname{grad}_{\xi_J} P_{\Gamma^{(k)}}(\eta) = 0 \tag{17}$$

hold, where $\Gamma^{(k)}$ is a senior face of $N(P)$ and $\xi_J = (\xi_{i_1}, \ldots, \xi_{i_s})$ is the set of essential variables of the face $\Gamma^{(k)}$. Select (by virtue of condition (A)) $q \in V_{(k)J} \setminus \partial V_{(k)}$ and set

$$\xi(t) = t^q \eta = (t^{q_1}\eta_1, \ldots, t^{q_n}\eta_n).$$

Then

$$\Xi_{\widehat{\delta}(P)}(\xi(t)) > \text{const}\, t^{\widehat{d}_P(q)}.$$

We shall show that if (17) is fulfilled, then for an appropriate choice of the polynomial Q we have

$$(P+Q)(\xi(t)) = o(t^{\widehat{d}_P(q)}) \quad \text{for} \quad t \to +\infty. \tag{18}$$

The resulting contradiction will prove the desired assertion.

According to (A), we have

$$P(\xi(t)) = t^{d_P(q)} P_{\Gamma^{(k)}}(\eta) + \omega t^{\widehat{d}_P(q)} + O(t^{\widehat{d}_P(q)-\varepsilon}).$$

Let us take the monomial $c\xi^\beta$ as $Q(\xi)$, where $\langle \beta, q\rangle = \widehat{d}_P(q)$. Selecting c from the condition $\omega + c\eta^\beta = 0$ (recall that $\eta_1, \ldots, \eta_n \neq 0$) and using the first condition (17) we obtain

$$P(\xi(t)) + Q(\xi(t)) = o(t^{\widehat{d}_P(q)}) \quad \text{for} \quad t \to +\infty.$$

Further, by the definition of the set $\widehat{V}_{(k),J}$, we have $q_{j_1} = \cdots = q_{j_s} < q_l$, $l \notin J$, and therefore

$$\sum_{\alpha>0} |P^{(\alpha)}(\xi(t)) + Q^{(\alpha)}(\xi(t))| = \sum_{i=1}^{s} \left|\frac{\partial P_{\Gamma^{(k)}}(\eta)}{\partial \xi_{j_i}}\right| t^{\hat{d}_P(q)} + o(t^{\hat{d}_P(q)}).$$

Using the second relation in (17) we prove (18).

We now assume that $\Gamma^{(k)}$ is an inessential face and that $\overline{d}(q) = \max_{\alpha\in\widehat{\delta}(P)} \langle \alpha, q\rangle$. According to condition (A), there is a vector q such that on the curve $\xi(t)$ defined above we have

$$\Xi_{\widehat{\delta}(P)}(\xi(t)) = O(t^{\overline{d}(q)}), \qquad (\widetilde{P+Q})(\xi(t)) = t^{d_P(q)} P_{\Gamma^{(k)}}(\eta) + o(t^{\overline{d}(q)}).$$

If $P_{\Gamma^{(k)}}(\eta) = 0$, then for $t \to \infty$ we arrive at a contradiction.

In §§2 and 3 we shall prove the following

Proposition. *Let a polynomial $P(\xi)$ satisfy the conditions of the definition. Then the inequality*

$$\Xi_{\widehat{\delta}(P)}(\xi) \leqslant c(|P(\xi)| + |\operatorname{grad} P(\xi)| + 1) \tag{19}$$

holds.

Since the definition of a polynomial of N-principal type involves only the various principal quasi-homogeneous parts $P_q(\xi)$, inequality (19) remains valid if P is replaced by $P+Q$, $N(Q) \subset \delta(P)$ (of course, the constant c in (19) should be replaced by $c(Q)$). Hence, the proposition implies (i)$\Longrightarrow$(ii).

Remarks. 1) If $N(P)$ is determined by an equation $\langle \alpha, q^0\rangle \leqslant m$, the above theorem goes into Theorem 4.1.3.

2) Inequality (16) implies

$$\Xi_{\delta(P)}(\xi) \leqslant c(\widetilde{P+Q})(\xi) \quad \forall \xi \in \mathbb{R}^n. \tag{16'}$$

However, inequality (16′) may hold not only for polynomials of N-principal type.

3) We note that condition (A) is not used in the proof of the proposition.

4) When proving the proposition we shall incidentally obtain a new proof of Theorem 5.1.2 in application to the case of polynomials P having regular Newton's polyhedra.

§2. Estimates for polynomials of *N*-principal type in regions of special form

When proving Proposition 1.5 in Chapter 4 for the case of two variables we first separated out special regions associated with the sides and vertices of Newton's polygon, proved estimates of the type (1.19) in these regions, and then established that the exterior of a circle of a sufficiently large radius can be covered by a finite number of regions where the estimate was proved. Following the same plan we shall prove inequality (1.19) for the multidimensional case as well. However, since for $n > 2$ a polyhedron in $\mathbb{R}^n$ has not only faces of zero dimension and dimension $n-1$ but also faces of intermediate dimensions k, $0 < k < n-1$, all the constructions are substantially complicated. In this section we shall define axiomatically those regions where estimates of the type (1.19) will be proved. In §3 we shall prove that these regions cover the complement to a ball in $\mathbb{R}^n$. In this way, along with Proposition 1.5, Theorem 1.5 will also be proved completely.

2.1. π-cones and π-cylinders. We associate with $q \in \mathbb{R}^n$ the one-parameter transformation group

$$T_q(\rho)\colon \mathbb{R}^n \setminus \{0\} \to \mathbb{R}^n \setminus \{0\} \quad (\xi \mapsto \rho^q \xi).$$

Let π be a subspace of $\mathbb{R}^n_q$. A set $W \subset \mathbb{R}^n \setminus \{0\}$ is called a π-cone if

$$T_q(\rho)W \subset W \quad \forall \rho > 0, \quad \forall q \in \pi.$$

We shall say that a subset $V \subset W$ generates a π-cone W if $\forall \xi \in W$ there are $\eta \in V$, $q \in \pi$, and $\rho \in \mathbb{R}_+$ such that $\xi = \rho^q \eta$.

Definition. A π-cone $W \subset \mathbb{R}^n_{(\xi)} \setminus \{0\}$ is said to be *regular* if it is generated by a compact set V not intersecting the coordinate planes, that is, more concisely, $V \subset (\mathbb{R} \setminus \{0\})^n$.

Remark. In contrast to the generating set, the π-cone itself may intersect coordinate planes. For instance, if $e_1 = (1, 0, \ldots, 0) \in \pi$, then $-e_1 \in \pi$, and W contains the curve $\xi(\rho) = (\eta_1/\rho, \eta_2, \ldots, \eta_n)$, $\eta \in W$, which intersects the plane $\{\xi_1 = 0\}$ as $\rho \to +\infty$.

A function $A(\xi)$ is said to be π-homogeneous if

$$A(\rho^q \xi) = \rho^{d(q)} A(\xi) \quad \forall q \in \pi,\ \forall \rho \in \mathbb{R}_+,\ \xi \in \mathbb{R}^n \setminus \{0\}. \tag{1}$$

The number $d(q)$ is called the q-order of the function.

If $N(P)$ is Newton's polyhedron of a polynomial P, $\Gamma^{(k)}$ is a face of $N(P)$, and $V_{(k)}$ is the polyhedral angle of normals to $\Gamma^{(k)}$, then $P_{\Gamma^{(k)}}(\xi)$ is a $\pi(V_{(k)})$-homogeneous polynomial.

Let π be a subspace of $\mathbb{R}^n$ and let $\dim \pi = n - k$. A set $w \subset \mathbb{R}^n$ is called a π-cylinder if

(a) w is invariant relative to translations along π, i.e. if $q' \in w$ and $q'' \in \pi$, then $q' + q'' \in w$;
(b) the section of w by a k-dimensional plane transversal to π is compact.

Remark. If $n = 3$ and $\dim \pi = 1$, we obtain an ordinary cylinder with compact base. If $\dim \pi = 2$, then we have a region in $\mathbb{R}^3$ bounded by two planes parallel to π.

We introduce the notation

$$\log \xi_+ = (\log |\xi_1|, \dots, \log |\xi_n|), \qquad |\xi^\alpha| = \exp\langle \alpha, \log \xi_+ \rangle \tag{2}$$

and consider the mappings

$$\begin{aligned} \text{Log}: &\quad \mathbb{R}^n_{(\xi)} \setminus \{0\} \to \mathbb{R}_q \quad (\xi \mapsto \log \xi_+), &(3)\\ \text{Exp}: &\quad \mathbb{R}^n_{(q)} \to \mathbb{R}^n_{(\xi)+} \quad (q \mapsto (e^{q_1}, \dots, e^{q_n})). &(3')\end{aligned}$$

These mapping establish a one-to-one correspondence between regular π-cones in $\mathbb{R}^n_{(\xi)}$ and π-cylinders in $\mathbb{R}^n_{(q)}$. More precisely, we have the following

Lemma. (i) *If $W \subset \mathbb{R}^n_{(\xi)}$ is a regular π-cone, then $w = \text{Log}\, W$ is a π-cylinder.*
(ii) *If $w \subset \mathbb{R}^n_{(q)}$ is a π-cylinder, then $W = \exp(w)$ is a regular π-cone.*

Proof. (i) According to the definition of a π-cone, if $\xi \in W$ then $T_q(\rho)\xi \in W$, whence $\log \xi_+ + q \log \rho \in w$ $\forall q \in \pi$, $\forall \rho \in \mathbb{R}_+$. Hence, it remains to show that the section of w by a plane λ transversal to π is a compact set.

If the π-cone W is regular, then it is generated by a compact set V not intersecting the coordinate planes. Therefore $v = \text{Log}\, V$ is a compact set in $\mathbb{R}^n_{(q)}$, and the projection of v on λ (coinciding with the section of w by the plane λ) is also a compact set.

(ii) Denote by v the section of w by a plane λ transversal to π. It is obvious that $W = \exp(w)$ is a π-cone generated by $V = \exp(v)$. If v is a compact set in $\mathbb{R}^n_{(q)}$, then V is a compact set in $(\mathbb{R} \setminus \{0\})^n$.

2.2. π-cones relating to Newton's polyhedron. If $\Gamma^{(k)}$ is a face of $N(P)$ and $V_{(k)}$ is its normal cone, then by $\pi(\Gamma^{(k)})$ and $\pi(V_{(k)})$ will be meant, respectively, the k-dimensional and $(n-k)$-dimensional hyperplanes containing $\Gamma^{(k)}$ and $V_{(k)}$.

Lemma 2.1 allows us to establish the following assertions that will play an important role in what follows.

Lemma. *Let $\Gamma^{(k)}$ be a face of $N(P)$ and let W be a regular $\pi(V_{(k)})$-cone. Then for every compact set $\Lambda \subset \pi(\Gamma^{(k)})$ there is $R = R(\Lambda) > 0$ such that*

$$|\xi^\alpha| < R|\xi^\beta| \quad \forall \alpha, \beta \in \Lambda, \quad \forall \xi \in W. \tag{4}$$

Proof. In view of (2), we have to select R such that

$$\langle \alpha - \beta, q \rangle < \log R, \quad \forall q \in w = \text{Log} W, \quad \forall \alpha, \beta \in \Lambda. \tag{4'}$$

By the definition of a normal cone, we have $\langle \alpha - \beta, q\rangle = 0$ if $\alpha, \beta \in \pi(\Gamma^{(k)})$ and $q \in \pi(V_{(k)})$. Therefore it suffices to select R such that (4′) is fulfilled for q belonging to the section of w by a plane λ transversal to $\pi(V_{(k)})$. According to the above lemma, this section is compact. Since $\alpha - \beta$ also runs over a compact set, R can always be selected in this way.

The definition of a regular π-cone implies the "local" form of the condition that P is a polynomial of N-principal type.

Proposition. *Let $\Gamma^{(k)}$ be a senior face of $N(P)$, let $V_{(k)}$ be the corresponding polyhedral angle of normals, and let W be a regular $\pi(V_{(k)})$-cone. If the conditions of Definition 1.5 hold, then W can be covered by a finite number of regular $\pi(V_{(k)})$-cones W_λ $(1 \leqslant \lambda \leqslant \Lambda)$ so that on the generating sets V_λ of each of the cones W_λ one of the following two conditions holds:*

(i) $P_{\Gamma^{(k)}}(\xi) \neq 0 \ \forall \xi \in V_\lambda$; (5)

(ii) *in each set ξ_J of essential variables of the face $\Gamma^{(k)}$*[1] *there is a variable $\xi_j (j \in J)$ such that*

$$\frac{\partial P_{\Gamma^{(k)}}}{\partial \xi_j} \neq 0 \quad \forall \xi \in V_\lambda. \tag{5'}$$

2.3. The property $G(\Gamma^{(k)}, \varepsilon)$. The basic estimate. Let $\Gamma^{(k)}$ be a senior face of $N(P)$. The convex set

$$T(\Gamma^{(k)}) = \{\alpha \in \mathbb{R}^n_{(\alpha)}, \langle \alpha, q\rangle \leqslant d(q), \forall q \in V_{(k)}\} \tag{6}$$

is called the polyhedral angle of $\Gamma^{(k)}$ relative to $N(P)$[2]. Denote by $T_+(\Gamma^{(k)})$ the intersection of this angle with the positive coordinate n-hedron. It is clear that $N(P) \subset T_+(\Gamma^{(k)})$ for any senior face $\Gamma^{(k)}$.

Definition. A region $U \subset \mathbb{R}^n_{(\xi)}$ is said to possess the property $G(\Gamma^{(k)}, \varepsilon)$ (more concisely, $U \in G(\Gamma^{(k)}, \varepsilon)$) if

(a) U is contained in a regular $\pi(V_{(k)})$-cone W;

(b) $\forall \alpha \in T_+(\Gamma^{(k)}) \cap \pi(\Gamma^{(k)})$ and any integral $\beta \in T_+(\Gamma^{(k)}) \setminus \pi(\Gamma^{(k)})$ the inequality

$$|\xi^\beta| < \varepsilon |\xi^\alpha| \tag{7}$$

holds.

The main result in this section is the following

---

[1] If the face $\Gamma^{(k)}$ is inessential, then according to Definition 1.5, condition (i) holds for it.

[2] The angle (6) is formed by the intersection of those supporting half-spaces to $N(P)$ which pass through the faces of maximum dimension on whose boundary $\Gamma^{(k)}$ lies.

Proposition. *Let $\Gamma^{(k)}$ be a senior face of $N(P)$ and let $U \in G(\Gamma^{(k)}, \varepsilon)$. Let the conditions of Definition 1.5 be fulfilled. Then for a sufficiently small ε there is c such that*

$$\Xi_{\widehat{\delta}(P)}(\xi) \leqslant c(|P(\xi)| + |\operatorname{grad} P(\xi)|). \tag{8}$$

Proof. 1. According to the above definition, U belongs to a regular $\pi(V_{(k)})$-cone W. By Proposition 2.2, there is a covering $\cup W_\lambda \supset W$ consisting of $\pi(V_{(k)})$-cones W_λ such that in each of them either (5) or (5′) holds. Replacing U by $U \cap W_\lambda$ we can assume, without loss of generality, that in U (or in the $\pi(V_{(k)})$-cone W containing U) one of the conditions of Proposition 2.2 holds.

2. We first assume that

$$P_{\Gamma^{(k)}} \neq 0 \quad \forall \xi \in V \subset W. \tag{9}$$

We shall prove that in this case the inequality

$$\Xi_{T_+(\Gamma^{(k)})}(\xi) < c|P(\xi)| \quad \forall \xi \in U \tag{10}$$

is fulfilled provided that ε is sufficiently small.

We first of all note that $\exists c_0 > 0$ such that $\forall \xi \in W$ we have

$$|\xi^\alpha| \leqslant c_0 |P_{\Gamma^{(k)}}| \quad \forall \alpha \in \pi(\Gamma^{(k)}) \cap T_+(\Gamma^{(k)}). \tag{11}$$

In view of the $\pi(V_{(k)})$-homogeneity, it suffices to find c_0 such that (11) holds on the generating set V of the cone W. By virtue of the compactness and condition (9), it is always possible to select such c_0. Combining (4), (7), and (11) we conclude that

$$\Xi_{T_+(\Gamma^{(k)})}(\xi) \leqslant c_1 |P_{\Gamma^{(k)}}(\xi)| \quad \forall \xi \in U.$$

We shall show that for a sufficiently small ε we have

$$\frac{1}{2}|P_{\Gamma^{(k)}}(\xi)| < |P(\xi)| < \frac{3}{2}|P_{\Gamma^{(k)}}(\xi)|, \tag{12}$$

whence (10) will follow.

Fixing $\alpha \in \pi(\Gamma^{(k)}) \cap T_+(\Gamma^{(k)})$ and applying Lemma 2.2 we find that

$$\begin{aligned} |P(\xi) - P_{\Gamma^{(k)}}(\xi)| &\leqslant \sum_{\beta \in N(P) \setminus \Gamma^{(k)}} |a_\beta||\xi^\beta| \leqslant \varepsilon\Big(\sum |a_\beta|\Big)|\xi^\alpha| \\ &\leqslant \varepsilon c_0 \Big(\sum |a_\beta|\Big)|P_{\Gamma^{(k)}}(\xi)| < \frac{1}{2}|P_{\Gamma^{(k)}}(\xi)| \end{aligned} \tag{12'}$$

provided that $\varepsilon c_0 \Sigma |a_\beta| < 1/2$. Inequality (12) is a trivial consequence of (12′).

3. Let (5′) hold in U and let J^l, $l = 1, \dots, s$, be the sets of indices of the essential variables of the face $\Gamma^{(k)}$. By the hypothesis, in every set there is an index, say l such that

$$\frac{\partial P_{\Gamma^{(k)}}}{\partial \xi_l} \neq 0 \quad \text{for} \quad \xi \in V \subset W. \tag{13}$$

We put $P^{(l)} = \partial P / \partial \xi_l$, and let $N(P^{(l)})$ be Newton's polyhedron of the polynomial $P^{(l)}$. It is the intersection of $\overline{\mathbb{R}^n_{(\alpha)+}}$ with the translation of $N(P)$ by the vector $-e_l$. The face $\Gamma^{(k)}$ goes into a face $\gamma^{(k)} \subset N(P^{(l)})$. The normal cone $v_{(k)}$ of the face $\gamma^{(k)}$ either coincides with the cone $V_{(k)}$ or is its extension, i.e. $v_{(k)} \supset V_{(k)}$[1]. Therefore, if $\pi(V_{(k)})$ is a cone with a compact generating set H, then it is contained in some $\pi(v_{(k)})$-cone w with the same generating set H. Since $P^{(l)}_{\gamma^{(k)}} = \partial P_{\Gamma^{(k)}} / \partial \xi_l$, by virtue of (13) we have

$$P^{(l)}_{\gamma_k}(\xi) \neq 0 \quad \text{for} \quad \xi \in H \subset w. \tag{13′}$$

It follows that (cf. the derivation of (11))

$$|\xi^\alpha| \leqslant c_0 |P^{(l)}_{\gamma^{(k)}}(\xi)| \quad \forall \xi \in w, \quad \forall \alpha \in \pi(\gamma^{(k)}) \cap \overline{\mathbb{R}^n_{(\alpha)+}}. \tag{14}$$

Denote by $T^{(l)}_+(\gamma^{(k)})$ the intersection of the polyhedral angle of the face $\gamma^{(k)}$ of $N(P^{(l)})$ with $\overline{\mathbb{R}^n_{(\alpha)+}}$. It follows from (7) that

$$|\xi^\beta| < \varepsilon |\xi^\alpha| \quad \forall \xi \in U, \quad \forall \alpha \in T^{(l)}_+(\gamma^{(k)}) \cap \pi(\gamma^{(k)})$$
$$\forall \beta \in (T^{(l)}_+(\gamma^{(k)}) \setminus \pi(\gamma^{(k)})) \cap \mathbb{Z}^n. \tag{15}$$

Similarly, (4) implies that $\exists R > 0$ such that

$$|\xi^\alpha| < R |\xi^\beta| \quad \forall \alpha, \beta \in \pi(\gamma^{(k)}) \cap T^{(l)}_+(\gamma^{(k)}), \quad \xi \in U. \tag{16}$$

It follows from (15) and (16) (cf. the derivation of (10)) that for a sufficiently small ε we have

$$\Xi_{T^{(l)}_+(\gamma^{(k)})}(\xi) \leqslant c \left| \frac{\partial P}{\partial \xi_l}(\xi) \right| \quad \forall \xi \in U. \tag{17}$$

Performing the summation of inequalities (17) over $l = 1, \dots, s$ we find

$$\Xi_{v(\Gamma^{(k)})}(\xi) \leqslant c |\operatorname{grad} P(\xi)| \quad \forall \xi \in U, \tag{18}$$

where

$$v(\Gamma^{(k)}) \text{ is the convex hull of } \bigcup_{l=1}^{s} T^{(l)}_+(\gamma^{(k)}). \tag{19}$$

[1] The extension of the normal cone can appear only in the case when $\gamma^{(k)}$ intersects a coordinate hyperplane which does not intersect the face $\Gamma^{(k)}$.

We now show that

$$\widehat{\delta}(P) \subset v(\Gamma^{(k)}), \tag{20}$$

where $\Gamma^{(k)}$ is an arbitrary essential face of $N(P)$. Inclusion (20) and inequality (18) imply that

$$\Xi_{\widehat{\delta}(P)}(\xi) \leqslant \text{const}\,|\operatorname{grad} P(\xi)| \quad \forall \xi \in U. \tag{21}$$

4. We proceed to the proof of inclusion (20). The definition of the polyhedrons $T_+^{(l)}(\gamma^{(k)})$ implies that

$$v(\Gamma^{(k)}) = \{\alpha \in \overline{\mathbb{R}^n_{(\alpha)+}}, \langle \alpha, q\rangle = d(q) - \min_{j\in J} q_j, \forall q \in V_{(k)}\}, \tag{22}$$

where $J = J^{(1)} \cup \cdots \cup J^{(s)}$. We begin with the case when the face $\Gamma^{(k)}$ does not lie in the coordinate planes. Here, according to Lemma 1.3, we have $V_k = \widehat{V}_k$ and $\min_{j\in J} q_j = \min_{1\leqslant j\leqslant n} q_j$. Using the definition of the function $\widehat{d}_P(q)$ we see that in this case we have

$$v(\Gamma^{(k)}) = \{\alpha \in \overline{\mathbb{R}^n_{(\alpha)+}}, \langle \alpha, q\rangle = \widehat{d}_P(q), \forall q \in V_{(k)}\}. \tag{22'}$$

It is clear that the polyhedron $\widehat{\delta}(P)$ determined by the inequalities

$$\langle \alpha, q\rangle \leqslant \widehat{d}_P(q) \quad \text{for} \quad q \in \mathbb{R}^n_+, \alpha \in \overline{\mathbb{R}^n_+}$$

is contained in (22′).

The case of faces lying in coordinate planes reduces to the above. Indeed, let a face $\Gamma^{(k)}$ be formed by the intersection of faces $\Gamma_l^{(n-1)}$, $l = 1, \ldots, L$, not lying in the coordinate planes and the coordinate plane $\{\alpha_i = 0, i \in I\}$. Denote by $\Gamma^{(k')}$ the face of $N(P)$ at the intersection of $\Gamma_l^{(n-1)}$, $1 \leqslant l \leqslant L$. Then

$$T_+(\Gamma^{(k)}) = T_+(\Gamma^{(k')}).$$

Since the face $\Gamma^{(k')}$ lies in none of the coordinate hyperplanes, what has been proved implies that

$$\widehat{\delta}(P) \subset v(\Gamma^{(k')}),$$

where $v(\Gamma^{(k')})$ is the polyhedron (19) corresponding to the face $\Gamma^{(k')}$. Inequality (20) will be proved if we show that

$$v(\Gamma^{(k')}) \subset v(\Gamma^{(k)}). \tag{23}$$

Inclusion (23) is implied by the definition of the polyhedrons $v(\Gamma^{(k)})$ and $v(\Gamma^{(k')})$ and the following geometrical

Lemma. *Let a face $\Gamma^{(k')} \subset N(P)$ lie in none of the coordinate hyperplanes and let*

$$\Gamma^{(k)} = \Gamma^{(k')} \cap \{\alpha_i = 0, i \in I\}. \tag{24}$$

Let ξ_l, $l \in I$ be an essential variable of the face $\Gamma^{(k)}$. Then ξ_l is an essential variable of $\Gamma^{(k')}$.

Proof. According to the definition of essential variables, $\exists q \in \mathbb{R}^n_+ \cap V_{(k)}$ such that $q_l = \min\limits_{1 \leqslant j \leqslant n} q_j$.

It follows from (24) that

$$V_{(k)} = \Big\{ q = q' + q'', q' = (q'_1, \dots, q'_n) \in V_{(k')},\ q'' = \sum_{i \in I} b_i e_i, b_i \leqslant 0 \Big\}.$$

If $q_l = \min q_j$ and $l \in I$, then $q'_l = q_l = \min q'_l$. Since $\Gamma^{(k')}$ does not lie in the coordinate planes, we have $\widehat{V}_{(k')} = V_{(k')}$. Once again using the definition of essential variables we conclude that ξ_l is an essential variable of the face $\Gamma^{(k')}$.

2.4. Concluding remarks. Since we have $\widetilde{P}(\xi) > \text{const}$, inequality (1.19) always holds in any bounded region in $\mathbb{R}^n$ and, in particular, in the cube

$$I(R) = \{\xi \in \mathbb{R}^n, |\xi_j| \leqslant R, j = 1, \dots, n\},$$

Hence, the desired estimate (1.19) is implied by the following

Proposition. *$\forall \varepsilon > 0$ there is $R = R(\varepsilon)$ such that $\mathbb{R}^n \setminus I(R)$ can be covered by a finite set of regions U_λ each of which possesses the property $G(\Gamma^{(k)}_\lambda, \varepsilon)$ for some senior face $\Gamma^{(k)}_\lambda \subset N(P)$.*

The next §3 is devoted to the proof of the proposition.

§3. The covering of $\mathbf{R}^n$ by special regions associated with Newton's polyhedron

In the foregoing section we associated with each face $\Gamma^{(k)}$ of Newton's polyhedron $N(P)$ of a polynomial P a class of regions possessing the property $G(\Gamma^{(k)}, \varepsilon)$. For a sufficiently small ε we established for polynomials of N-principal type an estimate of the form of

$$\Xi_{\widehat{\delta}(P)}(\xi) \leqslant c(\Omega)(|P(\xi)| + |\operatorname{grad} P(\xi)|), \qquad \xi \in \Omega, \tag{1}$$

in the regions $\Omega \subset G(\Gamma^{(k)}, \varepsilon)$. To complete the proof of Proposition 1.5 we must

(i) indicate an effective method for constructing regions possessing the property $G(\Gamma^{(k)}, \varepsilon)$;

(ii) show that by means of a finite set of such regions we can cover the complement of any compact set in $\mathbb{R}^n$, for instance, of the cube

$$I(R) = \{\xi \in \mathbb{R}^n, |\xi_j| < R, j = 1, \dots, n\},$$

where R depends on ε.

Then for an adequately chosen constant c we obtain the inequality

$$\Xi_{\widehat{\delta}(P)}(\xi) \leqslant c(|P(\xi)| + |\operatorname{grad} P(\xi)| + 1), \qquad \xi \in \mathbb{R}^n,$$

i.e. Proposition 1.5 together with Theorem 1.5 will be proved.

We shall in fact introduce a class of polyhedral regions in $\mathbb{R}^n_{(q)}$ possessing a property $g(\Gamma^{(k)}, \varepsilon)$ and going into regions possessing the property $G(\Gamma^{(k)}, \varepsilon)$ under the exponential mapping (2.3′). We shall show that by means of a finite number of such regions it is possible to cover the complement to the translated negative coordinate n-hedron

$$i_-(r) = \{q \in \mathbb{R}^n, q_j \leqslant r, j = 1, \dots, n\}. \tag{2}$$

3.1. The property $g(\Gamma^{(k)}, \varepsilon)$. In this section we shall deal only with polyhedral regions, i.e. regions that are intersections of a finite number of (closed) half-spaces; in what follows, when speaking of regions in $\mathbb{R}^n_{(q)}$ we shall mean polyhedral regions.

Let π be a subspace of $\mathbb{R}^n_{(q)}$, $\dim \pi = n - k$. As was noted in the foregoing section, a region $C \subset \mathbb{R}^n_{(q)}$ is called a π-cylinder if

(a) C is invariant relative to translations along π, i.e. coincides with its cylindrical hull

$$C = (C)_\pi \stackrel{\text{def}}{=} \{q = q' + q'', q' \in C, q'' \in \pi\}; \tag{3}$$

(b) the section of C by any k-dimensional hyperplane transversal to π is a compact set (i.e. a bounded polyhedron).

Let $\Gamma^{(k)}$ be a face of $N(P)$ and let $V_{(k)}$ be the polyhedral angle of normals to $\Gamma^{(k)}$. We shall say that a region $u \subset \mathbb{R}^n_{(q)}$ possesses the property $g(\Gamma^{(k)}, \varepsilon)$ if

(i) there is a $\pi(V_{(k)})$-cylinder C such that $u \subset C$;

(ii) $\forall \alpha \in T_+(\Gamma^{(k)}) \cap \pi(\Gamma^{(k)})$ and any integral $\beta \in T_+(\Gamma^{(k)}) \setminus \pi(\Gamma^{(k)})$ the following inequality holds:

$$\langle \alpha - \beta, q \rangle > \log(1/\varepsilon) \quad \forall q \in u. \tag{4}$$

3.2. Semi-cylinders. Let π be a subspace in $\mathbb{R}^n_{(q)}$ and let π_+ and π_- be the complementary half-spaces of π. A region $\mathcal{D}$ is called a π_+ semi-cylinder if

(a) $\mathcal{D} = (\mathcal{D})_{\pi+} \stackrel{\text{def}}{=} \{q = q' + q'', q' \in \mathcal{D}, q'' \in \pi_+\}$;

(b) $(\mathcal{D})_\pi$ (see notation (3)) is a π-cylinder (i.e. the section of $(\mathcal{D})_\pi$ by a plane transversal to π is compact).

(c) for any half-space $\widehat{\pi}_+ \subset \mathbb{R}^n_{(q)}$ transversal to π and containing π_+ there is a translation $S_{\widehat{\pi}_+}$ of the half-space such that $\mathcal{D} \subset S_{\widehat{\pi}_+}$.

Remark. Let π_+ and π_- be the complementary half-spaces of π and let $\partial\pi_+$ be the boundary of π_+ (and π_-). If $\mathcal{D}_\pm$ are $\pi_\pm$ semi-cylinders and $(\mathcal{D}_-)_\pi = (\mathcal{D}_+)_\pi$, then $\mathcal{D}_+ \cap \mathcal{D}_-$ is a $\partial\pi_+$-cylinder.

We indicate a natural method for constructing a semi-cylinder from a π-cylinder C. Let π_+ be a half-subspace of π and let R be an arbitrary half-space of $\mathbb{R}^n_{(q)}$ containing π_+ and transversal to π. Then the intersection of C with the translated half-space R is a π_+ semi-cylinder. This operation can be repeated several times. The intersection and union of π-cylinders and π_+ semi-cylindered (having the same name) do not fall outside these classes of sets.

The above definitions trivially imply the following

Lemma. *Let π be a subspace of $\mathbb{R}^n_{(q)}$ and let π^0 be a subspace of π of codimension 1; let μ_+ be a subspace of π transversal to π^0, let C be a π^0-cylinder, and let M be a μ_+ semi-cylinder. Then the intersection $C \cap M$ is a π^0_+ semi-cylinder where π^0_+ is a half-space of π^0 lying in μ_+.*

Let $V_{(l)}$ be a polyhedral angle in $\mathbb{R}^n_{(q)}$, $\dim V_{(l)} = n - l$, and let $V^1_{(l+1)}, \ldots, V^J_{(l+1)}$ be its faces of dimension $n - l - 1$. Denote by $\pi^1_+, \ldots, \pi^J_+$ the half-subspaces of $\pi(V_{(l)})$ corresponding to the faces $V^1_{(l+1)}$, etc. Then

$$V_{(l)} = \pi^1_+ \cap \cdots \cap \pi^J_+. \tag{5}$$

Definition. A region $\mathcal{D}$ is called a $V_{(l)}$ semi-cylinder if $\mathcal{D} = \mathcal{D}^1 \cap \cdots \cap \mathcal{D}^J$, where $\mathcal{D}^j$ are π^j_+ semi-cylinders (π^j_+ are those in (5)), all the cylinders $(\mathcal{D}^j)_{\pi(V_{(l)})}$ coinciding.

In the further presentation the following assertion plays an important role.

Proposition. *Let $\Gamma^{(l)}$ be a face of $N(P)$, let $V_{(l)}$ be the corresponding polyhedral angle of normals, and let C be a $\pi(V_{(l)})$-cylinder. Then $\forall \varepsilon > 0$ there is a $V_{(l)}$ semi-cylinder $\mathcal{D} \subset C$ such that*

$$\begin{aligned} &\langle \alpha - \beta, q \rangle > \log \frac{1}{\varepsilon} \quad \forall q \in \mathcal{D}, \\ &\forall \alpha \in T_+(\Gamma^{(l)}) \cap \pi(\Gamma^{(l)}), \\ &\forall \beta \in (T_+(\Gamma^{(l)}) \setminus \pi(\Gamma^{(l)})) \cap \mathbb{Z}^n. \end{aligned} \tag{6}$$

Proof. Let $\Gamma^{(l+1)}_j$, $j = 1, \ldots, J$, be those faces of $N(P)$ of dimension $l + 1$ on whose boundary $\Gamma^{(l)}$ lies,[1)] let $V^j_{(l+1)}$ be the corresponding polyhedral angles of normals lying on the boundary of $V_{(l)}$, and let π^j_+ be the half-subspaces of $\pi(V_{(l)})$ corresponding to $V^j_{(l+1)}$ so that (5) holds.

[1)] If $l = n - 1$, then $J = 1$ and $\Gamma^{(l+1)}_j = N(P)$.

In each plane $\pi(\Gamma_j^{(l+1)}) \cap T(\Gamma^{(l)})$ we take a set γ_j such that the following conditions are fulfilled:

1) γ_j $(j = 1, \ldots, J)$ do not intersect $\pi(\Gamma^{(l)})$;
2) the convex hull of $\gamma_1 \cup \cdots \cup \gamma_J$ contains

$$(T_+(\Gamma^{(l)}) \setminus \pi(\Gamma^{(l)})) \cap \mathbb{Z}^n.^{2)}$$

To prove the proposition it suffices to construct $\forall j$ a π_+^j semi-cylinder $\mathcal{D}_j \subset C$ such that

$$\langle \alpha - \beta, q \rangle > \log \frac{1}{\varepsilon} \quad \text{for} \quad \beta \in \gamma_j,\ q \in \mathcal{D}_j. \tag{6'}$$

Then $\mathcal{D} = \cap \mathcal{D}_j$ is the desired $V_{(l)}$ semi-cylinder.

To construct $\mathcal{D}_j$ we take an arbitrary element $q_0 \in \pi_+^j \setminus \partial\pi_+^j$. According to the definition of the half-space π_+^j, we have

$$\langle \alpha - \beta, q_0 \rangle > 0 \quad \forall \alpha \in \pi(\Gamma^{(l)}), \quad \forall \beta \in \pi(\Gamma_j^{(l+1)}) \setminus \pi(\Gamma^{(l)}).$$

We note that $\langle \alpha - \beta, q_0 \rangle$ does not depend on α (since $\langle \alpha, q_0 \rangle = \text{const}$ for $\alpha \in \pi(\Gamma^{(l)})$, $q_0 \in \pi_+^j \subset \pi(V_{(l)})$. By virtue of the compactness of γ_j, $\exists \delta > 0$ such that

$$\langle \alpha - \beta, q_0 \rangle > \delta \quad \forall \alpha \in \pi(\Gamma^{(l)}), \quad \forall \beta \in \gamma_j.$$

Let $C_j \subset C$ be an arbitrary π_+^j semi-cylinder and let $\mathcal{D}_j = C_j + Lq_0$. If $L = L(\varepsilon)$ is sufficiently large, then inequality (6′) holds for $q \in \mathcal{D}_j$. The proposition is proved.

3.3. Some additional remarks.

1) Let $V_{(l)}$ be determined by (5) and let $\hat{\pi}_+^j$ be the extension of the half-subspace π_+^j to a half-space of $\mathbb{R}^n$ transversal to $\pi(V_{(l)})$. Let R_+ be a translation of $\hat{\pi}_+^j$. The region $\hat{\mathcal{D}} = \mathcal{D} \cap R_+$ is called the truncation of the $V_{(l)}$ semi-cylinder $\mathcal{D}$ in the direction of π_+^j. As can easily be seen, $\hat{\mathcal{D}}$ is a $V_{(l)}$ semi-cylinder.

When proving Proposition 3.2 we have in fact established that by means of the operation of truncation in the directions of π_+^j it is possible to obtain from an arbitrary $V_{(l)}$ semi-cylinder $\mathcal{D}_0 \subset C$ (where C is a $\pi(V_{(l)})$-cylinder) a $V_{(l)}$ semi-cylinder $\mathcal{D}$ possessing the property $g(\Gamma^{(l)}, \varepsilon)$.

2) Let $\Gamma^{(l)} \subset \partial\Gamma_j^{(l+1)}$ and let $\Gamma_j^{(1)}$ be a one-dimensional face lying in $\Gamma_j^{(l+1)}$, transversal to $\Gamma^{(l)}$, and having a common vertex $\Gamma_j^{(0)}$ with $\Gamma^{(l)}$. Then as the above half-subspace π_+^j we can take the half-space bounded by a hyperplane Q_j orthogonal

---

[2)] We note that property 2) makes it necessary to take the points $\beta \in T(\Gamma^{(l)})$ lying outside the positive coordinate n-hedron.

to $\Gamma_j^{(1)}$. We note that on every such hyperplane Q_j there lies a face of maximum dimension of the polyhedral angle $V_{(0)}^j$ corresponding to $\Gamma_j^{(0)}$.

3) Thus, if $\Gamma^{(0)} \subset \Gamma^{(l)}$, then every $V_{(l)}$ semi-cylinder $\mathcal{D}$ belongs to appropriate translations of half-spaces supporting to those faces of maximum dimension of the polyhedral angle $V_{(0)}$ which do not contain $V_{(l)}$.

On the other hand, the definition of the π-cylinder C implies that it lies in appropriate translations of arbitrary half-spaces on whose boundary π lies. It follows that if $\Gamma^{(0)} \subset \Gamma^{(l)}$, then the $V_{(l)}$ semi-cylinder $\mathcal{D}$ belongs to appropriate translations of supporting half-spaces $V_{(0)}$ passing through the faces of maximum dimension containing $V_{(l)}$. We have thus proved the following

Lemma. *Let $\Gamma^{(l)}$ be a face of $N(P)$ and let $\Gamma_\lambda^{(0)}$, $\lambda = 1, \dots, \Lambda_0$, be the set of vertices of that face. Then each $V_{(l)}$ semi-cylinder $\mathcal{D}$ is contained in the intersection of translations of the polyhedral angles $V_{(0)}^\lambda$, $1 \leqslant \lambda \leqslant \Lambda_0$.*

Corollary. *Let $\Gamma^{(l)}$ be a minor face of a complete polyhedron $N(P)$. Then every $V_{(l)}$ semi-cylinder is contained in a translation of the coordinate n-hedron $\overline{\mathbb{R}_-^n}$:*

$$\mathbb{R}_-^n = \{q \in \mathbb{R}^n, q_j \leqslant 0, j = 1, \dots, n\}. \tag{7}$$

Proof. By virtue of the regularity of $N(P)$, the origin $\{0\} = \Gamma_0^{(0)}$ is the only minor vertex of $N(P)$. Since the coordinate hyperplanes $\{\alpha_j = 0\}$, $j = 1, \dots, n$, are the only faces of maximum dimension passing through $\Gamma_0^{(0)}$, we have $V_0^{(0)} = \mathbb{R}_-^n$. It remains to note that all minor faces of $N(P)$ lie in coordinate planes of various dimensions and pass through the origin.

3.4. Constructing covering by regions possessing the property $g(\Gamma^{(k)}, \varepsilon)$. The main result in this section is the following

Theorem. *For any $\varepsilon > 0$ there is $r = r(\varepsilon)$ such that the region $\mathbb{R}^n \setminus i_-(r)$*[1] *admits of a covering by a finite set of regions $\mathcal{D}_\lambda$ each of which satisfies the condition $g(\Gamma_\lambda^{(k)}, \varepsilon)$ for a certain senior face $\Gamma_\lambda^{(k)} \subset N(P)$.*

The theorem will be proved by induction. We shall proceed from the covering of $\mathbb{R}^n \setminus i_-(0)$ by the polyhedral angles of normals $V_{(0)}^j$ corresponding to the vertices of $N(P)$. At the lth step we shall obtain a covering of $\mathbb{R}^n$ by "final" $V_{(k)}^j$ semi-cylinders, $k \leqslant l - 1$ (they do not change in the further rearrangements) and "preliminary" $V_{(l)}^j$ semi-cylinders appearing at the foregoing step. Applying the operation of truncation in the directions corresponding to the faces $V_{(l)}^j$ of maximum dimension we cut out of the "preliminary" $V_{(l)}^j$ semi-cylinders the "final" $V_{(l)}$ semi-cylinders possessing the property $g(\Gamma_j^{(l)}, \varepsilon)$.

---

[1] For the definition of the region $i_-(r)$ see (2).

We now show (and this is the geometrical basis for the induction) that the part of a $V_{(l)}$ semi-cylinder that remains after a certain truncation of that semi-cylinder is removed is covered by the $V^j_{(l+1)}$ semi-cylinders corresponding to the $(l+1)$-dimensional faces on whose boundary $\Gamma^{(l)}$ lies.

Lemma. *Let $V_{(l)}$ be a polyhedral region (5), $\dim V_{(l)} = n - l$, and let $V^j_{(l+1)}$, $j = 1, \ldots, J$, be the faces of $V_{(l)}$ of maximum dimension corresponding to the half-spaces π^j_+. Let $\mathcal{D} = \cap \mathcal{D}^j$ be a $V_{(l)}$ semi-cylinder and let $\widehat{\mathcal{D}}$ be its truncation in the direction of a half-space R_+ which is a translation of the half-space $\widehat{\pi}^k_+$ passing through π^k_+ and transversal to $\pi(V_l)$. Then there is a $V^k_{(l+1)}$ semi-cylinder $\check{\mathcal{D}}$ such that*

$$\mathcal{D} \subset \widehat{\mathcal{D}} \cup \check{\mathcal{D}}. \tag{8}$$

Proof. Let $V^{i_1}_{(l+1)}, \ldots, V^{i_m}_{(l+1)}$ be those faces of $V_{(l)}$ of maximum dimension for which the dimension of their intersections with $V^k_{(l+1)}$ are maximum:

$$\dim(V^k_{(l+1)} \cap V^{i_s}_{(l+1)}) = n - l - 2, \qquad s = 1, \ldots, m.$$

Denote by R_- the complementary half-space to R_+. We set

$$\check{\mathcal{D}} = R_- \cap \mathcal{D}^k \cap \mathcal{D}^{i_1} \cap \cdots \cap \mathcal{D}^{i_m}.$$

Then inclusion (8) is obvious, and it remains to show that $\check{\mathcal{D}}$ is a $V^k_{(l+1)}$ semi-cylinder. To this end we denote by $\lambda^{i_1}_+, \ldots, \lambda^{i_m}_+$ the half-subspaces of $\partial\pi^k_+$ bounded by the intersections $\partial\pi^k_+ \cap V^{i_s}_{(l+1)}$, $s = 1, \ldots, m$. Then

$$V^k_{(l+1)} = \lambda^{i_1}_+ \cap \cdots \cap \lambda^{i_m}_+.$$

We shall show that $\check{\mathcal{D}}$ is representable as an intersection of $\lambda^{i_s}_+$ semi-cylinders. Indeed, we have

$$\check{\mathcal{D}} = (R_- \cap \mathcal{D}^k \cap \mathcal{D}^{i_1}) \cap \cdots \cap (R_- \cap \mathcal{D}^k \cap \mathcal{D}^{i_m}). \tag{9}$$

According to Remark 3.2, $R_- \cap \mathcal{D}^k$ is a $\partial\pi^k_+$-cylinder, and, by the condition of the lemma, $\mathcal{D}^{i_s}$ is a $\pi^{i_s}_+$ semi-cylinder. Applying now Lemma 3.2 to $\pi = \pi(V_{(l)})$, $\pi^0 = \partial\pi^k_+$, $\pi^0_+ = \lambda^{i_s}_+$, $\mu_+ = \pi^{i_s}$, $C = R_- \cap \mathcal{D}^k$, and $M = \mathcal{D}^{i_s}$ we find that $R_- \cap \mathcal{D}^k \cap \mathcal{D}^{i_s}$ is a $\lambda^{i_s}_+$ semi-cylinder. Therefore region (9) is a $V^k_{(l+1)}$ semi-cylinder. The lemma is proved.

The proof of the theorem. The general induction scheme has already been outlined, and it only remains to show that one can confine oneself to semi-cylinders satisfying the condition $g(\Gamma^{(k)}, \varepsilon)$ where $\Gamma^{(k)}$ runs over the set of the *senior* faces of $N(P)$.

So, let $\Gamma_\lambda^{(0)}$, $\lambda = 0, \dots, \Lambda_0$, be the vertices of $N(P)$, the vertex $\Gamma_0^{(0)} = \{0\}$ being minor and the other vertices being senior. It is clear that

$$\mathbb{R}^n_{(q)} = \bigcup_{\lambda=0}^{\Lambda_0} V^\lambda_{(0)}$$

($V_{(0)}$ are closed polyhedral angles). As has already been mentioned, $V^0_{(0)} = \overline{\mathbb{R}^n_-}$, whence

$$\mathbb{R}^n \setminus \overline{\mathbb{R}^n_-} \subset \bigcup_{\lambda=1}^{\Lambda_0} V^\lambda_{(0)}.$$

Based on Proposition 3.2, we cut out of each of the polyhedral angles $V^\lambda_{(0)}$ (which are $V^\lambda_{(0)}$ semi-cylinders) polyhedral regions $\mathcal{D}_0^{(\lambda)} \subset g(\Gamma_\lambda^{(0)}, \varepsilon)$ (by means of truncation in the directions corresponding to the faces of $V^\lambda_{(0)}$ of maximum dimension). According to the lemma, the set

$$\check{\mathcal{D}} \stackrel{\text{def}}{=} (V^1_{(0)} \setminus \mathcal{D}^1_{(0)}) \cup \dots \cup (V^{\Lambda_0}_{(0)} \setminus \mathcal{D}^{\Lambda_0}_{(0)})$$

is covered by a union of $V^\mu_{(1)}$ semi-cylinders, $\mu_1 = 1, \dots, M_1$. According to Corollary 3.3, the $V^\mu_{(1)}$ semi-cylinders corresponding to the minor faces $\Gamma_\mu^{(1)}$ (they lie on coordinate axes) are contained in translations of the n-hedron (7). Hence, after the first step there remains a finite set of $V^\mu_{(1)}$ semi-cylinders corresponding to the senior one-dimensional faces of $N(P)$. Continuing this process and using Corollary 3.3 at each step we arrive at the nth step at which there appears a $\{0\}$ semi-cylinder, i.e. a bounded polyhedron. Since the latter is always contained in region (2) (provided that r is sufficiently large), the theorem is proved completely.

Remarks. 1) Let $\Gamma^{(l)}$ be a face of $N(P)$ and let the corresponding polyhedral angle $V_{(l)}$ be the intersection of the half-subspaces $\pi^j_+ \subset \pi(V_{(l)})$, $j \in J$ (see (5)). Let $\widehat{J}$ be the set of those $j \in J$ for which $\partial\pi^j_+$ is not a coordinate plane. By the extended polyhedral angle of normals will be meant

$$\widehat{V}_{(l)} = \bigcap_{j \in J} \pi^j_+.$$

The geometrical meaning of the transition from $V_{(l)}$ to $\widehat{V}_{(l)}$ consists in discarding the faces of $V_{(l)}$ corresponding to the minor faces $\Gamma_k^{(l+1)}$ on whose boundary $\Gamma^{(l)}$ lies. The transition is nontrivial only for faces lying in coordinate planes.

It can be shown that the proof of Proposition 3.2 applies if $\widehat{V}_{(l)}$ is taken instead of $V_{(l)}$. Therefore the covering in the theorem can be constructed proceeding from

$\widehat{V}_{(l)}$. In this case no semi-cylinders corresponding to minor faces appear (and, accordingly, Corollary 3.3 is no longer needed).

2) In the theorem we have constructed a covering of $\mathbb{R}^n$ by closed polyhedral regions $\mathcal{D}_\lambda$. Examining the proof of the theorem one can readily see that the covering can be constructed so that the open parts of $\mathcal{D}_\lambda$ cover $\mathbb{R}^n$.

3) The regions $\mathcal{D}_\lambda$ constructed in the theorem are polyhedral, i.e. they are determined by a finite systems of inequalities of the form of

$$\langle l^{(\lambda,j)}, q\rangle \geqslant R_{\lambda j}, \qquad j = 1, \dots, J_\lambda, \tag{10}$$

where the vectors $l^{(\lambda,j)}$ and the numbers $R_{\lambda j}$ are determined up to within a normalizing factor, and $R_{\lambda j} = R_{\lambda j}(\varepsilon)$. In view of the remarks in Section 3.3, as the vectors $l^{(\lambda,j)} \subset \mathbb{R}^n_{(\alpha)}$ one can take only the direction vectors of the various one-dimensional faces (both senior and minor) of $N(P)$.

If $\mathcal{D}_\lambda$ corresponds to a face $\Gamma^{(l)}_\lambda$ of dimension $l > 0$, then among inequalities (10) there must be such that differ in sign. In other words, $\mathcal{D}_\lambda \subset g(\Gamma^{(l)}_\lambda, \varepsilon)$ is determined by a system of inequalities

$$R'_{\lambda,i} \leqslant \langle l^{(\lambda i)}, q\rangle \leqslant R''_{\lambda i}, \qquad i = 1, \dots, I_\lambda, \tag{11}$$

$$\langle l^{(\lambda,i)}, q\rangle \geqslant R_{\lambda j}, \qquad j = 1, \dots, J_\lambda. \tag{11$'$}$$

The vectors $l^{(\lambda,i)}$ correspond to the one-dimensional faces belonging to $\Gamma^{(l)}_\lambda$, and $l^{(\lambda,j)}$ correspond to the one-dimensional faces adjoining $\Gamma^{(l)}_\lambda$.

4) Since $N(P)$ is Newton's polyhedron of a polynomial, it has only integral vertices. Therefore as the direction vectors of one-dimensional faces of $N(P)$ one can also take vectors with integral and, which is more, even components. Here, we can assume, without loss of generality, that the vectors $l^{(\lambda,j)}$ in (10) (or in (11)) have only even components.

§4. Differential operators of *N*-principal type with variable coefficients

In this section we consider differential operators with coefficients belonging to C^∞:

$$P(x, D) = \sum a_\alpha(x) D^\alpha, \tag{1}$$

whose symbols $P(x;\xi)$ are polynomials of N-principal type at every fixed point $x \in \mathbb{R}^n$. Under the additional assumption that the coefficients in senior monomials of the polynomial $P(x;\xi)$ are real we shall prove a multidimensional analog of L_2 estimate (4.3.3). A simple modification of the argument in §4.4 makes it possible to extend the estimate to the scales of H^μ and to prove an analog of estimate (4.4.3) and a local solvability theorem generalizing Theorem 4.4.1. We shall not dwell on these questions and leave the proof of the indicated theorems to the reader as an exercise.

4.1. The statement of the basic result. As in §4.3, with a symbol $P(x;\xi)$ we associate the polyhedra $N(P(x))$, $\delta(P(x))$, and $\widehat{\delta}(P(x))$ at each point $x \in \mathbb{R}^n$ and denote by $N(P)$, $\delta(P)$, and $\widehat{\delta}(P)$ the convex hulls of the unions of the indicated polyhedra over $x \in \mathbb{R}^n$. Definition 4.3.1 is extended trivially to the case $n > 2$, i.e. operator (1) is called an operator of N-principal type if

(i) $N(P(x)) = N(P) \quad \forall x \in \mathbb{R}^n$;

(ii) $P(x^0;\xi)$ is a polynomial of N-principal type $\forall x^0 \in \mathbb{R}^n$.

Recall that condition (i) means, in particular, that if $\alpha^{(j)}(x)$, $j = 1,\dots,J$, are the senior vertices of $N(P)$, then $a_{\alpha^{(j)}}(x) \neq 0$, $x \in \mathbb{R}^n$. Condition (ii) means, in particular, that the polyhedron $N(P(x^0))$ is regular. By virtue of (i), the polyhedron $N(P)$ also possesses this property.

As in Chapter 4, we impose on the symbol $P(x;\xi)$ an additional condition, namely

Condition (R). *If an integral point $\alpha \in N(P)$ is not minor*[1]*, then $a_\alpha(x)$ is a real function.*

Theorem. *Let (1) be an operator of N-principal type and let the additional condition (R) be fulfilled. Then $\forall \varepsilon > 0\ \exists \omega(\varepsilon)$ such that in a region Ω of a sufficiently small diameter,* $\operatorname{diam}\Omega \leqslant \omega(\varepsilon)$, *the inequality (cf. (4.3.3))*

$$\|u\|_{\widehat{\delta}(P)} \leqslant \varepsilon \|P(x,D)u\| \quad \forall u \in \mathcal{D}(\Omega) \tag{2}$$

holds, where

$$\|u\|^2_{\widehat{\delta}(P)} = \int \Xi^2_{\widehat{\delta}(P)}(\xi)|\widehat{u}(\xi)|^2\, d\xi.$$

This theorem automatically implies the multidimensional generalizations of Corollaries 1 and 2 in Section 4.3.1, i.e. for any differential operator $Q(x;D)$ such that $N(Q(x;\xi)) \subset \widehat{\delta}(P)$, the inequality

$$\|u\|_{\widehat{\delta}(P)} \leqslant \varepsilon \|(P+Q)(x,D)u\| \quad \forall u \in \mathcal{D}(\Omega),\ \operatorname{diam}\Omega < \omega(\varepsilon, Q)$$

is fulfilled.

Since, by virtue of condition (R), the symbols of the operators P and P^* differ in minor terms, inequality (2) implies an analogous inequality for the adjoint operator P^* and, consequently, for the transposed operator ${}^tP$.

Remark. In the course of the proof we have incidentally established an estimate for N quasi-elliptic operators:

$$c \sum_{\alpha \in N(P)} \|D^\alpha u\| \leqslant \|P(x,D)u\| \quad \forall u \in \mathcal{D}(\Omega).$$

[1] That is integrally minor in the terminology of Chapter 4.

4.2. Microlocalization of estimate (2). With account of (i), the results of §§2 and 3 can be unified to state the following

Proposition 1. *Under the conditions of the theorem there exists a finite covering* $\{U_j\}$, $j = 1, \ldots, J$, *of the complement of the cube* $I_-(R)$:

$$\mathbb{R}^n_{(\xi)} \setminus I_-(R) \subset \bigcup_{j=1}^{J} U_j,$$

such that in each of the regions U_j *one of the following two conditions holds:*

$$P(x, \xi) > c_0 \Xi_{N(P)}(\xi), \quad \forall \xi \in U_j, \quad |x| < \delta_0, \tag{3}$$

$$|\operatorname{grad} P(x, \xi)| > c_1 \Xi_{\widehat{\delta}(P)}(\xi), \quad \forall \xi \in U_j, \quad |x| < \delta_0. \tag{4}$$

In the Appendix to the present chapter we prove

Proposition 2. *There exists a generalized partition of unity* $\{\psi_j(\xi)\}$, $0 \leqslant j \leqslant J$, *subordinate to the covering in Proposition* 1 (*cf. Section* 4.3.2), *i.e. satisfying the following conditions:*

(i) $\psi_j(\xi) \in C^\infty$;

(ii) $\psi_j(\xi) \geqslant 0$, $\operatorname{supp} \psi_j \subset U_j$, $U_0 = I_-(R)$;

(iii) *there is* $K > 0$ *such that*

$$K^{-1} \leqslant \sum_{j=0}^{J} \psi_j(\xi) \leqslant K; \tag{5}$$

(iv) *there is* ρ, $0 \leqslant \rho \leqslant 1$, *such that*

$$|\psi_j^{(\alpha)}(\xi)| < c_\alpha (1 + |\xi|)^{-\rho|\alpha|}, \qquad 0 < \rho \leqslant 1; \tag{6}$$

(v) *for any* $\alpha > 0$ *and* $\beta \in N(P)$ *we have*

$$|\xi^\beta \psi_j^{(\alpha)}(\xi)| < c_{j\alpha\beta} \Xi_{\widehat{\delta}(P)}(\xi). \tag{7}$$

Theorem. *Let the conditions of Theorem* 4.1 *be fulfilled. Let the covering* $\{U_j\}$ *and the generalized partition of unity* $\{\psi_j\}$ *satisfy the conditions of Propositions* 1 *and* 2. *Then* $\forall \varepsilon > 0$ $\exists \omega(\varepsilon)$ *such that* $\forall j = 1, \ldots, J$ *the inequality*

$$\|\psi_j u\|_{\widehat{\delta}(P)} \leqslant \varepsilon \|P(x, D)(\psi_j u)\| + c(\varepsilon) \|u\|_{(-t)} \tag{8}$$

holds in a region Ω, $\operatorname{diam} \Omega < \omega(\varepsilon)$, *where* $t > n/2$.

Remark. As in the remarks in Section 4.3.3, we assume that the coefficients of the operator P belong to C^∞ and are uniformly bounded on $\mathbb{R}^n$ together with all their derivatives.

The reduction of Theorem 4.1 to the above theorem is carried out by means of an almost literal repetition of the argument in Section 4.3.3, and we do not dwell on this here.

The proof of estimate (8) under condition (3) encounters no serious difficulties and is a simple modification of the argument in Section 4.3.4. Thus, the proof of the main Theorem 4.1 reduces to proving inequality (8) on condition that the symbol $P(x,\xi)$ satisfies inequality (4) in the region U_j.

4.3. The proof of estimate (8) under condition (4). To contract the notation we shall write ψ instead of ψ_j and also $v = \psi(D)u$. The general scheme for the derivation of the estimate is in many respects analogous to the case of real operators of principal type (see Hörmander [2] or Egorov [2]) and is based on an identity that will be derived below.

Lemma 1. *We have the identity*

$$\sum_{j=1}^{n} \|P^{(j)}(x,D)v\|^2 = \sum_{j=1}^{n} \operatorname{Im}[x_j P(x,D)v, P^{(j)}(x,D)v] - \sum_{j=1}^{n} \operatorname{Im}[x_j P^{(j)*}(x,D)v, P^*(x,D)v] + \sum_{j=1}^{n} \operatorname{Re}[P^{(j)*(j)}v, P^*v] + \sum_{j=1}^{n} \operatorname{Im}[x_j v, [P^{(j)}, P^*]v], \tag{9}$$

where $P^{(j)}$ is the operator with symbol $\partial P(x;\xi)/\partial\xi_j$, $P^{(j)*}$ is the adjoint operator of $P^{(j)}$, and the symbol $P^{(j)*(j)}$ is obtained by differentiating the symbol $P^{(j)*}$ with respect to ξ_j; $[\,,\,]$ designates the Hermitean scalar product in L_2.

Proof. For any differential operator we have

$$P(x,D)(x_j v) = x_j P(x,D)v - iP^{(j)}(x,D)v.$$

Multiplying scalarly by $P^{(j)}(x;D)v$ and taking the imaginary part we obtain

$$\|P^{(j)}(x,D)v\|^2 = \operatorname{Im}[x_j P(x,D)v, P^{(j)}(x,D)v] - \operatorname{Im}[P(x,D)(x_j v), P^{(j)}(x,D)v]. \tag{10}$$

Integrating by parts we transform the second term on the right-hand side in the following way:

$$\begin{aligned} -\operatorname{Im}[P(x_j v), P^{(j)}v] &= -\operatorname{Im}[x_j v, P^* P^{(j)}v] \\ &= -\operatorname{Im}[P^{(j)*}(x_j v), P^*v] + \operatorname{Im}[x_j v, [P^{(j)}, P^*]v] \\ &= -\operatorname{Im}[x_j P^{(j)*}v, P^*v] + \operatorname{Re}[P^{(j)*(j)}v, P^*v] + \operatorname{Im}[x_j v, [P^{(j)}, P^*]v]. \end{aligned}$$

Substituting into (10) and performing the summation over j we arrive at identity (9).

We now estimate separately the various terms in identity (9). The most difficult problem is to estimate the last term on the right-hand side of (9). We begin with estimating the left-hand side.

Lemma 2. *Let $v = \psi(D)u$, $u \in \mathcal{D}(\Omega)$, and let inequality* (4) *be fulfilled on the support of the function $\psi(\xi)$. Then $\forall t > n/2$ $\forall \omega$, K_1, and K_2 such that*

$$\sum_{j=1}^{n} \|P^{(j)}(x,D)v\|^2 > K_1\|v\|^2_{\delta(P)} - K_2\|u\|^2_{(-t)} \tag{11}$$

provided that Ω is a sufficiently small region, i.e. $\operatorname{diam}\Omega \leqslant \omega$.

Proof. Assume that the origin belongs to the region Ω (if otherwise, x_j in (9) should be replaced by $x_j - x_j^0$, $j = 1,\ldots,n$, where $x^0 = (x_1^0,\ldots,x_n^0) \in \Omega$). Setting $P(D) = P(0,D)$ and using inequality (4) obtain

$$\begin{aligned}\|v\|^2_{\delta(P)} &\leqslant c_1^{-2}\sum_{j=1}^{n}\|P^{(j)}(D)v\|^2\\ &\leqslant 2c_1^{-2}\sum_{j=1}^{n}\|P^{(j)}(x,D)v\|^2 + 2c_1^{-2}\sum_{j=1}^{n}\|(P^{(j)}(x,D)-P^{(j)}(D))v\|^2.\end{aligned} \tag{12}$$

Consider a truncating function $\chi(x) \in \mathcal{D}$ equal to 1 for $|x| \leqslant \omega_0$ and to zero for $|x| > 2\omega_0$. Then the last term on the right-hand side does not exceed

$$4c_1^{-2}\sum_{j=1}^{n}\|\chi(P^{(j)}(x,D)-P^{(j)}(D))v\|^2 + 4c_1^{-2}\sum_{j=1}^{n}\|(1-\chi)(P^{(j)}(x,D)-P^{(j)}(D))v\|^2. \tag{13}$$

We note that

$$\chi(P^{(j)}(x,D) - P^{(j)}(D)) = \sum_{\beta\in\delta(P)} c_\beta(x)D^\beta, \qquad \max|c_\beta(x)| = O(\omega_0).$$

Taking a sufficiently small ω_0 we estimate the first term in (13) from above by means of $\|v\|^2_{\delta(P)}/2$. Further, if $\omega_0 > \omega > \operatorname{diam}\Omega$, then the symbol $(1-\chi(x))(P^{(j)}(x;\xi) - P^{(j)}(\xi))\psi(\xi)$ is equal to zero on the support of the function $u \in \mathcal{D}(\Omega)$. Using the property of pseudo-locality of the corresponding operator (based on property (6) of the function $\psi = \psi_j$) we estimate the second term in (13) from above by means of $\operatorname{const}\|u\|_{(-t)}$, $t > n/2$. Substituting these estimates into (12) we arrive at (11).

Lemma 3. *Let the conditions of Theorem 4.2 hold. Then* $\forall\varepsilon > 0\ \exists\omega(\varepsilon)$ *such that in a region* Ω, $\operatorname{diam}\Omega \leqslant \omega(\varepsilon)$, *the right-hand side of identity* (9) *is estimated from above by means of*

$$\varepsilon\|P(x,D)v\|^2 + C\varepsilon\|v\|^2_{\widehat{\delta}(P)} + c(\varepsilon)\|u\|^2_{(-t)}, \tag{14}$$

where $t > n/2$, $v = \psi(D)u$, *and the function* ψ *satisfies conditions of the type* (6) *and* (7).

Comparing Lemmas 1, 2 and 3 and assuming that $C\varepsilon < K_1/2$ (where K_1 is the constant in (11)) we arrive at (8).

The proof of the Lemma 3.

1) *Estimation of the first term on the right-hand side of* (9). As in the proof of Lemma 1, we take the truncating function $\chi(x)$. We have

$$\begin{aligned}\operatorname{Im}[x_jP(x,D)v, P^{(j)}(x,D)v] &\leqslant \|P^{(j)}v\|\|x_jPv\|\\ &\leqslant c_3\|v\|_{\widehat{\delta}(P)}\|\chi x_iPv\| + c_3\|v\|_{\widehat{\delta}(P)}\|(1-\chi)x_jP\psi(D)v\|.\end{aligned} \tag{15}$$

Take $2\delta_0$, the diameter of the support of χ, so that $|x_j\chi| \leqslant \varepsilon$. Then the first term on the right-hand side of inequality (15) does not exceed

$$\varepsilon c_3\|v\|_{\widehat{\delta}(P)}\|Pv\| \leqslant \varepsilon\|Pv\|^2 + \varepsilon c_3^2\|v\|^2_{\widehat{\delta}(P)}. \tag{15$'$}$$

By virtue of the pseudo-locality of the operator $x_j(1-\chi)P\psi(D)$, the last term on the right-hand side of (15) does not exceed

$$c_tc_3\|v\|_{\widehat{\delta}(P)}\|u\|_{(-t)} \leqslant \varepsilon\|v\|^2_{\widehat{\delta}(P)} + c_3c_t\varepsilon^{-1}\|u\|^2_{(-t)}. \tag{15$''$}$$

Substituting (15′) and (15″) into (15) we estimate the left-hand side of (15) by means of expression (14).

2) *Estimation of the second term on the right-hand side of* (9). Repeating literally the above estimation we show that

$$-\operatorname{Im}[x_jP^{(j)*}v, P^*v] \leqslant \varepsilon\|P^*v\|^2 + c^*\varepsilon\|v\|^2_{\widehat{\delta}(P)} + c^*(\varepsilon)\|u\|_{(-t)}. \tag{16}$$

If condition (R)[1)] is fulfilled, then

$$P^*(x,\xi) = \overline{P(x,\xi)} + \sum_{\alpha>0}\overline{P}^{(\alpha)}_{(\alpha)}(x,\xi)/\alpha! = P(x,\xi) + Q(x,\xi), \tag{17}$$

where Q is a polynomial symbol in ξ, and we have $N(Q(x;\xi)) \subset \delta(P)$ for all x. Therefore

$$\|P^*v\| \leqslant \|Pv\| + \text{const}\,\|v\|_{\widehat{\delta}(P)}. \tag{17$'$}$$

---

1) Note that up to now we have not used this condition.

Substituting (17′) into (16) we estimate the left-hand side of (16) by means of expression (14).

3) *The estimation of the third term on the right-hand side of* (9). By Schwarz' inequality, we have

$$\mathrm{Re}[P^{(j)*(j)}v, P^*v] \leqslant \|P^{(j)*(j)}v\|\|P^*v\|. \tag{18}$$

The symbol of the operator $P^{(j)*(j)}$ is a linear combination of the monomials corresponding to the minor points of the polyhedron $N(P^{(j)*}) = N(P^{(j)}) \subset \widehat{\delta}(P)$. It follows that

$$\|P^{(j)*(j)}v\| \leqslant c_4 \sum_{\beta\in\delta(P^{(j)})} \|D^\beta v\|. \tag{19}$$

By virtue of the regularity of the polyhedron $\widehat{\delta}(P)$, $\forall\varepsilon$ and $\forall\beta \in \delta(P^{(j)})$ there is a constant $c_\beta(\varepsilon)$ such that

$$|\xi^\beta| < \varepsilon\Xi_{\widehat{\delta}(P)}(\xi) + c_\beta(\varepsilon).$$

It follows that

$$\|D^\beta v\| \leqslant \varepsilon\|v\|_{\widehat{\delta}(P)} + c_\beta(\varepsilon)\|v\| \leqslant 2\varepsilon\|v\|_{\widehat{\delta}(P)} + c'_\beta\|u\|_{(-t)}.$$

Substituting these inequalities into (19) we see that

$$\|P^{(j)*(j)}v\| \leqslant c_5\varepsilon\|v\|_{\widehat{\delta}(P)} + c_6(\varepsilon)\|u\|_{(-t)}. \tag{19′}$$

Substituting (19′) into (18) and estimating $\|P^*v\|$ by means of (17′) we obtain

$$\begin{aligned}\mathrm{Re}[P^{(j)*(j)}v, P^*v] &\leqslant c_5\varepsilon\|v\|_{\widehat{\delta}(P)}\|Pv\| \\ &\quad + c_7\varepsilon\|v\|^2_{\widehat{\delta}(P)} + c_6(\varepsilon)\|Pv\|\|u\|_{(-t)} + c_8(\varepsilon)\|v\|_{\widehat{\delta}(P)}\|u\|_{(-t)} \\ &\leqslant \varepsilon\|Pv\|^2 + \varepsilon(c_7 + c_5^2/2 + 1)\|v\|^2_{\widehat{\delta}(P)} + (\varepsilon^{-1}c_6^2(\varepsilon) + \varepsilon^{-1}c_8^2(\varepsilon))\|u\|^2_{(-t)},\end{aligned}$$

which proves the desired estimate.

4) Thus, to complete the proof of the lemma we have to establish the inequality

$$\mathrm{Im}[x_j v, [P^{(j)}, P^*v]v] \leqslant \varepsilon\|Pv\|^2 + c\varepsilon\|v\|_{\widehat{\delta}(P)} + c(\varepsilon)\|u\|_{(-t)}. \tag{20}$$

4.4. The proof of estimate (20). To begin with, we consider in detail the structure of the commutator of the operators $P^{(j)}$ and P^*.

Lemma 1. *The symbol $H_j(x;\xi)$ of the commutator $[P^{(j)},P^*]$ can be represented as*

$$H_j(x,\xi)=Q(x,\xi)+R(x,\xi), \tag{21}$$

where the symbol Q can be written in the form

$$Q(x,\xi)=\sum_{\substack{\alpha\in\delta(P)\\ \beta\in\delta(P)}} q_{\alpha\beta}(x)\xi^{\alpha+\beta}. \tag{22}$$

The symbol R is written

$$R(x,\xi)=\sum r_{\alpha\beta}(x)\xi^{\alpha+\beta}, \tag{23}$$

where $\beta\in N(P)$ and the (integral) multiindices α have the form

$$\alpha=\alpha'-\alpha''-\alpha''',\quad \alpha'\in N(P),\quad \alpha'',\alpha'''>0. \tag{23$'$}$$

Proof. We first of all note that if the condition (R) is fulfilled, then

$$P^*(x;D)=P(x;D)+Q(x;D),$$

where Q is a linear combination of the operators D^α, $\alpha\in\delta(P)$. Therefore we have

$$[P^{(j)},P^*]=[P^{(j)},P]+[P^{(j)},Q].$$

The symbol of the commutator $[P^{(j)},Q]$ has the form

$$\sum_{\gamma>0}(P^{(j)(\gamma)}Q_{(\gamma)}-P^{(j)}_{(\gamma)}Q^{(\gamma)})/\gamma!$$

and is obviously representable as (22). The symbol of the commutator $[P^{(j)},P]$ has the form

$$\sum_{\gamma>0}P^{(j)(\gamma)}P_{(\gamma)}/\gamma!-\sum_{\gamma>0}P^{(j)}_{(\gamma)}P^{(\gamma)}/\gamma!$$

Here the second sum is of the form of (22), and the first sum has the form (23), where in (23$'$) we have $\alpha'\in N(P)$, $\alpha''=e_j$, and $\alpha'''=\gamma$.

By virtue of the above lemma, the proof of inequality (20) is implied by the following lemmas.

Lemma 2. *$\forall\varepsilon > 0$ there is $\omega(\varepsilon)$ such that in a region Ω, $\operatorname{diam}\Omega < \omega(\varepsilon)$, the inequality*

$$[x_j q(x)v, D^{\alpha+\beta}v] \leqslant c_1\varepsilon\|v\|^2_{\widehat{\delta}(P)} + c_2(\varepsilon)\|u\|^2_{(-t)},$$
$$\alpha,\beta \in \delta(P), \quad v = \psi(D)u, \quad u \in \mathcal{D}(\Omega), \tag{24}$$

holds.

Proof. If the operation of differentiation D^α is transferred to $x_j qv$ by means of integration by parts, then the left-hand side of (24) takes the form

$$[x_j q(x)D^\alpha v, D^\beta v] + \sum_{\gamma>0}\binom{\alpha}{\gamma}[D^\gamma(x_j q)D^{\alpha-\gamma}v, D^\beta v].$$

As has already been done many times, we take a truncating function $\chi(x)$ such that it is equal to 1 on the support of $u(x)$ and the inequality $|x_j q(x)\chi(x)| < \varepsilon/2$ holds. Then we have

$$\begin{aligned}[x_j q D^\alpha v, D^\beta v] &\leqslant \|D^\beta v\|(\|x_j\chi q D^\alpha v\| + \|x_j(1-\chi)qD^\alpha v\|)\\ &\leqslant \frac{\varepsilon}{2}\|v\|^2_{\widehat{\delta}(P)} + c(t)\|D^\beta v\|\|u\|_{(-t)} \leqslant \varepsilon\|v\|^2_{\widehat{\delta}(P)} + c(\varepsilon)\|u\|^2_{(-t)}.\end{aligned}$$

As to the terms in the sum corresponding to $\gamma > 0$, we can use in the case the same argument as in the proof of estimate (19′).

Lemma 3. *Under the conditions of Theorem 4.2 $\forall\varepsilon > 0$ $\exists\omega(\varepsilon)$ such that in a region Ω, $\operatorname{diam}\Omega < \omega(\varepsilon)$, we have*

$$[x_j rv, D^{\alpha+\beta}v] \leqslant c_3\varepsilon\|v\|^2_{\widehat{\delta}(P)} + c_4(\varepsilon)\|u\|^2_{(-t)}, \tag{25}$$

where the multiindices α and β are of the form of (23′).

Proof. If the vector $\alpha+\beta$ can be represented as a sum $\alpha'+\beta'$, $\alpha',\beta' \in \delta(P)$, then the desired assertion reduces to the foregoing lemma. Therefore we shall consider the case when no such representation is possible and the ordinary integration by parts does not help.

1) Put

$$h(\xi) = \Big(1 + \sum_{j=1}^n |P^{(j)}(x^0,\xi)|^2\Big)^{1/2}, \qquad x^0 \in \Omega. \tag{26}$$

It is clear that

$$h(\xi) \leqslant \operatorname{const}\Xi_{\widehat{\delta}(P)}(\xi) \quad \forall\xi \in \mathbb{R}^n. \tag{27}$$

On the other hand, according to (4), we have

$$h(\xi) > \operatorname{const}\Xi_{\widehat{\delta}(P)}(\xi) \quad \text{for} \quad \xi \in \operatorname{supp}\psi. \tag{27′}$$

Let us show that there is ρ, $0 < \rho \leqslant 1$, such that $\forall \gamma > 0$ we have

$$|\partial^\gamma h(\xi)| < c_\gamma (1 + |\xi|)^{-\rho|\gamma|} \Xi_{\widehat{\delta}(P)}(\xi). \tag{28}$$

To prove (28) we note that the regularity of the polyhedron $2\widehat{\delta}(P)$ implies that there always exists $\rho > 0$ such that

$$|\partial^\gamma h^2(\xi)| < c \Xi^2_{\widehat{\delta}(P)}(\xi)(1 + |\xi|)^{-\rho|\alpha|}.$$

With account of inequality (27′), we conclude that

$$|\partial^\gamma h^2(\xi)| < c' h^2(\xi)(1 + |\xi|)^{-\rho|\gamma|} \quad \text{for} \quad \xi \in \operatorname{supp} \psi.$$

It follows that the function $h(\xi) = \sqrt{h^2(\xi)}$ satisfies the inequality (cf. the proof of Lemma 2 in the Appendix to §1.4)

$$|\partial^\gamma h(\xi)| < c'' h(\xi)(1 + |\xi|)^{-\rho|\gamma|} \quad \text{for} \quad \xi \in \operatorname{supp} \psi.$$

By virtue of (27), we obtain (28).

2) Take a function $\varphi(\xi)$ possessing the following properties:

(i) $\varphi(\xi) \equiv 1$ for $\xi \in \operatorname{supp} \psi$,

(ii) $h(\xi) > \text{const}\, \Xi_{\widehat{\delta}(P)}(\xi)$, $\xi \in \operatorname{supp} \varphi$,

(iii) $|\partial^\gamma \varphi(\xi)| < c_\gamma (1 + |\xi|)^{-\rho|\gamma|}$.

The construction of this function is carried out by means of the same procedure as in the case of the function $\psi = \psi_j$ using the original partition of unity (see the Appendix to the present section).

3) We come back to the proof of inequality (25). Setting $h_1(\xi) = \varphi(\xi)h(\xi)$ we have

$$\begin{aligned} [x_j r v, D^{\alpha+\beta} v] &= [x_j r v, \varphi(D)h(D)h^{-1}(D)D^{\alpha+\beta} v] \\ &= [h_1(D)(x_j r v), h^{-1}(D)D^{\alpha+\beta} v] \leqslant \|h_1(D)(x_j r v)\| \|h^{-1}(D)D^{\alpha+\beta} v\|. \end{aligned} \tag{29}$$

4) Let us estimate the second factor on the right-hand side of (29). We note that if (23′) holds. then

$$\alpha + \beta \in 2\widehat{\delta}(P) \tag{30}$$

Indeed, if $q \in \mathbb{R}^n_+$ and $\min q_i = 1$, then

$$\langle q, \alpha + \beta \rangle = \langle q, \alpha' \rangle + \langle q, \beta \rangle - \langle q, \alpha'' \rangle - \langle q, \alpha''' \rangle \leqslant 2d_P(q) - 2 = 2\widehat{d}_P(q).$$

Recalling the definition of $\widehat{\delta}(P)$ we obtain (30). This inclusion implies that

$$|\xi^{\alpha+\beta}| \leqslant \Xi_{2\widehat{\delta}(P)} \leqslant (\Xi_{\widehat{\delta}(P)}(\xi))^2.$$

With account of (27), we obtain

$$|h^{-1}(\xi)\xi^{\alpha+\beta}| < \text{const}\, \Xi_{\widehat{\delta}(P)}(\xi).$$

Thus, if $v = \psi(D)u$, then

$$\|h^{-1}(D)D^{\alpha+\beta}v\| \leqslant \text{const}\, \|v\|_{\widehat{\delta}(P)}. \tag{31}$$

5) We now pass to the estimation of the first factor in (29). We have

$$\|h_1(D)(x_j rv)\| \leqslant \|x_j r h_1(D)v\| + \|[h_1(D), x_j r]v\|. \tag{32}$$

Using the smallness of diam Ω and the truncating function χ, $\chi(x) = 1$, $x \in \Omega$, we estimate the first term on the right-hand side:

$$\|x_j r h_1(D)v\| \leqslant \varepsilon\|v\|_{\widehat{\delta}(P)} + c(\varepsilon)\|u\|_{(-t)}. \tag{33}$$

Using the commutation formula we write

$$[h_1(D), x_j r]v = \sum_{\gamma>0}^{N-1} D^\gamma(x_j r)h_1^{(\gamma)}(D)v/\gamma! + R_N v.$$

Comparing (28) and property (iii) of the function $\varphi(\xi)$ we find that for some $\rho > 0$ and $\forall\varepsilon > 0$ we have

$$|h_1^{(\gamma)}(\xi)| < c_\gamma \Xi_{\widehat{\delta}(P)}(1+|\xi|)^{-\rho} \leqslant \varepsilon\Xi_{\widehat{\delta}(P)} + K(\varepsilon).$$

This inequality implies that

$$\|D^\gamma(x_j r)h_1^{(\gamma)}(D)v\| \leqslant C\varepsilon\|v\|_{\widehat{\delta}(P)}. \tag{34}$$

Comparing again (28) and property (iii) of the function $\varphi(\xi)$ we see that for some M the inequality

$$|\partial^\gamma h_1(\xi)| < c_\gamma(1+|\xi|)^{M-\rho|\alpha|}$$

holds. Therefore, according to Hörmander [4], we have

$$\|T_N v\| \leqslant \text{const}\, \|v\|_{(M-\rho N)} \leqslant \text{const}\, \|u\|_{(-t)} \tag{35}$$

provided that the chosen number N is sufficiently large. Comparing (31)–(35) we arrive at (25). The lemma is proved.

Appendix

Here we present the proof of Proposition 2 in Section 4.2 and construct a system of functions $\{\psi_j(\xi)\}$ satisfying conditions (4.5), (4.6), and (4.7). We shall use the construction that was presented in Section 4.3.3 and in the Appendix to §4.3 for the two-dimensional case. As in the two-dimensional case, the central point of the whole construction relates to the regions corresponding to the zero-dimensional faces (vertices). The case of the faces $\Gamma^{(k)}$, $k > 0$ readily reduced to the case $k = 0$.

1. By virtue of Remark 3, in Section 3.4, we can assume that the image of U_j under the logarithmic mapping (3) is determined by the equations

$$\langle l^{j\mu}, q\rangle > R_{j\mu}, \qquad \mu - 1, \ldots, M_j, \tag{1}$$

where the vectors $l^{j\mu} = (l_1^{j\mu}, \ldots, l_n^{j\mu})$ have even components. As in (4.3.10), we set

$$\psi_j(\xi) = \theta(\xi^2 - R^2) \prod_j \theta(\exp\langle l^{j\mu}, \log \xi_+\rangle - R_{j\mu}), \tag{2}$$

where $\theta(t) \in C^\infty$, $\theta(t) \geqslant 0$, $\theta(t) = 0$ for $t < 0$, and $\theta(t) = 1$ for $t \geqslant \varkappa$. The set of functions (2) is supplemented with the function

$$\psi_0(\xi) = \theta(R^2 - \xi^2). \tag{2'}$$

As was noted in Section 4.3.3, a special analysis is required for the verification of the fulfilment of the smoothness condition on the coordinate hyperplanes and the verification of conditions (4.6) and (4.7). In what follows we shall assume that $|\xi| > R$, and therefore the first factor in (2) will be dropped.

2. As in Item 1 in Appendix to §4.3, we first verify inequality (4.7) for $\beta = \gamma$ i.e. the inequality

$$|\xi^\beta \psi_j^{(\beta)}(\xi)| < c_{j\beta} \quad \text{for} \quad \xi \in \mathbb{R}^n. \tag{3}$$

Its derivation is a literal repetition of the derivation of inequality (1) in the indicated Appendix.

It follows from (3) that if there are $\omega > 0$ and ρ, $0 < \rho \leqslant 1$, such that

$$|\xi_k| > \omega(1 + |\xi|)^\rho \quad \text{for } \xi \in \operatorname{supp} \psi_j, \tag{4}$$

then (3) implies (4.6).

3. Consider a region U_j corresponding to a vertex $\Gamma_\lambda^{(0)}$ not lying on the coordinate axes. The normal cone of this vertex consists of positive vectors $q \in \mathbb{R}_+^n$, i.e. there is ρ such that

$$\min_{j,k}(q_j/q_k) > \rho > 0 \quad \forall q \in V_{(0)}^\lambda.$$

Since we have $\operatorname{supp} \psi_j \in U_j$ and the image of U_j under the logarithmic mapping $\xi \mapsto (\log|\xi_1|, \ldots, \log|\xi_n|)$ is contained in the translation of $V_{(0)}^\lambda$, we have

$$\log|\xi_k| > \rho \log|\xi_l| - \log \omega_0 \quad \forall \xi \in \operatorname{supp} \psi_j, \quad k, l = 1, \ldots, n. \tag{4'}$$

Inequalities (4′) imply (4) and hence condition (4.6) for the functions ψ_j.

We now verify condition (4.7). According to the hypothesis, we have $\operatorname{supp}\psi_j \in U_j$, where the region U_j possesses the property $G(\Gamma_\lambda^{(0)}, \varepsilon)$, which means (see Section 2.3) that if $\alpha_{(\lambda)}$ is a coordinate of the vertex $\Gamma_\lambda^{(0)}$, then

$$|\xi^\beta| < \text{const}\, |\xi^{\alpha(\lambda)}| \quad \forall \xi \in U_j, \quad \forall \beta \in N(P) \setminus \Gamma_\lambda^{(0)}.$$

Hence, it suffices to verify (4.7) for $\alpha = \alpha_{(\lambda)}$. Since $\Gamma_\lambda^{(0)}$ does not lie on the coordinate hyperplanes, all the coordinates $(\alpha_{(\lambda)1}, \dots, \alpha_{(\lambda)n})$ are positive, and consequently are no less than 1. Using inequality (3) for $\beta = (\beta_1, \dots, \beta_n)$, $\beta_1 \geqslant 1$, we obtain

$$|\psi_j^{(\beta)}(\xi)| < c_{j\beta}|\xi^{-\beta}| = (c_{i\beta}|\xi_1|^{-\beta_1+1} \dots |\xi_n|^{-\beta_n})|\xi_1|^{-1}.$$

By virtue of condition (4), the expression in the parenthesis does not exceed a constant. Multiplying both side of the inequality by $|\xi^{\alpha(\lambda)}|$ we obtain

$$|\xi^{\alpha(\lambda)}\psi_j^{(\beta)}(\xi)| < \text{const}\, |\xi_1^{\alpha(\lambda)1 - 1} \dots \xi_n^{\alpha(\lambda)}| \leqslant \text{const}\, \Xi_{\widehat{\delta}(P)}(\xi).$$

4. We now consider the case when $\Gamma_\lambda^{(0)}$ lies in a coordinate hyperplane of codimension $n-m$. Expand $\mathbb{R}^n_{(\alpha)}$ as a direct sum: $\alpha = (\beta, \gamma)$, $\beta \in \mathbb{R}^m_{(\beta)}$, $\gamma \in \mathbb{R}^{n-m}_{(\gamma)}$, and assume, to simplify the notation, that if $\alpha = (\alpha_1, \dots, \alpha_n)$, then $\beta = (\alpha_1, \dots, \alpha_m)$ and $\gamma = (\alpha_{m+1}, \dots, \alpha_n)$. Accordingly, the variables ξ are divided into two groups: $\xi = (\eta, \zeta)$, $\eta \in \mathbb{R}^m$, $\zeta \in \mathbb{R}^{n-m}$.

Thus, let the vertex $\Gamma_\lambda^{(0)}$ belong to the plane $\{\alpha_{m+1} = \dots = \alpha_n = 0\}$ and not belong to a coordinate plane of a higher dimension. Let us divide the one-dimensional faces passing through $\Gamma_\lambda^{(0)}$ into two groups, namely

$$\Gamma_\mu^{(1)}, \mu = 1, \dots, \mu_1, \text{ belonging to } \{\alpha_{m+1} = \dots = \alpha_n = 0\}; \tag{5}$$

$$\Gamma_\mu^{(1)}, \mu = \mu_1 + 1, \dots, \mu_2, \text{ transversal to } \{\alpha_{m+1} = \dots = \alpha_n = 0\}. \tag{5′}$$

In accordance with this division of one-dimensional faces we write:

$$\psi_j(\xi) = \psi_j^{(1)}(\xi)\psi_j^{(2)}(\xi) = \prod_{\mu=1}^{\mu_1}(|\xi^{l^{j\mu}}| - R_{j\mu}) \prod_{\mu>\mu_1}(|\xi^{l^{j\mu}}| - R_{j\mu}). \tag{6}$$

We note that the function $\psi_j^{(1)}$ depends only on the variables η (and does not depend on the variables ζ). Denote by N_m the section of $N(P)$ by the coordinate plane $\{\alpha_{m+1} = \dots = \alpha_n = 0\}$. Obviously, N_m is a regular polyhedron in $\mathbb{R}^m$ and $\Gamma_\lambda^{(0)}$ is a senior vertex of that polyhedron not lying in the coordinate planes. Therefore the argument in Item 3 implies that

$$|\eta_l| > \omega_1|\eta_k|^{\rho_1}, \quad \eta \in \operatorname{supp}\psi_j^{(1)}, \quad l, k = 1, \dots, m, \quad 0 < \rho_1 \leqslant 1. \tag{7}$$

We now consider the function $\psi_j^{(2)}$. Let $(\beta_{(\lambda)}, 0)$ be the coordinates of the vertex $\Gamma_\lambda^{(0)}$ and let $(0, \gamma_{(\mu)})$, $\mu = \mu_1 + 1, \dots, \mu_l$, be the coordinates of the intersections of the straight lines passing through the one-dimensional faces (5′) with the subspace $\{\alpha_1 = \dots = \alpha_m = 0\}$. Denote by N_{n-m} the convex hull of $\gamma_{(\mu)}$, $\mu = \mu_1 + 1, \dots, \mu_2$, and the origin. The polyhedron N_{n-m} is the section of the regular polyhedron $T_+(\Gamma_\lambda^{(0)})$ (see (2.6)) by the plane $\{\alpha_1 = \dots = \alpha_m = 0\}$. It is clear that N_{n-m} is a regular polyhedron in $\mathbb{R}^{n-m}$.

For an appropriate normalization of the direction vectors l^μ of the one-dimensional faces (5′) we have

$$l^\mu = (2\beta_{(\lambda)}, 2\gamma_{(\mu)}), \qquad \mu = \mu_1 + 1, \dots, \mu_2.$$

Consequently, the function $\psi_j^{(2)}(\xi)$ in (6) has the form

$$\psi_j^{(2)}(\eta, \zeta) = \prod_{\mu=\mu_1+1}^{\mu_2} \theta(\eta^{2\beta_{(\lambda)}} \zeta^{-2\gamma_{(\mu)}} - R_{\mu j}). \tag{8}$$

It follows from the definition of the function θ that

$$|\zeta^{\gamma_{(\mu)}}| < R_{\mu_j}^{-1/2} |\eta^{\beta_{(\lambda)}}| \quad \text{for} \quad (\eta, \zeta) \in \operatorname{supp} \psi_j^{(2)}, \quad \mu_1 < \mu < \mu_2. \tag{9}$$

Among the vertices $\gamma_{(\mu)}$ there are vertices $\gamma_{(\mu_l)} = \gamma_l e_l$, $l = m+1, \dots, n$, lying on coordinate axes. Hence, (9) implies

$$|\eta^{\beta_{(\lambda)}}| > c_l |\xi_l|^{\gamma_l}, \qquad l = m+1, \dots, n.$$

With account of (7), we have

$$|\eta_k| > \omega_2 |\xi_l|^{\rho_k}, \quad k = 1, \dots, m;\ l = 1, \dots, n,\ \xi \in \operatorname{supp} \psi_j. \tag{10}$$

We now show that

$$|\zeta_s| > \omega_3 |\xi_l|^{\rho_3}, \quad s = m+1, \dots, n;\ l = 1, \dots, n,\ \zeta \in \operatorname{supp} \partial\psi_j / \partial\xi_s. \tag{11}$$

For definiteness, we shall assume that $s = n$. Differentiating (8) with respect to ξ_k and using the property that $\theta'(t)$ is identically equal to zero for $t < 0$ and $t > \varkappa$, we see that for every point $\xi \in \operatorname{supp} \partial\psi_j / \partial\xi_n$ there is ν, $\mu_1 + 1 \leqslant \nu \leqslant \mu_2$, such that

$$|\zeta^{\gamma_{(\nu)}}| > (R_\nu + \varkappa)^{1/2} |\eta^{\beta_{(\lambda)}}|.$$

Comparing this inequality with (9) we conclude that

$$|\zeta^{\gamma)_{(\nu)}}| > \text{const}\, |\zeta^{\gamma_{(\mu)}}| \quad \forall \mu = \mu_1 + 1, \dots, \mu_2.$$

Since $\gamma_{(\mu)}$ are the vertices of the polyhedron N_{n-m}, we have

$$|\zeta^{\gamma)(\nu)}| > \text{const}\,|\zeta^{\gamma}| \quad \gamma \in N_{n-m}, \quad (\eta,\zeta) \in \operatorname{supp} \partial\psi_j^{(2)}/\partial\xi_n.$$

Since all the coordinates of the vertex $\gamma_{(\nu)}$ are positive, we have $\gamma_{(\nu)n} > 0$. We select a point $\gamma = (\gamma_{m+1}, \dots, \gamma_n)$ so that $\gamma_n < \gamma_{(\nu)n}$ and for some s the condition $\gamma_s > \gamma_{(\nu)s}$ holds. Then we see that

$$|\xi_n| > \omega_n |\xi_s|^{\rho_s}, \qquad m+1 \leqslant s \leqslant n. \tag{12}$$

Hence, we have proved inequality (4.6). We now proceed to the verification of inequality (4.7). Without loss of generality, we can assume that $\alpha = (\beta_{(\lambda)}, 0)$ (this follows from (9)). If among the components of β in (4.7) there is $\beta_l \geqslant 1$, $1 \leqslant l \leqslant m$, then the desired estimate is proved using the argument in Item 3. Therefore we shall assume that $\beta = (0, \dots, 0, \beta_{m+1}, \dots, \beta_n)$. It suffices to consider the case $|\beta| = 1$, say $\beta_n = e_n$.

Then we have

$$\left|\frac{\partial\psi_j}{\partial\xi_n}\right| \leqslant \sum_{\nu=\mu_1+1}^{\mu_2} \text{const}\, |\theta'(\eta^{2\beta_{(\lambda)}}\zeta^{-2\gamma_{(\lambda)}} - R_\nu)||\xi_n|^{-1},$$

whence

$$\left|\eta^{\beta_{(\lambda)}}\frac{\partial\psi_j}{\partial\xi_n}\right| < \text{const} \sum \left[|\theta'(\eta^{2\beta_{(\lambda)}}\zeta^{-2\gamma_{(\lambda)}} - R_{(\nu)}||\eta^{\beta_{(\lambda)}}\zeta^{-\gamma_{(\nu)}}|\right] |\zeta^{\gamma_{(\nu)}-e_n}|.$$

The factor in square brackets is bounded and the other factor is estimated by means of $\Xi_{\widehat{\delta}(P)}(\xi)$.

5. We now consider the case of the region corresponding to the face $\Gamma^{(k)}$, $k > 0$. In the covering constructed in §3 to this region there corresponds a $V_{(k)}^{\lambda}$ semi-cylinder $\mathcal{D}_\lambda$. The region $\mathcal{D}_\lambda$ is contained in the intersection of translations of the polyhedral angles $V_{(0)}^{\mu}$, $\mu = 1, \dots, M$, corresponding to the vertices $\Gamma_\mu^{(0)} \subset \Gamma_\lambda^{(k)}$. In this case the region $\mathcal{D}_\lambda$ is determined by M groups of inequalities, each of the groups determining a translation of some $V_\mu^{(0)}$. We now represent the function ψ_j as a product $\psi_j^{(1)} \dots \psi_j^{(M)}$ where each of the factors is a function considered in Sections 1 to 4 and, consequently, satisfies conditions of the type of (4.6) and (4.7). It remains to note that the functions satisfying these conditions form a ring relative to multiplication.

CHAPTER VII

THE METHOD OF ENERGY ESTIMATES IN CAUCHY'S PROBLEM

§1. Introduction. The functional scheme of the proof of the solvability of Cauchy's problem

1.1. The present chapter is devoted to sufficient conditions for correctness of Cauchy's problem for differential operators with variable coefficients. In Chapter 2 (see Theorem 2.5.5) a general result on the solvability of Cauchy's problem was established. Under the assumption that the symbol satisfied the conditions of constant strength and exponential correctness we obtained a priori estimates in the $H^{(s)}_{[\gamma]}$ norms for the original operator P and its formal adjoint operator P^* (see (2.5.31) and (2.5.32)):

$$\sum_{\alpha>0} \|P^{(\alpha)}(y;D)u\|_{(s),\gamma} \leqslant \varepsilon_s(\gamma)\|P(y;D)u\|_{(s),\gamma}, \qquad \gamma \leqslant \gamma_0(s), \tag{1}$$

$$\sum_{\alpha>0} \|\overline{P}^{(\alpha)}(y;D)v\|_{(s),\rho} \leqslant \varepsilon^*_s(\rho)\|P^*(y;D)v\|_{(s),\rho}, \qquad \rho \geqslant \rho_0(s), \tag{1$'$}$$

where $y = (x,t)$, $x \in \mathbb{R}^n$, $t \in \mathbb{R}$, and $\varepsilon_s(\gamma)$, $\varepsilon^*_s(\rho) \to 0$ as $\gamma \to -\infty$, $\rho \to +\infty$. Although the existence and uniqueness theorems in §2.5 were proved by constructing the parametrix for the operator P, they can be in fact deduced directly from inequalities (1) and (1$'$).

The condition of constant strength of the symbol used in the derivation of the above estimates is very stringent and is not fulfilled for strictly hyperbolic and, the more so, dominantly correct differential operators with variable coefficients. In this chapter we present a scheme for studying Cauchy's problem which does not use the condition of constant strength of the symbol and makes it possible to prove the correctness of Cauchy's problem for dominantly correct differential operators with variable coefficients.

The approach in this chapter is based on a rather simple observation that to prove the correctness of Cauchy's problem in the spaces $H^{(s)}_{[\gamma]+}$ it suffices to have weaker estimates (as compared to (1) and (1$'$)).

For a natural l we set

$$P^{(l)}(y;\xi,\tau) \stackrel{\text{def}}{=} P^{(0,\dots,0,l)}(y;\xi,\tau) = \partial^l P(y;\xi,\tau)/\partial\tau^l.$$

By the solution to homogeneous Cauchy's problem

$$P(y,D)u = f \tag{2}$$

will be meant a distribution u with support in the half-space $t \geqslant 0$ satisfying (2) in the sense of distributions:

$$(u, {}^tPv) = (f,v) \quad \forall v \in \mathcal{D} \tag{2$'$}$$

We have

Theorem A. *Let a differential operator $P(y;D)$ and its formal adjoint operator $P^*(y;D)$ satisfy the a priori estimates*

$$c_s \sum_{l=1}^{m} (\gamma_0 - \gamma)^l \|P^{(l)}(y;D)u\|_{(s),\gamma} \leqslant \|P(y;D)u\|_{(s),\gamma}, \quad \forall u \in H^{(\infty)}_{[\gamma]}, \tag{3}$$

$$c_s^* \sum_{l=1}^{m} (\gamma_0 - \gamma) \|P^{*(l)}(y;D)v\|_{(-s),-\gamma} \leqslant \|P^*(y;D)v\|_{(-s),-\gamma},$$

$$\forall u \in H^{(\infty)}_{[-\gamma]}, \ -\gamma \geqslant -\gamma_0, \tag{3'}$$

where $|s| \leqslant M$ and $\gamma_0 = \gamma_0(M)$.

Then $\forall \overline{M} > 0$ there is $\gamma(\overline{M})$ such that $\forall f \in H^{(s)}_{[\gamma]+}$, $|s| \leqslant \overline{M}$, and $\gamma < \gamma(\overline{M})$ the homogeneous Cauchy problem (2) has a unique solution $u \in H^{(s)}_{(\gamma)+}$. Moreover, we have $P^{(l)}(y;D)u \in H^{(s)}_{[\gamma]+}$, $l = 1, \ldots, m$, and the solution satisfies inequality (3).

Before presenting the proof of the theorem we discuss the general scheme for derivation of estimates (3) and (3′).

1.2. Estimates (3) and (3′) are a consequence of more exact inequalities obtained by Leray's method of "separating operator" [1]. The method is based on the following elementary

Proposition. *Let $P(\xi,\tau)$ be a polynomial in the variables $\xi \in \mathbb{R}^n$, $\tau \in \mathbb{C}^1$ solved with respect to the highest power of τ:*

$$P(\xi,\tau) = \tau^m + \sum_{j\geqslant 1,\alpha} a_{\alpha j} \xi^\alpha \tau^{m-j}, \tag{4}$$

and let

$$H_P(\xi,\sigma,\gamma) = -\operatorname{Im}(P(\xi,\tau)\overline{P^{(1)}(\xi,\tau)}), \qquad \tau = \sigma + i\gamma. \tag{5}$$

Then the conditions below are equivalent.

(I) *Petrovskiĭ's correctness condition is fulfilled, i.e. there is γ_0 such that*

$$P(\xi,\tau) \neq 0 \quad \text{for} \quad \operatorname{Im}\tau \leqslant \gamma_0, \quad (\xi, \operatorname{Re}\tau) \in \mathbb{R}^{m+1}. \tag{6}$$

(II) *There is c, depending only on n and m (where m is the degree of P with respect to τ), and a constant γ_0 such that*

$$c \sum_{l=1}^{m} (\gamma_0 - \gamma)^{2l-1} |P^{(l)}(\xi, \sigma + i\gamma)|^2 \leqslant H_P(\xi,\sigma,\gamma), \qquad \gamma \leqslant \gamma_0. \tag{7}$$

(III) *There is γ_0 such that*

$$H_P(\xi,\sigma,\gamma) > 0 \quad \text{for} \quad \gamma \leqslant \gamma_0, \quad (\xi,\sigma) \in \mathbb{R}^{n+1}. \tag{8}$$

Proof. (II)$\Longrightarrow$(III)$\Longrightarrow$(I). The first implication follows from the fact that $P^{(m)}(\xi,\tau)\equiv m!$, and the second implication is based on the property that $H_P=0$ when $P=0$.

(I)$\Longrightarrow$(II) For any fixed ξ we factorize polynomial (4):

$$P(\xi,\tau)=\prod_{j=1}^{m}(\tau-\tau_j(\xi)). \tag{9}$$

Differentiating (9) with respect to τ we obtain

$$P^{(1)}(\xi,\tau)=\sum_{k=1}^{m}\prod_{j\neq k}(\tau-\tau_j(\xi)). \tag{10}$$

Substituting (9) and (10) into (5) we find

$$H_P(\xi,\sigma,\gamma)=\sum_{k=1}^{m}(-\gamma+\operatorname{Im}\tau_k(\xi))\prod_{j\neq k}|\sigma+i\gamma-\tau_j(\xi)|^2. \tag{11}$$

If (6) is fulfilled, then $\operatorname{Im}\tau_k(\xi)\geqslant\gamma_0$, whence

$$|\gamma_0-\gamma||P^{(1)}(\xi,\sigma+i\gamma)|^2\leqslant m(\gamma_0-\gamma)\sum_{k=1}^{m}\prod_{j\neq k}|\sigma+i\gamma-\tau_j(\xi)|^2\leqslant mH_P(\xi,\sigma,\gamma).$$

Differentiating consecutively (10) with respect to τ we readily prove inequality (7) to full extent.

Corollary. *If polynomial* (4) *satisfies Petrovskiĭ's correctness condition, then there are* γ_0 *and* $\varkappa>0$ *such that*

$$(1+|\xi|)^{-2\varkappa}(|\gamma|+|\sigma|)^{2m-2}\leqslant c_\varepsilon(\gamma_0-\gamma)^{-1}H_P(\xi,\sigma,\gamma), \tag{12}$$

where $\gamma\leqslant\gamma_0-\varepsilon$, $\varepsilon>0$.

Proof. Differentiating (4) consecutively we obtain the triangular system

$$P^{(l)}(\xi,\tau)=m(m-1)\dots(m-l+1)\tau^{m-l}$$
$$+\sum(m-j)\dots(m-j-l+1)a_{\alpha j}\xi^\alpha\tau^{m-j-l},\quad l=1,\dots,m,$$

making it possible to express τ^j, $j=0,\dots,m-1$, in terms of $P^{(l)}$, $l=1,\dots,m$:

$$\tau^j=\sum_k a_{jk}(\xi)P^{(j+k)}(\xi,\tau),$$

where $a_{jk}(\xi)$ are polynomials in ξ. Taking the maximum degree of these polynomials as $\varkappa$ and using (7) we derive inequality (12).

1.3. If $P(D_x, D_t)$ is a differential operator with constant coefficients, then passing to the Fourier transform we find

$$- \operatorname{Im}[\exp(\gamma t)P(D)u, \exp(\gamma t)P^{(1)}(D)u] = \iint H_P(\xi, \sigma, \gamma)|\widehat{u}(\xi, \sigma + i\gamma)|^2 \, d\xi \, d\sigma.$$

By virtue of Proposition 1.2, the right-hand side can be estimated from below by means of

$$\text{const} \sum_{l=1}^{m} (\gamma_0 - \gamma)^{2l-1} \|P^{(l)}(D)u\|_\gamma^2.$$

According to Schwarz' inequality, the left-hand side can be estimated from above by means of

$$\|Pu\|_\gamma \|P^{(1)}u\|_\gamma \leqslant (\gamma_0 - \gamma)^{-1/2} \|Pu\|_\gamma \Big(\sum_{l \geqslant 1} (\gamma_0 - \gamma)^{2l-1} \|P^{(l)}u\|_\gamma^2 \Big)^{1/2}.$$

Comparing these estimates we obtain inequality (3) with $s = 0$ for the operator $P(D)$. Replacing u by $\lambda_s^+(D)u = (iD_t + \gamma_0 + \sqrt{1 + |D_x|^2})^s u$ in this inequality we obtain (3) with an arbitrary $s \in \mathbb{R}$.

If the operator $P(D_x, D_t)$ is correct in Petrovskiĭ's sense, then the operator $\overline{P}(D_x, -D_t)$ possesses the same property. Writing down an inequality of the type (3) for this operator and making change of variable $t \to -t$ we obtain (3′).

The aim of the present chapter is to develop the indicated approach in application to operators with variable coefficients. We consider the quadratic form

$$- \operatorname{Im}[\exp(\gamma t)\lambda_s^+(D)P(y; D)u, \exp(\gamma t)\lambda_s^+(D)P^{(1)}(y; D)u].$$

Under some (rather cumbersome) conditions on the symbol $P(y; \xi, \tau)$ presented in §2 the form can be estimated from above and below by means of

$$\iint H_P(\xi, \sigma, \gamma)(1 + |\xi|^2 + \sigma^2 + \gamma^2)^s |\widehat{u}(\xi, \sigma + i\gamma)|^2 \, d\xi \, d\sigma,$$

where $H_P(\xi, \sigma, \gamma)$ denotes the value of the symbol

$$H_P(y; \xi, \sigma, \gamma) = - \operatorname{Im}(P(y; \xi, \tau)\overline{P^{(1)}(y; \xi, \tau)}) \tag{13}$$

at a fixed point $y = y^0$, whence inequality (3) is obtained comparatively simply. Inequality (3′) is deduced in like manner. An analysis of sufficient conditions for the validity of inequalities (3) and (3′) leads to the following assertion.

Theorem B. *Let a symbol $P(y;\xi,\tau)$, solved with respect to the highest power of τ:*

$$P(y;\xi,\tau)=\tau^m+\sum_{j\geqslant 1}P_j(y;\xi)\tau^{m-j},$$

satisfy the following conditions:

(I) $\forall y^0\in\mathbb{R}^{n+1}$ *the polynomial $P(y^0;\xi,\tau)$ is correct in Petrovskiĭ's sense;*

(II) *symbol (13) satisfies the condition of constant strength, i.e. there are γ_0 and $A>0$ such that*

$$A^{-1}\leqslant H_P(y';\xi,\sigma,\gamma)/H_P(y'';\xi,\sigma,\gamma)\leqslant A,\quad \forall y',y''\in\mathbb{R}^{n+1},\quad \forall\gamma\leqslant\gamma_0; \tag{14}$$

(III) $\forall y',y''\in\mathbb{R}^{n+1}$ *and* $\forall y\in\mathbb{R}^{n+1}$ *we have*

$$|P^{(\beta)}|(y';\xi,\tau)\leqslant\varepsilon_\beta(\operatorname{Im}\tau)|P(y'';\xi,\tau)|,\quad \beta>0, \tag{15}$$

$$|P_{(\alpha)}(y;\xi,\tau)|\leqslant\varepsilon_\alpha(\operatorname{Im}\tau)|P(y;\xi,\tau)|(1+|\xi|),\quad \alpha>0, \tag{16}$$

where $\varepsilon_\beta(\operatorname{Im}\tau)$, $\varepsilon_\alpha(\operatorname{Im}\tau)\to 0$ as $\operatorname{Im}\tau\to-\infty$.

Then the differential operator $P(y;D)$ satisfies a priori estimates (3) *and* (3′), *and, consequently, homogeneous Cauchy's problem is uniquely solvable.*

The theorem will be proved in §3.

In the case of constant coefficients conditions (I), (II), and (III) go into a single condition, namely the operator $P(\xi,\tau)$ should be exponentially correct.

Conditions (15) and (16) are fulfilled automatically for exponentially correct symbols of constant strength. Later (see §3) we shall show that the condition of constant strength for P implies an analogous condition for H_P, i.e. Theorems A and B imply the correctness theorem for Cauchy's problem for exponentially correct operators of constant strength.

In §§3 and 4 we shall present meaningful examples of (pluriparabolic and dominantly correct) differential operators with variable coefficients for which the condition of constant strength does not hold but all conditions of Theorem B are fulfilled.

1.4. The proof of Theorem A.

1) The uniqueness of the solution to problem (2) is the simpler assertion in the theorem. This property and the stronger assertion

$$\{u\in H^{(-\infty)}_{[\gamma]},Pu=0\}\Rightarrow\{u\equiv 0\} \tag{17}$$

follow from the inequality

$$\|u\|_{(s),\gamma}\leqslant c_s\|Pu\|_{(s),\gamma}\quad \forall u\in H^{(\infty)}_{[\gamma]}.$$

In view of the continuity, this inequality is extended to the space $H^{(s+N)}_{[\gamma]}$ for any $s\in\mathbb{R}$ and $N\geqslant\deg P$:

$$\|u\|_{(s),\gamma}\leqslant c_s\|Pu\|_{(s),\gamma}\quad \forall u\in H^{(s+N)}_{[\gamma]}. \tag{18}$$

Further, let $u \in H^{(-\infty)}_{[\gamma]}$ and let $Pu = 0$. The definition of $H^{(-\infty)}_{[\gamma]}$ implies that $u \in H^{(s_1)}_{[\gamma]}$ for some s_1. Therefore, if a sequence $u_j \in H^{(\infty)}_{[\gamma]}$ converges to u in $H^{(s_1)}_{[\gamma]}$, then the sequence Pu_j converges to $Pu = 0$ in $H^{(s_2)}_{[\gamma]}$, where $s_2 \leqslant s_1 - \deg P$. Applying (18) we conclude that $u \equiv 0$, i.e. (17) holds.

2) In view of the duality relation

$$(H^{(s)}_{[\gamma]})' = H^{(-s)}_{[-\gamma]},$$

it follows from the inequality

$$\|v\|_{(-s),-\gamma} \leqslant c^*_s \|P^* v\|_{(-s),-\gamma} \quad \forall v \in H^{(\infty)}_{[-\gamma]} \tag{18'}$$

that for any given right-hand side $f \in H^{(s)}_{[\gamma]}$ there exists a solution $u \in H^{(s)}_{[\gamma]}$ to Equation (2). Indeed (cf. Hörmander [3, §8.7]), consider the linear functional $L(\psi) = (f, \varphi)$ on the linear manifold $\{\psi, \psi = P^*\varphi, \varphi \in H^{(\infty)}_{[-\gamma]}\}$. By virtue of (18′), we have

$$|L(\psi)| \leqslant \|f\|_{(s),\gamma} \|\varphi\|_{(-s),-\gamma} \leqslant c_s \|f\|_{(s),\gamma} \|P^*\varphi\|_{(-s),-\gamma}.$$

Since $P^*\varphi = \psi$, the norm of the functional does not exceed $c_s\|f\|_{(s),\gamma}$. Therefore, by the Hahn-Banach theorem, there exists an element $u \in H^{(s)}_{[\gamma]}$, $\|u\|_{(s),\gamma} \leqslant c_s \|f\|_{(s),\gamma}$, such that

$$(u, \psi) = L(\psi), \qquad \text{that is} \qquad (u, P^*\varphi) = (f, \varphi).$$

Thus, the proof of the theorem has reduced to the proof of the fact that u belongs to $H^{(s)}_{[\gamma]+}$ for $f \in H^{(s)}_{[\gamma]+}$. It is this property for whose proof the stronger estimate (3′) (as compared to (18′)) is applied. For this proof we need some estimates in the more general spaces $H^{(s)}_{[\gamma',\gamma'']}$ [1)] whose special cases are the spaces $H^{(s)}_{[\gamma]}$ and $H^{(s)}_{[\gamma]+}$ and their conjugate spaces.

3) We define the norm in $H^{(s)}_{[\gamma]}$ by means of the PDO with symbol

$$\lambda^+_s(\xi, \tau) = (i\tau + \gamma_0 + \sqrt{1 + |\xi|^2})^s, \tag{19}$$

where γ_0 is selected so that symbol (19) does not vanish for the values of $\operatorname{Im}\tau$ in question. So, we set

$$\|u\|_{(s),\gamma} = \|\exp(\gamma t)\lambda^+_s(D_x, D_t)u\|. \tag{20}$$

Let $H^{(s)}_{[\gamma',\gamma'']}$ with $\gamma' \leqslant \gamma''$ denote the set of those u belonging $H^{(s)}_{[\gamma]}$ with $\gamma' \leqslant \gamma \leqslant \gamma''$ for which the norm

$$\|u\|_{(s),\gamma',\gamma''} = \sup_{\gamma' \leqslant \gamma \leqslant \gamma''} \|u\|_{(s),\gamma} \tag{21}$$

[1)] For a detailed presentation of the theory of these spaces see Volevich and Gindikin [1, 7].

is finite. If γ' and γ'' are finite numbers, then $H^{(s)}_{[\gamma',\gamma'']}$ coincides with the intersection $H^{(s)}_{[\gamma']}\cap H^{(s)}_{[\gamma'']}$ and the norm (21) is equivalent to the natural norm in the intersection of spaces:

$$\|u\|_{(s),\gamma'}+\|u\|_{(s),\gamma''}. \tag{21'}$$

In case $\gamma'=-\infty$, we obtain the subspace $H^{(s)}_{[\gamma_1,\gamma_2]}$.

4) We now show that the proof of the existence of a solution $u\in H^{(s)}_{[\gamma]+}$ reduces to the proof of the existence of a solution belonging to $H^{(s)}_{[\gamma_1',\gamma_2]}$ for arbitrary $\gamma_1<\gamma_2$, where $-\gamma_2$ is sufficiently large.

Indeed, let $f\in H^{(s)}_{[\gamma]+}$. Then $f\in H^{(s)}_{[\rho]}$ for all $\rho<\gamma$. If inequalities (3) and (3′) are fulfilled, then estimates (18) and (18′) are sure to hold, and consequently for every $\rho\leqslant\gamma$ there exists a unique solution $u_\rho\in H^{(s)}_{[\rho]}$ to Equation (2), and we have

$$\|u_\rho\|_{(s),\rho}\leqslant c_s\|f\|_{(s),\rho}\leqslant c_s\|f\|_{(s),\gamma}, \tag{22}$$

where the constant c_s does not depend on ρ. If we prove for at least sufficiently large ρ that the function u_ρ does not in fact depend on ρ, i.e. $u_\rho=u$, $\rho<\rho_0$, then, taking the supremum of the left-hand side of (22) over all $\rho\leqslant\rho_0$, we obtain (for $\rho_0<\gamma$) the inequality

$$\|u\|_{(s),-\infty,\rho_0}\leqslant c_s\|f\|_{(s),\gamma}.$$

It follows (see Volevich and Gindikin [1, §2.5]) that $u=0$ for $t<0$.

Let $\gamma_1<\gamma_2\leqslant\rho_0$ and let $-\rho_0$ be sufficiently large. Then, by the hypothesis, for any right-hand side $f\in H^{(s)}_{[\gamma_1,\gamma_2]}\subset H^{(s)}_{[\gamma]+}$ there exists a unique solution $u_{\gamma_1\gamma_2}\in H^{(s)}_{[\gamma_1,\gamma_2]}$ to Equation (2). However, the uniqueness in the spaces $H^{(s)}_{[\rho]}$ implies that $u_{\gamma_1}=u_{\gamma_1\gamma_2}=u_{\gamma_2}$.

5) We now deduce the existence of a solution in $H^{(s)}_{[\gamma_1,\gamma_2]}$ with $\gamma_1<\gamma_2<\rho_0$ from an a priori estimate in the conjugate spaces $(H^{(s)}_{[\gamma_1,\gamma_2]})'$. We state the necessary definitions. We set

$$H^{(-s)}_{[-\gamma_1,-\gamma_2]}=(H^{(s)}_{[\gamma_1,\gamma_2]})' \tag{23}$$

and endow the left-hand space with the natural norm of the Banach conjugate space of $H^{(s)}_{[\gamma_1,\gamma_2]}$. It is proved that this space coincides with the linear hull of $H^{(-s)}_{[-\gamma_1]}+H^{(-s)}_{[-\gamma_2]}$, and as the norm in the space (23) the expression

$$\|v\|_{(-s),-\gamma_1,-\gamma_2}=\|\chi v\|_{(-s),-\gamma_2}+\|(1-\chi)v\|_{(-s),-\gamma_1} \tag{24}$$

can be taken, where $\chi(t)\in C^\infty(\mathbb{R}^1)$, $\chi(t)=1$ for $t\geqslant 0$, $\chi(t)=0$ for $t\leqslant -1$. The central point of the proof of Theorem A is the following

Lemma. *Let $\gamma_2 > \gamma_1$, let $\gamma_2 - \gamma_1 \leqslant \delta$, and let the conditions of the theorem hold. Then $\forall M > 0$ and $\forall \delta > 0$ there is $\gamma_0(M,\delta)$ such that for $|s| \leqslant M$ and $\gamma_1 < \gamma_2 < \gamma_0(M,\delta)$ the inequalities*

$$\|u\|_{(s),\gamma_2,\gamma_1} \leqslant K_s \|Pu\|_{(s),\gamma_2,\gamma_1}, \tag{25}$$

$$\|v\|_{(-s),-\gamma_1,-\gamma_2} \leqslant K_s^* \|Pu\|_{(-s),-\gamma_1,-\gamma_2} \tag{25'}$$

hold.

In view of the duality, inequality (25′) implies the existence of a solution belonging to $H^{(s)}_{[\gamma_1,\gamma_2]}$. Therefore the assertion of the theorem follows from what has been said in 4).

6) *The proof of the lemma.* To simplify the notation, we confine ourselves to the proof of (25) (inequality (25′) is proved in a similar way). By definition, for $\gamma_2 > \gamma_1$ we have

$$\|Pu\|_{(s),\gamma_2,\gamma_1} = \|\chi Pu\|_{(s),\gamma_1} + \|(1-\chi)Pu\|_{(s),\gamma_2}.$$

By Leibniz's formula,

$$\chi Pu = P(\chi u) - \sum_{l=1}^{m} \chi_l P^{(l)} u, \qquad \chi_l = D_t^l \chi/\alpha!$$

On writing down an analogous relation for $(1-\chi)Pu$ we obtain

$$\begin{aligned}
\|Pu\|_{(s),\gamma_1,\gamma_1} &\geqslant \|P(\chi u)\|_{(s),\gamma_1} + \|P(1-\chi)u\|_{(s),\gamma_2} \\
&\quad - \sum_{l=1}^{m} (\|\chi_l P^{(l)} u\|_{(s),\gamma_1} + \|\chi_l P^{(l)} u\|_{(s),\gamma_2}) \\
&\geqslant \sum_{l=1}^{m} \Big[(\gamma_0-\gamma_1)^l \|P^{(l)}(\chi u)\|_{(s),\gamma_1} + (\gamma_0-\gamma_2)^l \|P^{(l)}(1-\chi)u\|_{(s),\gamma_2} \\
&\quad - \|\chi_l P^{(l)} u\|_{(s),\gamma_1} - \|\chi_l P^{(l)} u\|_{(s),\gamma_2}\Big].
\end{aligned} \tag{26}$$

We have

$$\|\chi_l P^{(l)} u\|_{(s),\gamma_1} \leqslant \|\chi_l P^{(l)}(\chi u)\|_{(s),\gamma_1} + \|\chi_l P^{(l)}(1-\chi)u\|_{(s),\gamma_1}. \tag{27}$$

Using the fact that $\chi_l(t)$, $l \geqslant 1$, is a function of compact support we show that

$$\|\chi_l P^{(l)}(1-\chi)u\|_{(s),\gamma_1} \leqslant c(s,\gamma_2-\gamma_1)\|P^{(l)}(1-\chi)u\|_{(s),\gamma_2}. \tag{28}$$

Comparing (27) and (28) we find

$$\|\chi_l P^{(l)} u\|_{(s),\gamma_1} \leqslant c'(s,\gamma_2-\gamma_1)\big[\|P^{(l)}(\chi u)\|_{(s),\gamma_1} + \|P^{(l)}(1-\chi)u\|_{(s),\gamma_2}\big]. \tag{29}$$

It is shown in a similar way that $\|\chi_l P^{(l)} u\|_{(s),\gamma_2}$ is also estimated by means of the right-hand side of (29). Substituting these inequalities into (27) and taking a sufficiently large $-\gamma_2 < -\gamma_1$ we estimate the right-hand side of (26) from below by means of

$$\frac{1}{2}\sum_{l=1}^{m}\left[(\gamma_0-\gamma_1)^l\|P^{(l)}(\chi u)\|_{(s),\gamma_1}+(\gamma_0-\gamma_2)^l\|P^{(l)}(1-\chi)u\|_{(s),\gamma_2}\right]$$
$$\geqslant \frac{m!}{2}(\gamma_0-\gamma_2)^m\left[\|\chi u\|_{(s),\gamma_1}+\|(1-\chi)u\|_{(s),\gamma_2}\right]=\frac{m!}{2}(\gamma_0-\gamma_2)^m\|u\|_{(s),\gamma_2,\gamma_1},$$

i.e. we arrive at inequality (25).

7) Thus, to complete the proof of the theorem we have to establish (28). We write

$$\begin{aligned}\|\chi_l P^{(l)}(1-\chi)u\|_{(s),\gamma_1} &\overset{\text{def}}{=} \|\exp(\gamma_1 t)\lambda_s(D_x,D_t)\chi_l P^{(l)}(1-\chi)u\|\\ &= \|\lambda_s(D_x,D_t+i\gamma_1)\exp(\gamma_1 t)\chi_l P^{(l)}(1-\chi)u\|\\ &= \left\|\frac{\lambda_s(D_x,D_t+i\gamma_1)}{\lambda_s(D_x,D_t+i\gamma_2()}\lambda_s(D_x,D_t+i\gamma_2)[\exp(\gamma_1 t-\gamma_2 t)\chi_l]\right.\\ &\qquad \left.\times\exp(\gamma_2 t)P^{(l)}(1-\chi)u\right\|. \end{aligned} \tag{30}$$

By virtue of the elementary inequality

$$|\lambda_s(\xi,\sigma+i\gamma_1)\lambda_s^{-1}(\xi,\sigma+i\gamma_2)|\leqslant(1+|\gamma_2-\gamma_1|)^s,$$

the expression $\lambda_s(D_x,D_t+i\gamma_1)\lambda_{-s}(D_x,D_t+i\gamma_2)$ is a bounded operator in L_2, and therefore the right-hand side of (30) is estimated from above by means of

$$c(\gamma_2-\gamma_1)\left\|\lambda_s(D_x,D_t+i\gamma_2)[\exp(\gamma_2 t-\gamma_1 t)\chi_l]\exp(\gamma_2 t)P^{(l)}(1-\chi)u\right\|. \tag{31}$$

The expression in square brackets is a function of compact support, and its derivatives are estimated by means of constants depending on the difference $\gamma_2-\gamma_1$ solely. It can be shown that (31) is estimated by means of

$$c'(\gamma_2-\gamma_1)\|\lambda_s(D_x,D_t+i\gamma_2)\exp(\gamma_2 t)P^{(l)}(1-\chi)u\| \overset{\text{def}}{=} c'(\gamma_2-\gamma_1)\|P^{(l)}(1-\chi)u\|_{(s),\gamma_2}$$

(this inequality is proved particularly simply in the case of integral values of $s \geqslant 0$).

§2. Sufficient conditions for the existence of energy estimates

In this section we present a set of rather cumbersome conditions making it possible to estimate from above and below the forms

$$-\operatorname{Im}[Q(y;D)u,Q^{(1)}(y;D)u]_{(s),\gamma},\quad Q=P,P^*, \tag{1}$$

by means of analogous forms corresponding to an operator Q with constant coefficients (frozen at a point $y=y^0$). From the estimate for forms (1) we derive inequalities (1.3) and (1.3′) and thus establish the solvability of Cauchy's problem. As consequences of the conditions in the present section, we shall obtain in §3 some easily verifiable conditions guaranteeing the solvability of Cauchy's problem. The results in this section are taken from the paper by Volevich [1].

2.1. Formulation of the main results. We shall deal with differential operators $P(y;D)$, $y=(t,x)\in\mathbb{R}^{n+1}$, solved with respect to the highest derivative with respect to t, i.e. the symbol $P(y;\xi,\tau)$ has the form

$$P(y;\xi,\tau)=\tau^m+\sum_{j>0}P_j(y;\xi)\tau^{m-j}. \tag{2}$$

As in Chapters 2 and 3, we shall assume, without a special stipulation, that the coefficients of the polynomial symbol (2) belong to C^∞ and do not depend on y for large $|y|$. It will also be assumed that conditions (I) and (II) in the foregoing section are fulfilled.

When estimating the forms (1) we shall use the norm

$$\{u\}_{(s),\gamma}=\Big(\iint H_P(y^0;\xi,\sigma,\gamma)(1+|\xi|^2+\sigma^2+\gamma^2)^s|\widehat{u}(\xi,\sigma+i\gamma)|^2\,d\xi\,d\sigma\Big)^{1/2}, \tag{3}$$

where y^0 is a fixed point. By the condition of constant strength for symbol (2), the replacement of y_0 by any other point results in an equivalent norm.

If we set, as in §1,

$$\lambda_s^{\pm}=(\pm iD_t+\gamma_0+\sqrt{1+|D_x|^2})^s, \tag{4}$$

$$^{\pm}[u,v]_{(s),\gamma}=[\exp(\gamma t)\lambda^{\pm}(D)u,\exp(\gamma t)\lambda_s^{\pm}(D)v], \tag{5}$$

then, as was noted in §1, the norm (3) is equivalent to

$$-\operatorname{Im}[P(y^0;D)u,P^{(1)}(y^0;D)u]_{(s),\gamma},\qquad \gamma\leqslant\gamma_0. \tag{3'}$$

We introduce the following convenient notation. If $Q(y;\xi,\tau)$ and $R(y;\xi,\tau)$ are two polynomial functions in τ, we set

$$\{Q,R\}(y;\xi,\sigma,\gamma)=\frac{1}{2i}\left[\frac{\partial R}{\partial\tau}(y;\xi,\tau)\overline{Q(y;\xi,\tau)}-R(y;\xi,\tau)\overline{\frac{\partial Q(y;\xi,\tau)}{\partial\tau}}\right],\qquad \tau=\sigma+i\gamma.$$

This definition implies that

$$\{P,P\}(y;\xi,\sigma,\gamma)=H_P(y;\xi,\sigma,\gamma),\qquad \{Q,R\}=\overline{\{R,Q\}}. \tag{6}$$

Proposition 1. *Let symbol* (2) *satisfy conditions* (I) *and* (II) *in §1 and, moreover, let there be* γ_0 *such that* $\forall y\in\mathbb{R}^n$ *and* $\forall\gamma<\gamma_0$ *we have*

$$|H_P^{(\beta)}(y;\xi,\sigma,\gamma)|<\varepsilon_\beta(\gamma)H_P(y;\xi,\sigma,\gamma),\qquad \beta>0, \tag{7}$$

$$|\{P,P^{(\beta)}\}_{(\beta)}(y;\xi,\sigma,\gamma)|<\varepsilon_\beta(\gamma)H_P(y;\xi,\sigma,\gamma),\qquad \beta>0, \tag{8}$$

$$|\{P_{(\alpha)},P^{(\beta)}\}_{(\beta)}(y;\xi,\sigma,\gamma)|<\varepsilon_{\alpha\beta}(\gamma)H_P(y;\xi,\sigma,\gamma)(1+|\xi|)^{|\alpha|},\quad |\beta|\geqslant 0,\ |\alpha|>0, \tag{9}$$

where $\varepsilon_\beta(\gamma)$, $\varepsilon_{\alpha\beta}(\gamma)\to 0$ *as* $\gamma\to-\infty$.

Then $\forall M>0\ \exists\gamma_0(M)$ *such that* $\forall s\in\mathbb{R}$, $|s|\leqslant M$, *and* $\gamma\leqslant\gamma_0(M)$ *the two-sided estimate*

$$c_s^{-1}\{u\}^2_{(s),\gamma}\leqslant-\operatorname{Im}{}^{+}[P(y;D)u,P^{(1)}(y;D)u]_{(s),\gamma}\leqslant c_s\{u\}^2_{(s),\gamma}\quad\forall u\in H^{(\infty)}_{[\gamma]} \tag{10}$$

holds.

Proposition 2. *Let conditions* (I) *and* (II) *in §1 be fulfilled, let condition* (7) *hold, and, besides, let*

$$|\{P, P_{(\beta)}\}^{(\beta)}(y;\xi,\sigma,\gamma)| < \varepsilon_\beta(\gamma) H_P(y;\xi,\sigma,\gamma), \quad \beta > 0, \tag{11}$$

$$|\{P, P_{(\alpha+\beta)}\}^{(\beta)}(y;\xi,\sigma,\gamma)| < \varepsilon_{\alpha\beta}(\gamma) H_P(y;\xi,\sigma,\gamma)(1+|\xi|)^{|\alpha|}, \quad \beta \geqslant 0,\ \alpha > 0. \tag{12}$$

Then $\forall M > 0\ \exists \gamma_0^*(M)$ *such that* $\forall s \in \mathbb{R}$, $|s| \leqslant M$, *and* $\gamma \leqslant \gamma_0^*(M)$ *the two-sided estimate*

$$c_s^{*-1}\{v\}^2_{(-s),-\gamma} \leqslant \operatorname{Im}{}^-[P^*(y;D)v, P^{*(1)}(y;D)v]_{(-s),-\gamma} \leqslant c_s^*\{v\}_{(-s),-\gamma}, \quad \forall v \in H^{(\infty)}_{[-\gamma]} \tag{13}$$

holds.

Remark. Formally, (9) and (12) are infinite sets of conditions. However, if $|\alpha| > \varkappa + \deg H_P + 1$, where $\varkappa$ is the constant in Corollary 1.2, then inequalities (9) and (12) are fulfilled automatically.

2.2.

Theorem. *Let symbol* (2) *satisfy all conditions of Propositions* 1 *and* 2 *in the foregoing section. Then inequalities* (1.3) *and* (1.3′) *take place.*

Proof. If conditions (I) and (II) hold, then, by virtue of Proposition 1.2 (II), $\forall y' \in \mathbb{R}^{n+1}$ we can write the inequality

$$c_s' \sum_{l=1}^{m} (\gamma_0 - \gamma)^{2l-1} \|P^{(l)}(y';D)u\|^2_{(s),\gamma} \leqslant \{u\}^2_{(s),\gamma}, \tag{14}$$

where the constant γ_0 can be selected so that inequalities (14) are fulfilled for any $y' \in \mathbb{R}^{n+1}$ and $\gamma \leqslant \gamma_0$. These inequalities readily imply an analogous inequality for the operator with variable coefficients:

$$c_s'' \sum_{l=1}^{m} (\gamma_0 - \gamma)^{2l-1} \|P^{(l)}(y;D)u\|^2_{(s),\gamma} \leqslant \{u\}^2_{(s),\gamma}. \tag{15}$$

Indeed, since the degrees of all polynomials $P(y;\xi,\tau)$ are uniformly bounded with respect to $y \in \mathbb{R}^{n+1}$, they form a finite-dimensional space, and among them there are a finite number of linearly independent elements. Therefore we can write

$$P(y;\xi,\tau) = \sum_{j=1}^{J} c_j(y) P(y^j;\xi,\tau), \tag{16}$$

where the functions $c_j(y)$ possess the same smoothness properties as the coefficients of the original polynomial (2). Differentiating (16) with respect to τ we obtain

$$P^{(l)}(y;\xi,\tau)=\sum_{j=1}^{J}c_j(y)P^{(l)}(y^j;\xi,\tau). \tag{16'}$$

Applying inequality (14) with $y'=y^j$ to each of the operators on the right-hand side of (16′) we obtain (15).

Comparing (15) and (10) we find (cf. Section 1.3)

$$c_s''\sum_{l=1}^{m}(\gamma_0-\gamma)^{2l-1}\|P^{(l)}u\|^2_{(s),\gamma}\leqslant -\operatorname{Im}{}^{+}[Pu,P^{(1)}u]_{(s),\gamma}\leqslant \|Pu\|_{(s),\gamma}\|P^{(1)}u\|_{(s),\gamma}$$
$$\leqslant(\gamma_0-\gamma)^{-1/2}\|Pu\|_{(s),\gamma}\Big(\sum_{l=1}^{m}(\gamma_0-\gamma)^{2l-1}\|P^{(l)}u\|^2_{(s),\gamma}\Big)^{1/2},$$

whence follows inequality (1.3).

Similarly, to prove (1.3′) it suffices to show that

$$c_s'^{*}\sum_{l-1}^{m}(\gamma_0-\gamma)^{2l-1}\|P^{*(l)}v\|^2_{(-s),-\gamma}\leqslant\{v\}^2_{(-s),-\gamma}. \tag{15'}$$

According to (16), we have

$$P^*(y;\xi,\tau)=\sum\overline{P}^{(\beta)}_{(\beta)}(y;\xi,\tau)/\beta!=\sum\frac{1}{\beta!}D^\beta c_j(y)\overline{P}^{(\beta)}(y^j;\xi,\tau).$$

Hence, to prove (15′) it suffices to establish the inequalities

$$\text{const}\sum_{l=1}^{m}(\gamma_0-\gamma)^{2l-1}\sum_{\beta\geqslant 0}\|\overline{P}^{(l)(\beta)}(y^j;D)v\|_{(-s),-\gamma}\leqslant\{v\}^2_{(-s),-\gamma},$$

which are equivalent to the set of inequalities

$$\text{const}(\gamma_0-\gamma)^{2l-1}|\partial^\beta\overline{P}^{(l)}(y';\xi,\sigma+i\gamma)|\leqslant H_P(y';\xi,\sigma,\gamma),\qquad \gamma\leqslant\gamma_0. \tag{17}$$

We put $\eta=(\xi,\sigma)$. For any polynomial $Q(\eta)$ we have the inequality

$$c|Q^{(\beta)}(\eta)|\leqslant\sup_{|\theta|\leqslant 1}|Q(\eta+\theta)|, \tag{18}$$

where the constant c depends only on the degree of Q and the dimension of the space. In view of inequality (18) (applied to the polynomial $Q(\eta)=P^{(l)}(y';\xi,\sigma+i\gamma)$), we have

$$c_1(\gamma_0-\gamma)^{2l-1}|P^{(l)(\beta)}(y';\xi,\sigma+i\gamma)|^2\leqslant\sup_{|\theta|\leqslant 1}H_P(y';\eta+\theta,\gamma).$$

We now show that if $-\gamma$ is sufficiently large, then

$$H_P(\eta+\theta,\gamma) \leqslant 2H_P(\eta,\gamma), \qquad |\theta| \leqslant 1,\ \gamma \leqslant \gamma_1.$$

Indeed, expanding the left-hand side by Taylor's formula we obtain

$$H_P(\eta+\theta,\gamma) = H_P(\eta,\gamma) + \sum_{\beta>0} \frac{\theta^\beta}{\beta!} H_P^{(\beta)}(\eta,\gamma).$$

It now remains to apply condition (7).

2.3. The plan of the proof of the propositions in Section 2.1. We set $w = \exp(\gamma t)\lambda_s^+(D)u$. If u runs over $H^{(\infty)}_{[\gamma]+}$, then the function w runs over $H_+^{(\infty)}$. Setting $D_\gamma = (D_x, D_t + i\gamma)$ we can rewrite the middle term in (10) in the form

$$\begin{aligned} &-\operatorname{Im}[\lambda_s^+(D_\gamma)P(y;D_\gamma)\lambda_{-s}^+(D_\gamma)w, \lambda_s^+(D_\gamma)P^{(1)}(y;D_\gamma)\lambda_{-s}^+(D_\gamma)w] \\ &\quad = \Big(\frac{i}{2}\lambda_{-s}^-(D_{-\gamma})\{P^{(1)*}(y;D_{-\gamma})\lambda_s^+(D_\gamma)\lambda_s^-(D_{-\gamma})P(y;D_\gamma) \\ &\qquad - P^*(y;D_{-\gamma})\lambda_s^+(D_\gamma)\lambda_s^-(D_{-\gamma})P^{(1)}(y;D_\gamma)\}\lambda_{-s}^+(D_\gamma)w, \overline{w}\Big). \end{aligned} \tag{19}$$

Here we used the fact that

$$\overline{(\lambda_s^+(D_\gamma))^*} = \lambda_s^-(D_{-\gamma}), \qquad (Q(y;D_\gamma))^* = Q^*(y;D_{-\gamma}).$$

If the coefficients of P were constant (cf. Section 1.3), we would obtain the quadratic form $(H_P(D,\gamma)w,\overline{w})$. In the case of variable coefficients we separate out the Hermitian form corresponding to the differential operator $H_P(y;D,\gamma)$, i.e. rewrite (19) in the form

$$\operatorname{Re}(H_P(y;D,\gamma)w,\overline{w}) + (Q_{s\gamma}w,\overline{w}), \tag{20}$$

where

$$\begin{aligned} Q_{s\gamma} = \frac{i}{2}\lambda_{-s}^-(D_{-\gamma})\Big\{&P^{(1)*}(y;D_{-\gamma})\lambda_s^+(D_\gamma)\lambda_s^-(D_{-\gamma})P(y;D_\gamma) \\ &- P^*(y;D_{-\gamma})\lambda_s^+(D_\gamma)\lambda_s(D_{-\gamma})P^{(1)}(y;D_\gamma)\Big\}\lambda_{-s}^+(D_\gamma) \\ -\frac{1}{2}H_P(y;D,\gamma) - \frac{1}{2}H_P^*(y;D,\gamma).& \end{aligned} \tag{21}$$

We shall prove the following inequalities:

$$c_1^{-1}(H_P(y^0;D,\gamma)w,\overline{w}) \leqslant \operatorname{Re}(H_P(y;D,\gamma)w,\overline{w}) \leqslant c_1(H_P(y^0,D)w,\overline{w}) \tag{22}$$

$$|(Q_{s\gamma}w,\overline{w})| \leqslant \varepsilon(\gamma)(H_P(y^0;D,\gamma)w,\overline{w}), \quad \varepsilon(\gamma)\to 0, \quad \gamma\to-\infty. \tag{23}$$

Noting that

$$(H_P(y^0;D,\gamma)w,\overline{w}) = {}^+[H_P(y^0;D)u,u]_{(s),\gamma} = \{u\}^2_{(s),\gamma},$$

we obtain the proof of Proposition 1 in Section 2.1.

We note that (22) is an analog of Gårding's inequality for inhomogeneous quadratic forms.

As to Proposition 2 in Section 2.1, after the substitution of

$$z = \exp(-\gamma t)\lambda_s^-(D)v$$

the middle term in (13) can be rewritten as

$$\operatorname{Re}(H_P(y;D,\gamma)z,\overline{z}) + (R_{s\gamma}z,\overline{z}), \tag{24}$$

where

$$\begin{aligned} R_{s\gamma} = \frac{1}{2i}\lambda^+_{-s}(D_\gamma)\Big\{&P^{(1)}(y;D_\gamma)\lambda^+_{-s}(D_\gamma)\lambda^-_s(D_{-\gamma})P^*(y;D_{-\gamma}) \\ &- P(y;D_\gamma)\lambda^+_{-s}(D_\gamma)\lambda^-_{-s}(D_{-\gamma})P^{(1)*}(y;D_{-\gamma})\Big\}\lambda^-_s(D_{-\gamma}) \\ &- \frac{1}{2}H_P(y;D,\gamma) - \frac{1}{2}H_P^*(y;D,\gamma). \end{aligned} \tag{25}$$

Proposition 2 in Section 2.1 follows from (22) and the inequality

$$|(R_{s\gamma}z,\overline{z})| \leqslant \varepsilon^*(\gamma)(H_P(y^0;D,\gamma)z,\overline{z}), \quad \varepsilon^*(\gamma)\to 0,\ \gamma\to -\infty. \tag{26}$$

2.4. The proof of Inequality (23). If $Q_1(y;D)$ and $Q_2(y;D)$ are two PDO's then for any natural N we set

$$R_N(Q_1,Q_2) = Q_1(y;D)\cdot Q_2(y;D) - \sum_{|\alpha|\leqslant N-1}\frac{1}{\alpha!}(Q_1^{(\alpha)}Q_{2(\alpha)})(y;D).$$

This relation is usually called the commutation formula. Setting

$$\lambda_s^-(D_{-\gamma})\lambda_s^+(D_\gamma) = (|D_t|^2 + (\gamma_0 - \gamma + \sqrt{1+|D_x|^2})^2)^s \stackrel{\text{def}}{=} \Delta_{2s}(D,\gamma) \tag{27}$$

we write

$$\begin{aligned} P^{(1)*}\Delta_{2s}P - P^*\Delta_{2s}P^{(1)} &= \sum\frac{1}{\alpha!}\Big(P^{(1)*}P_{(\alpha)} - P^*P^{(1)}_{(\alpha)}\Big)\Delta_{2s}^{(\alpha)} \\ &\quad + P^{(1)*}R_N(\Delta_{2s},P) - P^*R_N(\Delta_{2s},P^{(1)}). \end{aligned} \tag{28}$$

The expression under the summation sign in (28) is a PDO. To calculate its symbol we note that if Q_1 and Q_2 are two differential operators, then, in view of Leibnitz's formula, the symbol of $Q_1^* Q_2$ is equal to

$$\sum \frac{1}{\alpha!\,\delta!} \overline{Q}_{1(\delta)}^{(\alpha+\delta)} Q_{2(\alpha)} = \sum \frac{1}{\beta!} (\overline{Q_1}^{(\beta)} Q_2)_{(\beta)}.$$

Therefore the symbol of the PDO under the summation sign in (28) is equal to

$$\sum \frac{1}{\alpha!\,\beta!} \left(\overline{P^{(1)(\beta)}(y;\xi,\tau)} P_{(\alpha)}(y;\xi,\tau) - \overline{P^{(\beta)}(y;\xi,\tau)} P^{(1)}_{(\alpha)}(y;\xi,\tau) \right)_{(\beta)} \Delta^{(\alpha)}(\eta,\gamma)$$
$$= -2i \sum \frac{1}{\alpha!\,\beta!} \{P_{(\alpha)}, P^{(\beta)}\}_{(\beta)}(y;\xi,\sigma,\gamma) \Delta_{2s}^{(\alpha)}(\eta,\gamma).$$

Applying the commutation formula once again we write operator (21) in the form

$$Q_{s\gamma} = Q_{s\gamma N} + T_{s\gamma N}, \tag{29}$$

where $Q_{s\gamma N}$ is a PDO with symbol

$$\sum_{\alpha,\beta,\theta} \frac{1}{\alpha!\,\beta!\,\theta!} \{P_{(\alpha)}, P^{(\beta)}\}_{(\beta+\theta)} \Delta_{2s}^{(\alpha)} \lambda_{-s}^{-(\theta)} \lambda_{-s}^{+} - H_P(y;\eta,\gamma) - \frac{1}{2} \sum_{\beta>0} H_{P(\beta)}^{(\beta)}(y;\eta,\gamma). \tag{30}$$

Noting that $\{P, P\} = H_P$ we rewrite the symbol in the form

$$\sum_{|\alpha+\beta+\theta|>0} \frac{1}{\alpha!\,\beta!\,\theta!} \{P_{(\alpha)}, P^{(\beta)}\}_{(\beta+\theta)} \Delta_{2s}^{(\alpha)} \lambda_{-s}^{-(\theta)} \lambda_{-s}^{+} - \frac{1}{2} \sum_{\beta>0} H_{P(\beta)}^{(\beta)}. \tag{30'}$$

The operator $T_{s\gamma N}$ is written

$$\frac{i}{2} \lambda_{-s}^{-} (P^{(1)} R_N(\Delta_{2s}, P) - P^* R_N(\Delta_{2s}, P^{(1)})_{-s}^{+})$$
$$+ \sum_{|\alpha+\beta+\theta|>0} \frac{1}{\alpha!\,\beta!} R_N(\lambda_{-s}^{-}, \{P_{(\alpha)}, P^{(\beta)}\}_{(\beta+\theta)}) \Delta_{2s}^{(\alpha)} \lambda_{-s}^{-(\theta)} \lambda_{-s}^{+}. \tag{31}$$

Lemma 1. *The symbol of the operator $Q_{s\gamma N}$ is represented as*

$$Q_{s\gamma N}(y;\eta) = \sum_{j=1} c_j(y) b_j(\eta,\gamma), \tag{32}$$

where

$$|b_j(\eta,\gamma)| < \varepsilon_j(\gamma) H_P(y^0;\gamma,\eta), \quad \varepsilon_j(\gamma) \to 0, \quad \gamma \to -\infty. \tag{33}$$

Proof. We now show that, by virtue of Proposition 1 in Section 1.2, we have

$$|Q_{s\gamma N}(y;\eta)| < \varepsilon(\gamma)H_P(y^0;\eta,\gamma), \quad \varepsilon(\gamma)\to 0, \quad \gamma\to-\infty. \tag{34}$$

Writing the symbol $P(y;\xi,\tau)$ in the form of (16) we arrive at representation (32).

The inequalities

$$|H^{(\beta)}_{P(\beta)}(y;\eta,\gamma)| < \varepsilon_\beta(\gamma)H_P(y^0;\eta,\gamma)$$

follow from (7) and the condition of constant strength. We have to estimate the first sum in (30′). As can easily be seen, for $\gamma < \gamma_0$ we have

$$|\Delta^{(\alpha)}_{2s}(\eta,\gamma)\lambda^{-(\theta)}_{-s}(\eta,\gamma)\lambda^{+}_{-s}(\eta,\gamma)| \leqslant K_{\alpha\theta}(1+|\gamma-\gamma_0|+|\eta|)^{-1}$$
$$(1+|\xi|)^{-|\alpha+\theta|+1} \leqslant K'_{\alpha\theta}(1+|\xi|)^{-|\alpha+\theta|}.$$

With account of (8) and (9), we arrive at (34).

To estimate the operator $Q_{s\gamma N}$ we need

Lemma 2. *Let $\mu(\eta)$ be a positive function and let*

$$\mu(\eta')\mu^{-1}(\eta'') \leqslant K(1+|\eta'-\eta''|)^l.$$

Let $a(y) = a + a'(y)$, let $a'(y)\in\mathcal{D}$, and let

$$|b(\eta)| \leqslant \delta\mu^2(\eta).$$

Then

$$|(a(y)b(D)v,\bar v)| \leqslant \delta K(a)\|\mu(D)v\|^2.$$

Proof. By Schwarz' inequality, we have

$$|(a(y)b(D)v,\bar v)| \leqslant \|\mu(D)v\|\|\mu^{-1}(D)a(y)b(D)v\|.$$

Using the inequality (see Volevich and Paneyakh [1, (1.10)])

$$\|\mu^{-1}(D)a(y)w\| \leqslant K\Big(|a| + \int|\tilde a(\eta)|(1+|\eta|)^l\,d\eta\Big)\|\mu^{-1}(D)w\|$$

we prove the assertion of the lemma.

To apply Lemma 2 to estimate the operator $Q_{s\gamma N}$ we put

$$h(\eta) = H_P^{1/2}(y^0;\eta,\gamma).$$

According to condition (7), we have

$$\frac{h(\eta')}{h(\eta'')} = \Big[1+\sum_{\beta>0}\frac{(\eta'-\eta'')^\beta}{\beta!}\frac{H_P^{(\beta)}(y^0;\eta'',\gamma)}{H_P(y^0;\eta'',\gamma)}\Big]^{1/2} \leqslant \text{const}(1+|\eta''-\eta'|)^l,$$

where $2l \geqslant \deg H_P$. Applying Lemma 2 to the forms $(c_j(y)b_j(D,y)w,\overline{w})$ we prove the inequality

$$|(Q_{s\gamma N}(y;D)w,\overline{w})| \leqslant K(H_P(y^0;D,\gamma)w,\overline{w}). \tag{23'}$$

We now proceed to the estimation of the form corresponding to the operator $T_{s\gamma N}$ in (29).

Lemma 3. *The operator $T_{s\gamma N}$ (see (3.1)) is represented in the form*

$$T_{s\gamma N} = \sum P_{jk}\gamma^j D_t^k, \qquad j + k \leqslant 2m - 2, \tag{35}$$

where P_{jk} are operators of the $(2\mu - N)$th order with respect to the variables x, i.e.

$$\|D_x^\beta(P_{jk}w)\| \leqslant \text{const}\,\|w\|, \qquad |\beta| \leqslant 2N - 2\mu. \tag{35'}$$

Here μ is the maximum degree of the operators P_j in (2), and the constant in (35′) does not depend on γ.

Proof. By the linearity, we have

$$R_N(\Delta_{2s}, P) = \sum_{j>0} R_N(\Delta_{2s}, a_{\alpha j}(y))D_x^\alpha(D_t + i\gamma)^{m-j}$$

(we have used the fact that the coefficient in the operator D_t^m is identically equal to 1). The expression $R_N(\Delta_{2s}, a_{\alpha j})$ is an operator of the $(2s - N)$th order with respect to x. It follows that the order of the operator $\lambda_{-s}^- P^{(1)*} R_N(\Delta_{2s}, a_{\alpha j})\lambda_{-s}^+$ with respect to x is equal to $2\mu - N$. The remaining terms in (31) are considered in like manner.

We now proceed to the estimation of the quadratic form

$$(P_{jk}\gamma^j D_t^k w, \overline{w})$$

corresponding to operator (35). Applying integration by parts with respect to t we can rewrite the form thus:

$$(P'_{jk}\gamma^{j'} D_t^{k'}, \gamma^{j''} D_t^{k''}), \qquad j' + k' \leqslant m - 1,\ j'' + k'' \leqslant m - 1.$$

The absolute value of the latter form does not exceed

$$\text{const}\,\|(1 + |D_x|)^{-\varkappa}\gamma^j D_t^k w\|^2, \qquad \varkappa = \frac{N}{2} - \mu.$$

If N is sufficiently large, then by virtue of Corollary 1.2, this expression can be estimated from above by means of

$$(\gamma_0 - \gamma)^{-1} \int H_P(y^0; D, \gamma)|\widehat{w}(\eta)|^2\, d\eta.$$

We have thus proved that

$$|(T_{s\gamma N}w, \overline{w})| \leqslant \varepsilon(\gamma)(H_P(y^0; D, \gamma)w, \overline{w}). \tag{23''}$$

Comparing (23′) and (23″) we obtain (23).

2.5. The proof of Inequality (26). The proof of (26) is carried out according to the same plan as the proof of (23). If Q_1 and Q_2 are two differential operators, then the symbol of $Q_1 Q_2^*$ is equal to

$$\sum \frac{1}{\alpha!\,\delta!} Q_1^{(\alpha)} \overline{Q}_{2(\alpha+\delta)}^{(\delta)} = \sum \frac{1}{\beta!} (Q_1 \overline{Q_{2(\beta)}})^{(\beta)}.$$

Using the commutation formula we can represent operator (25) as a sum a PDO with symbol

$$\sum_{\substack{|\alpha+\beta+\theta|>0 \\ |\alpha|,|\theta|\leqslant N-1}} \frac{1}{\alpha!\,\beta!\,\theta!} \{P, P_{(\beta+\alpha)}\}_{(\theta)}^{(\beta)} \Delta_{-2s}^{(\alpha)} \lambda_{-s}^{+(\theta)} \lambda_{-s}^{-} - \frac{1}{2} \sum_{\beta>0} H_{P(\beta)}^{(\beta)}$$

and operators representable in the form of (35). Applying the conditions of Proposition 2 in Section 2.1 and Lemma 2 we prove (26).

2.6. Gårding's inequality for inhomogeneous quadratic forms (the proof of inequality (22)). We shall prove the following

Proposition. *Let $H(y;\eta)$ be a positive polynomial symbol in η having smooth stabilized coefficients*[1] *and satisfying the following conditions:*

(i) $\quad H(y';\eta)H^{-1}(y'';\eta) < c_0 \quad \forall y', y'' \in \mathbb{R}^{n+1}, \quad \forall \eta \in \mathbb{R}^{n+1},$

(ii) $\quad |H^{(\beta)}(y;\eta)| < \varepsilon H(y;\eta) \quad \forall y \in \mathbb{R}^{n+1}, \quad \forall \eta \in \mathbb{R}^{n+1}.$

Assume that the constant ε in (ii) *is sufficiently small. Then the two-sided estimate*

$$c^{-1}(H(D)w, \overline{w}) < \operatorname{Re}(H(y;D)w, \overline{w}) < c(H(D)w, \overline{w}) \tag{36}$$

holds, where $H(D) = H(y^0; D)$, $y^0 \in \mathbb{R}^{n+1}$.

The proof is carried out according to the same scheme as the ordinary Garding inequality.

1) We first consider inequality (36) on the functions w with support in a ball of a sufficiently small radius δ and center at y_0. Write the operator $H(y;D)$ in the form

$$H(y;D) = H(y^0;D) + \sum_{j=1}^{J} h_j(y) H(y^j;D).$$

We take a truncating function $\psi(y) \in \mathcal{D}$ equal to 1 for $|y - y^0| \leqslant \delta$ and to zero for $|y - y^0| \geqslant 2\delta$. Then we have

$$H(y;D)w = H(D)w + \sum (\psi h_j)(y) H(y^j;D)w.$$

[1] This condition can be dropped (see Volevich [1, Section 3.4].

We now write

$$
\begin{aligned}
|\operatorname{Re}(H(y;D)w,\overline{w}) - (H(D)w,\overline{w})| &\leqslant \sum_{j=1}^{J} |(\psi h_j H(y^j;D)w,\overline{w})| \\
&\leqslant \|H^{1/2}(D)w\| \sum_{j=1}^{J} \|H^{-1/2}(D)\psi h_j H(y^j;D)w\| \\
&\leqslant \|H^{1/2}(D)w\| \sum_{j=1}^{J} \Big(\|\psi h_j H^{-1/2}(D)H(y^j;D)w\| \\
&\qquad + \|(H^{-1/2}(D)\psi h_j - \psi h_j H^{-1/2}(D)H(y^j;D)w)\|\Big).
\end{aligned} \tag{37}
$$

Since $h_j(y^0) = 0$, the maximum of the function $(\psi h_j)(y)$ does not exceed $c\delta$, and therefore

$$
\sum_{j=1}^{J} \|\psi h_j H^{-1/2}(D)H(y^j;D)w\| \leqslant c\delta c_0^{1/2}\|H^{1/2}(D)w\| = c_1\delta\|H^{1/2}w\|.
$$

The Fourier transform of the operator $(H^{-1/2}(D)\psi h_j - \psi h_j H^{-1/2}(D))H(y^j;D)$ is an integral operator with kernel

$$
\widehat{\psi h_j}(\eta' - \eta'')[H^{-1/2}(\eta') - H^{-1/2}(\eta'')]H(y^j;\eta''). \tag{38}
$$

Condition (ii) implies that

$$
|H^{-1/2}(\eta') - H^{-1/2}(\eta'')| < c_2\varepsilon H^{-1/2}(\eta'').
$$

Since the function $\widehat{\psi h_j}(\eta)$ decreases faster than any power of $|\eta|$, the modulus of function (38) does not exceed

$$
\varepsilon c(\delta)(1 + |\eta' - \eta''|)^{-n-2}H^{1/2}(\eta'').
$$

Therefore inequality (37) assumes the form

$$
|\operatorname{Re}(H(y;D)w,\overline{w}) - (H(D)w,\overline{w})| < (c_1\delta + c_2(\delta)\varepsilon)(H(D)w,\overline{w}).
$$

Taking δ satisfying the condition $c_1\delta < 1/4$ and ε satisfying the inequality $c_1(\delta)\varepsilon < 1/4$ we conclude that

$$
\frac{1}{2}(H(D)w,\overline{w}) \leqslant Re(H(y;D)w,\overline{w}) \leqslant \frac{3}{2}(H(D)w,\overline{w}).
$$

2) *Localization.* Take a system of nonnegative functions $\{\varphi_j(y)\}$ possessing the following properties:

(a) $\varphi_j \in \mathcal{D}$ and $\operatorname{supp}\varphi_j$ belongs to a ball of radius δ;

(b) $\sum \varphi_j^2(y) \equiv 1$;

(c) every point $y \in \mathbb{R}^{n+1}$ belongs to at most J supports of the functions $\varphi_j(y)$;

(d) $\max|D^\alpha \varphi_j|$, $\int |D^\alpha \varphi_j(y)|\,dy < c_\alpha(\delta) \quad \forall \alpha, j$.

Then we have

$$\operatorname{Re}(H(y;D)w, \overline{w}) = \sum_j \operatorname{Re}(\varphi_j H(y;D)w, \overline{\varphi_j w})$$
$$= \sum_j \operatorname{Re}(H(y;D)\varphi_j w, \overline{\varphi_j w}) - \sum_j \sum_{\beta>0} (D^\beta \varphi_j H^{(\beta)}(y;D)w, \overline{\varphi_j w})/\beta! \tag{39}$$

By what has been proved, we have

$$\frac{1}{2}\sum (H(D)(\varphi_j w), \overline{\varphi_j w}) \leqslant \sum \operatorname{Re}(H(y;D)(\varphi_j w), \overline{\varphi_j w})$$
$$\leqslant \frac{3}{2}\sum (H(D)(\varphi_j w), \overline{\varphi_j w}). \tag{40}$$

To estimate the second term on the right-hand side of (39) we take the functions $\psi_j(y) \in \mathcal{D}$ satisfying conditions (a), (b), and (c) and such that $\varphi_j(y)\psi_j(y) \equiv 1$. Then the second term on the right-hand side of (39) can be rewritten as

$$\sum_{\beta>0}\sum_j (\varphi_{j\beta} H^{(\beta)}(y;D)(\psi_j w), \overline{\psi_j w}), \qquad \varphi_{j\beta} = \varphi_j D^\beta \varphi_j/\beta!$$

Using Lemma 2 in the foregoing section we estimate this expression from above by means of

$$c\varepsilon \sum \|H^{1/2}(\psi_j w)\|^2.$$

Let us show that

$$\sum \|H^{1/2}(\psi_j w)\|^2 \leqslant c_1 \|H^{1/2}(\varphi_j w)\|^2. \tag{41}$$

To this end we note that

$$\psi_j w = \sum_k \psi_j \varphi_k^2 w = \sum_{k\in U_j} \psi_j \varphi_k(\varphi_k w),$$

where U_j is a finite set with the number of elements J_1 not depending on j. Therefore (cf. Lemma 2 in Section 2.4) we have

$$\|H^{1/2}\psi_j w\|^2 \leqslant J_1 \sum_{k\in U_j} \|H^{1/2}(\psi_j\varphi_k)(\varphi_k w)\|^2 \leqslant J_1 c_2 \sum_{k\in U_j} \|H^{1/2}\varphi_k w\|^2. \tag{42}$$

Performing the summation of inequalities (42) over j we obtain (41). Comparing (39) with (40) and (41) we conclude that if ε is sufficiently small, then the left-hand side of (39) is estimated from above and below by means of

$$\sum_j (H(D)(\varphi_j w), (\overline{\varphi_j w})).$$

Further, repeating in fact the already performed calculations we find

$$\begin{aligned}\left|\sum (H(D)(\varphi_j w), \overline{\varphi_j w}) - (H(D)w, \overline{w})\right| &\leqslant \sum \frac{1}{\beta!} |(D^\beta \varphi_j H^{(\beta)} w, \overline{\varphi_j w})| \\ &\leqslant c_3 \varepsilon \sum \|H^{1/2}(\psi_j w)\| \leqslant c_4 \varepsilon \sum \|H^{1/2}(\varphi_j w)\|.\end{aligned}$$

If ε is sufficiently small, then

$$\frac{1}{2}(H(D)w, \overline{w}) \leqslant \sum \|H^{1/2}(D)\varphi_j w\|^2 \leqslant \frac{3}{2}(H(D)w, \overline{w})$$

which proves the proposition.

§3. An analysis of conditions for the existence of energy estimates

It was shown in §2 that the conditions of Propositions 1 and 2 in Section 2.1 guarantee the existence of estimates (1.3) and (1.3′) under which Cauchy's problem is uniquely solvable. However, as was mentioned above, it is difficult to verify the fulfilment of these conditions for given classes of differential operators. In the present section we shall present cruder sufficient conditions for the existence of energy estimates and, in particular, prove Theorem B stated in §1. Examples of differential operators satisfying the conditions of this theorem will also be given. A more complicated example of dominantly correct operators will be considered in §4. The results in this section are taken from the papers by Volevich [1] and Gindikin [1] (see the Appendix in the latter paper).

3.1. Some immediate consequences of Theorem 2.1. As in §1, we shall consider a symbol $P(y;\xi,\tau)$ with smooth stabilized coefficients solved with respect to the highest power τ^m, the coefficient in τ^m being identically equal to 1. In what follows these conditions on the symbol will not be stipulated.

Proposition 1. *Let a symbol $P(y;\xi,\tau)$ satisfy the correctness conditions $\forall y \in \mathbb{R}^{n+1}$ and let for the function $H_P(y;\xi,\sigma,\gamma)$ the condition of constant strength be fulfilled.*

Let, additionally, the conditions below hold:

$$|\{P^{(\beta)}, P_{(\alpha)}\}(y;\xi,\sigma,\gamma)| < \varepsilon_{\alpha\beta}(\gamma) H_P(\xi,\sigma,\gamma), \quad \beta > 0,\ \alpha \geqslant 0, \tag{1}$$

$$|\{P, P^{(\beta)}_{(\alpha)}\}(y;\xi,\sigma,\gamma)| < \varepsilon_{\alpha\beta}(\gamma) H_P(\xi,\sigma,\gamma), \quad \beta > 0,\ \alpha \geqslant 0, \tag{2}$$

$$|\{P, P_{(\alpha)}\}(y;\xi,\sigma,\gamma)| < \varepsilon_\alpha(\gamma) H_P(\xi,\sigma,\gamma)(1+|\xi|), \quad \alpha > 0, \tag{3}$$

where $\varepsilon_{\alpha\beta}(\gamma)$, $\varepsilon_\alpha(\gamma) \to 0$ as $\gamma \to -\infty$. Then the symbol $P(y;\xi,\tau)$ satisfies all conditions of Propositions 1 *and* 2 *in Section* 1.2.

Remark. In the case of constant coefficients all conditions of the proposition reduce to the condition

$$|\{P^{(\beta)}, P\}(\xi,\sigma,\gamma)| < \varepsilon_\beta(\gamma) H_P(\xi,\sigma,\gamma), \quad \beta > 0. \tag{1'}$$

This condition is equivalent to the exponential correctness of the polynomial P.

The proof of Proposition 1. 1) We first verify the conditions

$$|H_P^{(\beta)}| < \varepsilon_\beta(\gamma) H_P, \qquad \beta > 0. \tag{4}$$

For $|\beta| = 1$ condition (4) follows from (1) with $\alpha = 0$ since in the case we have

$$H_P^{(\beta)} \stackrel{\text{def}}{=} \frac{1}{2}(P\overline{P}^{(1)} - \overline{P}P^{(1)})^{(\beta)} = \{P^{(\beta)}, P\} + \{P, P^{(\beta)}\}.$$

We now show that condition (4) for $|\beta| > 1$ follows from the analogous condition for $|\beta| = 1$. We use the simple argument in the paper by Shilov [1]. By Lagrange's formula, we have

$$\begin{aligned}\left|\log\frac{H_P(\eta+\theta,\gamma)}{H_P(\eta,\gamma)}\right| &= |\log H_P(\eta+\theta,\gamma) - \log H_P(\eta,\gamma)| \\ &= H_P^{-1}(\eta+\theta^*,\gamma)\left|\sum \theta_k \partial_k H(\eta+\theta^*,\gamma)\right| \leqslant \varepsilon(\gamma) \to \infty, \quad \gamma \to -\infty,\end{aligned}$$

where $\theta \in \mathbb{R}^{n+1}$, $|\theta| \leqslant 1$, and θ^* is a point on the line segment joining 0 and θ. It follows that

$$\left|\frac{H_P(\eta+\theta,\gamma)}{H_P(\eta,\gamma)} - 1\right| = \left|\sum_{\beta>0} \frac{\theta^\beta}{\beta!}\frac{H_P^{(\beta)}(\eta,\gamma)}{H_P(\eta,\gamma)}\right| < \varepsilon(\gamma)e^{\varepsilon(\gamma)} = \varepsilon_1(\gamma),$$

where $\varepsilon_1(\gamma) \to 0$, $\gamma \to -\infty$. Since the monomials θ^β are linerally independent as functions of $\theta_1, \dots, \theta_{n+1}$, the above inequality implies that (4) holds for any values of $|\beta|$.

2) Let us verify conditions (2.8). Since the space of polynomials is finite-dimensional, we write

$$\begin{aligned}&\overline{P^{(1)(\beta)}(y,\eta+i\gamma)}P(y;\eta+i\gamma) - \overline{P^{(\beta)}(y;\eta+i\gamma)}P^{(1)}(y;\eta+i\gamma) \\ &= \sum d_j(y)\Big[\overline{P^{(1)(\beta)}(y^j;\eta+i\gamma)}P(y^j;\eta+i\gamma) \\ &\qquad - \overline{P^{(\beta)}(y^j;\eta+i\gamma)}P^{(1)}(y^j;\eta+i\gamma)\Big],\end{aligned}$$

that is

$$\{P^{(\beta)}(y), P(y)\} = \sum d_j(y)\{P^{(\beta)}(y^j), P(y^j)\}. \tag{5}$$

Applying the operator D_y^β to both sides and using inequalities (1) for $\alpha = 0$ we obtain (2.8).

3) The fulfilment of conditions (2.9) is verified in like manner with replacement of P in (5) by $P_{(\alpha)}$.

4) We verify (2.11) and (2.12). Conditions (2.12) for $\beta = 0$ go into (3), and therefore they should be verified for $\beta > 0$. In this case (2.11) and (2.12) are special cases of the more general condition

$$|\{P, P_{(\alpha)}\}^{(\beta)}| < \varepsilon_\beta(\gamma) H_P(y; \xi, \sigma, \gamma), \qquad \beta > 0. \tag{6}$$

The proof of (6) is analogous to the above verification of (4). For $|\beta| = 1$ the left-hand side of (6) is equal to

$$|\{P^{(\beta)}, P_{(\alpha)}\} + \{P, P_{(\alpha)}^{(\beta)}\}|,$$

and the required estimate is a consequence of (1) and (2). If $|\beta| > 1$, then the vector β is representable as a sum of integral vectors: $\beta = \beta' + \beta''$, $|\beta'| = 1$. Applying inequality (2.18) with $\beta = \beta''$ to the symbol $Q = \{P, P_{(\alpha)}\}^{(\beta')}$ and using the fact that, by virtue of the already proved inequalities (4), we have

$$H_P(\eta + \theta, \gamma) H_P^{-1}(\eta, \gamma) \to 0 \quad \text{for } \gamma \to \infty \text{ (uniformly with respect to } \eta),$$

we arrive at (6).

A direct consequence of Proposition 1 is

Proposition 2. *Let a symbol $P(y; \xi, \eta)$ satisfy Petrovskiĭ's correctness condition $\forall y \in \mathbb{R}^{n+1}$, the condition of constant strength for H_P, and condition (3), and, moreover, let for any pair of points $y', y'' \in \mathbb{R}^{n+1}$ the inequality*

$$|\{P^{(\beta)}(y'), P(y'')\}(\xi, \sigma, \gamma)| < \varepsilon_\beta(\gamma) H_P(y; \xi, \sigma, \gamma), \qquad \beta > 0, \tag{7}$$

hold, where $\varepsilon_\beta(\gamma) \to 0$ for $\gamma \to -\infty$. Then all conditions of Proposition 1 and, consequently, of Theorem 2.1 are fulfilled.

Proof. Using the representation

$$P(y; \xi, \tau) = \sum d_j(y) P(y^j; \xi, \tau),$$

we write

$$\{P^{(\beta)}(y'), P(y'')\} = \sum d_j(y') d_k(y'') \{P^{(\beta)}(y^j), P(y^k)\}.$$

We now apply the operator $D_{y''}^\alpha$ and set $y' = y'' = y$ to obtain, by virtue of (7), relation (1). Applying the operator $D_{y'}^\alpha$ and putting $y = y' = y''$ we derive (2).

3.2. The proof of Theorem B in Section 1.3. We shall need some auxiliary assertions.

Let $Q(\eta,\gamma)$ and $R(\eta,\gamma)$ be some functions of the variables $\gamma\in(-\infty,\gamma_0)$ and η, where η runs over the same set for the two functions, say (for definiteness) over $\mathbb{R}^{n+1}$.

Lemma 1. *Let*

$$Q(\eta,\gamma)\neq 0\quad\forall\gamma\leqslant\gamma_0,\quad\forall\eta\in\mathbb{R}^{n+1}.$$

Then the following conditions are equivalent:

(i) $R(\eta,\gamma)Q^{-1}(\eta,\gamma)\to 0$ *for* $\gamma\to-\infty$ *uniformly with respect to* η;

(ii) $\forall a\in\mathbb{C}$ *there is* $\gamma_0(a)$ *such that*

$$Q(\eta,\gamma)+aR(\eta,\gamma)\neq 0,\quad\gamma<\gamma_0(a),\quad\eta\in\mathbb{R}^{n+1},$$

where the function $\gamma_0(a)$ *can be chosen so that it is bounded from below on any compact set in* $\mathbb{C}$.

Proof. (i)$\Longrightarrow$(ii) is obvious.

(ii)$\Longrightarrow$(i). Assume that (i) does not hold. Then for some $\varepsilon>0$ there is a sequence (η_j,γ_j), $\gamma_j\to-\infty$, such that

$$|R(\eta_j,\gamma_j)Q^{-1}(\eta_j,\gamma_j)|\geqslant\varepsilon.$$

We set $a_j=R(\eta_j,\gamma_j)/Q(\eta_j,\gamma_j)$. Then $|a_j|<\varepsilon^{-1}$, and, by virtue of (ii), the sequence $\gamma_0(a_j)$ is bounded from below by a constant $\overline{\gamma_0}$. Since $\gamma_j\to-\infty$, we have $\gamma_j<\overline{\gamma_0}$ for sufficiently large j, and we hence

$$Q(\eta_j,\gamma_j)+a_jR(\eta_j,\gamma_j)=0,\qquad\gamma_j<\overline{\gamma_0}.$$

The resulting contradiction proves the lemma.

Lemma 2. *Let polynomials* $P_j(\xi,\tau)$, $j=1,2$, *satisfy Petrovskiĭ's condition, that is* $\exists\gamma_0$ *such that*

$$P_j(\xi,\tau)\neq 0,\quad\operatorname{Im}\tau\leqslant\gamma_0,\quad\xi\in\mathbb{R}^n,\quad\tau=\sigma+i\gamma. \tag{8}$$

Then

$$\{P_1,P_2\}(\xi,\sigma,\gamma)\neq 0,\quad\gamma\leqslant\gamma_0,\quad(\xi,\sigma)\in\mathbb{R}^{n+1},\quad\tau=\sigma+i\gamma. \tag{9}$$

Proof. If $\tau_{1k}(\xi)$ and $\tau_{2k}(\xi)$ are the roots of the polynomials P_1 and P_2, then replacing P_1 and P_2 by their factorizations we derive

$$\begin{aligned}\{P_1,P_2\}(\xi,\sigma,\gamma)&=2iP_1(\xi,\tau)\overline{P_2(\xi,\tau)}\left[\frac{P_1^{(1)}(\xi,\tau)}{P_1(\xi,\tau)}-\overline{\frac{P_2^{(1)}(\xi,\tau)}{P_2(\xi,\tau)}}\right]\\&=2iP_1(\xi,\tau)\overline{P_2(\xi,\tau)}\sum_{k=1}^{m}\left[\frac{1}{\tau-\tau_{1k}(\xi)}-\frac{1}{\overline{\tau}-\overline{\tau}_{2k}(\xi)}\right].\end{aligned}$$

If $\operatorname{Im}\tau \leqslant \gamma_0$, then, according to (8), we have $P_1(\xi,\tau)\overline{P_2(\xi,\tau)} \neq 0$. Further, since $\operatorname{Im}(\tau - \tau_{jk}(\xi)) < 0$, $j = 1, 2$, the two expressions in the square brackets have positive imaginary parts, and consequently (9) holds.

If two functions $Q(\eta,\gamma)$ and $R(\eta,\gamma)$ are related by conditions of Lemma 1, we shall write $R \prec Q$. The same notation will be retained for $Q(\xi,\tau)$ and $R(\xi,\tau)$ regarded as functions of the variables $\eta = (\xi, \operatorname{Re}\tau)$ and $\gamma = \operatorname{Im}\tau$.

Lemma 3. *If P is a polynomial correct in Petrovskiĭ's sense and $R(\xi,\tau) \prec P(\xi,\tau)$, then*

$$\{R,P\}(\xi,\sigma,\gamma) \prec H_P(\xi,\sigma,\gamma). \tag{10}$$

Proof. According to Lemma 1, if $R \prec P$, then $\forall \overline{a} \in \mathbb{C}\ \exists\gamma(\overline{a})$ such that

$$(P + \overline{a}R)(\xi,\tau) \neq 0, \qquad \operatorname{Im}\tau < \gamma_0(\overline{a}),$$

i.e. the polynomial $P + \overline{a}R$ is also correct in Petrovskiĭ's sense. By Lemma 2, we have

$$\{P + \overline{a}R, P\} = \{P,P\} + a\{R,P\} = H_P + a\{R,P\} \neq 0, \quad \operatorname{Im}\tau \leqslant \gamma_0(\overline{a}).$$

Applying Lemma 1 once again we obtain (10).

We now can prove Theorem B. We have in fact to show that condition (1.15) of the theorem implies (7) and that (1.16) implies (3). Indeed, take arbitrary $y', y'' \in \mathbb{R}^{n+1}$, $\beta > 0$. According to (1.15), we have

$$R \stackrel{\text{def}}{=} P^{(\beta)}(y';\xi,\tau) \prec P(y'';\xi,\tau) \stackrel{\text{def}}{=} P.$$

Applying condition (10) of Lemma 3 to the polynomials P and R be obtain (7).

We note that in the assertions of Lemmas 2 and 3 the polynomial dependence of the functions involved on τ is essential while the polynomial dependence on ξ is inessential. In view of this fact, Lemma 3 can be applied to $R = P_{(\alpha)}(y;\xi,\tau)$ and $P = P(y;\xi,\tau)(1 + |\xi|)$. Therefore

$$\{P_{(\alpha)}(y), P(y)\}(1 + |\xi|) \prec H_P(y)(1 + |\xi|)^2,$$

whence follows (3). Theorem B is proved completely.

3.3. Remarks on the condition of the constant strength for H_P. Condition (1.15) means that the symbol $P(y;\xi,\tau)$ is exponentially correct for every fixed $y \in \mathbb{R}^{n+1}$. If the condition of constant strength is imposed on the symbol P, that is $\exists\gamma_0$, $A > 0$ such that $\forall y', y'' \in \mathbb{R}^{n+1}$ the inequalities

$$A^{-1} \leqslant |P(y';\xi,\tau)P^{-1}(y'';\xi,\tau)| \leqslant A, \qquad \operatorname{Im}\tau \leqslant \gamma_0,$$

hold, then conditions (1.15) and (1.16) are fulfilled automatically. Moreover, as will be shown, the exponential correctness and the condition of constant strength for P imply the condition of constant strength for H_P, i.e. Theorems A and B imply a non-trivial generalization of Theorem 2.5.5.

Proposition. *Let $P(y;\xi,\tau)$ be a polynomial correct in Petrovskiĭ's sense at every point and let the condition of constant strength be fulfilled for it. Then the function $H_P(y;\xi,\sigma,\gamma)$ satisfies the condition of constant strength.*

To prove this assertion we first of all note that the lemma below is readily proved by repeating of the argument in Lemma 1 in the foregoing section.

Lemma 1′. *Let*

$$Q(\eta,\gamma)\neq 0 \quad \forall\gamma\leqslant\gamma_0, \quad \forall\eta\in\mathbb{R}^{n+1}.$$

Then the conditions below are equivalent:

(i) $R(\eta,\gamma)Q^{-1}(\eta,\gamma)<c$ *for* $\gamma\leqslant\gamma_0$ *and some* c;
(ii) $(Q+aR)(\eta,\gamma)\neq 0$ *for* $|a|<c$ *and* $\gamma<\gamma_0$.

If the conditions of Lemma 1′ are fulfilled, we shall say that R is *weakly subordinate to* Q. The following lemma is a modification of Lemma 3 in the foregoing section.

Lemma 2. *If P is a polynomial correct in Petrovskiĭ's sense and R is weakly subordinate to P, then $\{R,R\}$ is weakly subordinate to $\{P,P\}$.*

Applying Lemma 2 to the polynomials $R=P(y';\xi,\tau)$ and $P=P(y'';\xi,\tau)$ we conclude that $\{P(y'),P(y')\}$ is weakly subordinate to $\{P(y''),P(y'')\}$ for any y' and y'', which exactly means that the assertion of the proposition is true.
We outline the scheme of the proof of Lemma 2. We shall also need

Lemma 3. *If P_1 and P_2 are polynomials correct in Petrovskiĭ's sense and R is weakly subordinate to P_1, then $\{R,P_2\}$ is weakly subordinate to $\{P_1,P_2\}$.*

Proof. If $|a|$ is sufficiently small, then P_1+aR is a polynomial correct in Petrovskiĭ's sense, and, by Lemma 2 in the foregoing section, we have

$$\{P_1,P_2\}+a\{R,P_2\}=\{P_1+aR,P_2\}\neq 0.$$

The proof of Lemma 2. If a is sufficiently small, then the polynomial $\{P+aR,R\}$ is weakly subordinate to $\{P+aR,P\}=\{P,P\}+a\{R,P\}$. Since, according to Lemma 3, $\{R,P\}$ is weakly subordinate to $\{P,P\}$, we see that $\{P,P\}+a\{R,R\}$ is weakly subordinate to $\{P,P\}$. We have thus proved that for sufficiently small $|a|$

$$\{P,R\}+a\{R,R\} \text{ is weakly subordinate to } \{P,P\}.$$

With account of the fact that $\{P,R\}$ is weakly subordinate to $\{P,P\}$, it follows that the assertion of the lemma holds.

The condition of constant strength for H_P is the main condition of Theorem B, and its verification is most difficult. However, there are classes of symbols for which

H_P is estimated from below by means of the sums of the moduli of the monomials involved, and Newton's polyhedron of H_P does for depend on y. In this case the condition of constant strength for H_P is fulfilled automatically.

Let a symbol $P(y;\xi,\tau)$ be given. We denote by $N(P(y))$ and $\delta^0(P(y))$ Newton's polyhedron of the polynomial $P(y;\dots)$ and the polyhedron of the integrally minor terms of the polynomial, respectively. Regarding $\{P(y),P(y)\}$ as a polynomial in $n+2$ variables (ξ,σ,γ) we denote by $N(H_{P(y)})$ the corresponding Newton polyhedron. Finally, let $N(P)$, $\delta^0(P)$, and $N(H_P)$ be the convex hulls of the unions of $N(P(y))$, etc. over all $y\in\mathbb{R}^{n+1}$.

Theorem. *Let the following conditions be fulfilled for the symbol* $P(y;\xi,\tau)$:

(i) $N(H_{P(y)})=N(H_P)\ \forall y\in\mathbb{R}^{n+1}$;

(ii) *the symbol* $H_P(y;\xi,\sigma,\gamma)$ *is estimated from below in terms of the moduli of the monomials involved in it*:

$$H_P(y;\xi,\sigma,\gamma)>\text{const}\sum_{(\alpha,\beta,r)\in N(H_P)}|\xi^\alpha\sigma^\beta\gamma^r|;$$

(iii) $\forall y\in\mathbb{R}^{n+1}$ *and any polynomials* Q_1 *and* Q_2, $N(Q_1)\in N(P)$, $N(Q_2)\subset\delta^0(P)$, *we have*

$$\{Q_1,Q_2\}(\xi,\sigma,\gamma)\prec H_P(y;\xi,\sigma,\gamma).$$

Then the symbol P satisfies all conditions of Proposition 2 in Section 3.1.

Proof. It is obvious that (i) and (ii) imply the condition of constant strength for H_P, and hence we have to verify conditions (3) and (7).

Applying condition (iii) to $Q_2(\xi,\tau)=P^{(\beta)}(y';\xi,\tau)$, $\beta>0$, and $Q_1(\xi,\tau)=P(y'';\xi,\tau)$ we obtain (7).

To prove (3) we write P in the form

$$P=\tau^m+\sum_{j=1}^n\xi_jP_j+P_0,\quad N(P_j)\subset\delta^0(P),\quad j=0,\dots,n.$$

Differentiating this relation with respect to y (it is this place where use is made of the fact that the coefficient in the highest power of τ is identically equal to a constant) we find

$$\begin{aligned}|\{P,P_{(\alpha)}\}|&=\Big|\sum\xi_j\{P,P_{j(\alpha)}\}+\{P,P_{0(\alpha)}\}\Big|\\&\leqslant\varepsilon_\alpha(\gamma)H_P(y;\xi,\sigma,\gamma)(1+|\xi|),\quad\varepsilon(\gamma)\to0,\quad\gamma\to-\infty.\end{aligned}$$

3.4. Strictly pluriparabolic differential operators. Here we shall present a class of differential operators whose symbols satisfy the conditions of Theorem B. These operators include as special cases the strictly hyperbolic and q-parabolic operators. We first give the definition and description of these operators for the case of constant coefficients, i.e. for polynomials.

Represent the space $\mathbb{R}^{n+1}$ as a direct sum of the subspace $\mathbb{R}^k$ of the variables $\sigma = (\sigma_1, \dots, \sigma_k)$ and the subspace $\mathbb{R}^l$ of the variables $\zeta = (\zeta_1, \dots, \zeta_l)$, $l + k = n + 1$.

We separate out the variable σ_1, and let $\sigma' = (\sigma_2, \dots, \sigma_k)$; in Cauchy's problem σ_1 plays the role of a variable dual to time.

Let $q = 2b$ be an even positive integer. In what follows we shall assign the weights q and 1 to the variables σ and ζ, respectively.

Definition 1. A $(q, \dots, q, 1, \dots, 1)$-homogeneous polynomial $P_0(\sigma, \zeta)$ is said to be *strictly pluriparabolic* (see Gindikin [2], and Volevich and Gindikin [5]) if

(i) the polynomial $P_0(\sigma, 0)$ is strictly hyperbolic;

(ii) there is $\lambda > 0$ such that

$$\operatorname{Im} \tau_{0j}(\sigma', \zeta) \geqslant \lambda |\zeta|^q, \quad j = 1, \dots, m,$$

where $\tau_{0j}(\sigma', \zeta)$ are the roots of the polynomial P with respect to σ_1.

Definition 2. A polynomial $P(\sigma, \zeta)$ is said to be *strictly pluriparabolic* if its principal $(q, \dots, q, 1, \dots, 1)$-homogeneous part possesses this property.

The strict hyperbolicity of $P(\sigma, 0)$ implies that this polynomial and, consequently, the polynomial $P(\sigma, \zeta)$ as well can be solved with respect to the highest power σ_1^m (the coefficient in σ_1^m is assumed to be equal to 1):

$$P(\sigma, \zeta) = \sigma_1^m + \sum_{j \geqslant 1} a_{j\varkappa\beta} \sigma_1^{m-j} \sigma'^{\varkappa} \zeta^\beta. \tag{11}$$

Proposition. *Let $q > 0$ be even. Then for polynomial* (11) *the following conditions are equivalent:*

(I) *polynomial* (11) *is strictly pluriparabolic;*

(II) *there are γ_0 and $c > 0$ such that for $\gamma \leqslant \gamma_0$ we have*

$$c^{-1} \leqslant H_P(\zeta, \sigma, \gamma)/(|\gamma| + |\eta|^q)(|\gamma| + |\sigma| + |\eta|^q)^{2(m-1)} \leqslant c, \tag{12}$$

where the notation $\eta = (\sigma', \xi)$

$$H_P(\zeta, \sigma, \gamma) = -\operatorname{Im}(P(\sigma_1 + i\gamma, \sigma', \zeta)\overline{\partial P(\sigma_1 + i\gamma, \sigma', \zeta)/\partial \sigma_1}) \tag{13}$$

is used;

(III) *there are γ_0 and $c_1 > 0$ such that*

$$c_1(|\gamma| + |\zeta|^q)(|\gamma| + |\sigma| + |\zeta|^q)^{m-1} \leqslant |P(\sigma_1 + i\gamma, \sigma', \zeta)|, \qquad \gamma \leqslant \gamma_0. \tag{14}$$

Proof. (I)$\Longrightarrow$(II). 1) We first assume that P is a $(q,\dots,q,1,\dots,1)$-homogeneous polynomial, and let $\tau_{0j}(\sigma',\zeta)$, $j=1,\dots,m$, be its roots. According to (1.11), for $\gamma<0$ we have

$$\begin{aligned} H_P(\zeta,\sigma,\gamma) &= \sum_{k=1}^{m}(-\gamma+\operatorname{Im}\tau_{0k}(\sigma',\zeta))\prod_{j\neq k}|\sigma_1+i\gamma-\tau_{0k}(\sigma',\xi)|^2 \\ &\geqslant (|\gamma|+\lambda|\zeta|^q)H(\zeta,\sigma,\gamma), \end{aligned}$$

where

$$H(\zeta,\sigma,\gamma)=\sum_{k=1}^{n}\prod_{j\neq k}[(\sigma_1-\operatorname{Re}\tau_{0k}(\sigma',\xi))^2+(-\gamma+\operatorname{Im}\tau_{0k}(\sigma',\xi))^2] \tag{15}$$

is a $(q,\dots,q,1,\dots,1)$-homogeneous function of degree $2(m-1)q$. To prove (12) in the quasi-homogeneous case it suffices to show that

$$H(\zeta,\sigma,\gamma)\neq 0 \quad \text{for} \quad \gamma\leqslant 0, \quad |\sigma|^2+|\zeta|^{2q}+\gamma^2=1. \tag{16}$$

Since $-\gamma+\operatorname{Im}\tau_{0k}(\sigma',\zeta)\geqslant|\gamma|+\lambda|\zeta|^q$, it suffices to verify (16) only for $\gamma=0$, $\zeta=0$ (i.e. for the strictly hyperbolic polynomial $P(\sigma,0)$). Since the roots $\tau_{0k}(\sigma',0)$ are real and are distinct for $|\sigma'|\neq 0$, one of the numbers $\sigma_1-\operatorname{Re}\tau_{0k}(\sigma',0)$ is nonzero, whence follows (16). In case $\sigma'=0$, we have

$$H(0;\sigma_1,0,0)=m\sigma_1^{2m-2}\neq 0 \quad \text{for} \quad \sigma_1\neq 0.$$

2) Now let $P(\sigma,\zeta)=P_0(\sigma,\zeta)+Q(\sigma,\zeta)$, where P_0 is a $(q,\dots,q,1,\dots,1)$-homogeneous polynomial and the $(q,\dots,q,1,\dots,1)$-degree of Q does not exceed $mq-1$. By what was proved, for $P=P_0$ inequality (12) has already been proved. To prove it in the general case we show that

$$|(H_{P_0+Q}-H_{P_0})(\zeta,\sigma,\gamma)|\leqslant\varepsilon(\gamma)(|\gamma|+|\zeta|^q)(|\gamma|+|\sigma|+|\zeta|^q)^{2m-2}. \tag{17}$$

To prove (17) we note that $H_P-H_{P_0}=\{P_0,Q\}+\{Q,P_0\}+\{Q,Q\}$ is a polynomial in ξ, σ, and γ of $(q,1,\dots,1)$-degree no higher than $2mq-q-1$, i.e. is a linear combination of monomials of the form of

$$(\gamma,\sigma)^\alpha\zeta^\beta, \quad |\alpha|\leqslant 2m-2, \quad q|\alpha|+|\beta|\leqslant 2mq-q-1.$$

These monomials can be represented as the expressions

$$(\gamma,\sigma)^\alpha\zeta^{\beta'}\zeta^{\beta''}, \quad |\alpha|q+|\beta'|\leqslant 2mq-2q, \quad |\beta''|\leqslant q-1,$$

which are obviously estimated by means of the right-hand side of (17) with $\varepsilon(\gamma)=\text{const}\,|\gamma|^{-1/q}$.

(II)$\Longrightarrow$(III). Since $\partial P/\partial\sigma_1$ is a polynomial of $(q,\dots,q,1,\dots,1)$-degree no higher than $(m-1)q$, we have

$$H_P(\zeta,\sigma,\gamma) \leqslant \text{const}\,|P(\sigma_1+i\gamma,\sigma',\zeta)|(1+|\sigma|+|\gamma|+|\zeta|^q)^{m-1},$$

whence for large $-\gamma$ follows (14).

(III)$\Longrightarrow$(I). Inequality (14) for the polynomial P implies an analogous inequality with $\gamma_0 = 0$ for its $(q,\dots,q,1,\dots,1)$-homogeneous part. As was already done many times, to show this one should replace (σ,ζ) by $(t^q\sigma, t\zeta)$ and pass to the limit for $t\to+\infty$. In what follows we assume that the polynomial P is $(q,1,\dots,1)$-homogeneous.

If we set $\zeta = 0$ in (14), this results in

$$c_1|\gamma|(|\gamma|+|\sigma|) \leqslant |P(\sigma_1+i\gamma,\sigma',0)|,$$

whence it follows that $P(\sigma,0)$ is strictly hyperbolic.

It now remains to verify condition (ii) in Definition 1. In view of the quasi-homogeneity, it suffices to show that there is $\lambda > 0$ such that

$$P(\sigma_1+i\gamma,\sigma',\zeta)\neq 0 \quad \text{for} \quad \gamma\leqslant\lambda, \quad |\zeta| = 1, \quad \sigma\in\mathbb{R}^k. \tag{18}$$

By virtue of (14), we have to consider only the case $\gamma \geqslant 0$. Setting $\gamma = 0$ in (14) and assuming that $|\zeta| = 1$ we find

$$|P(\sigma,\zeta)| > c_1(1+|\sigma|)^{m-1},$$

whence

$$\begin{aligned}|P(\sigma_1+i\gamma,\sigma',\zeta)| &> |P(\sigma,\zeta)| - |P(\sigma_1+i\gamma,\sigma',\zeta)-P(\sigma,\zeta)|\\ &\geqslant c_1(1+\sigma)^{m-1} - c_2\gamma[(1+|\sigma|)^{m-1}+\gamma^{m-1}].\end{aligned}$$

Taking $\lambda < c_1/4c_2$ and $\gamma \leqslant \lambda \leqslant 1$ we obtain (18).

Remark 1. Let $P(\sigma,\zeta)$ be a strictly pluriparabolic polynomial and let $\tau_{0j}(\sigma',\zeta)$ be the roots of its principal $(q,\dots,q,1,\dots,1)$-homogeneous part. Then, by virtue of condition (i), there is $\delta > 0$ such that

$$|\tau_{0j}(\sigma',0)-\tau_{0k}(\sigma',0)| > \delta|\sigma'|, \quad j\neq k. \tag{19}$$

A careful examination of the proof of the proposition shows that the constant c in (12) depends on δ (this follows form (19)), λ (by the condition (11)), and the maximum of the moduli of the coefficients of the polynomial P.

We now consider a symbol $P(y;\xi,\zeta)$ with smooth stabilized coefficients, solved with respect to σ_1^m (the highest power of σ_1), the coefficient in σ_1^m being equal to 1.

Definition 3. A symbol $P(y;\xi,\zeta)$ is said to be *strictly pluriparabolic* if the polynomial $P(y^0;\sigma,\zeta)$ is strictly pluriparabolic for each y^0 and, moreover, the roots $\tau_{0j}(y';\sigma',\zeta)$ of the principal $(q,\dots,q,1,\dots,1)$-homogeneous part of P satisfy for some λ, $\delta>0$ the inequalities

$$|\tau_{0j}(y;\sigma',0)-\tau_{0k}(y;\sigma',0)|>\delta|\sigma'|,\quad j\neq k, \tag{20}$$

$$\operatorname{Im}\tau_{0j}(y;\sigma',\zeta)>\lambda|\zeta|^q. \tag{21}$$

Theorem. *A symbol P satisfying the conditions of Definition* 3 *satisfies conditions of Theorem* B.

Proof. With account of Remark 1, inequality (12) holds for $P=P(y;\sigma,\zeta)$ with a unified constant c. It follows that the condition of constant strength holds for H_P. Condition (1.15) is a direct consequence of inequality (14) for $P=P(y;\sigma,\zeta)$ (recall that, by virtue of Remark 1, the constant c_1 in (14) does not depend on y).

To prove (3) we note that the symbol $P_{(\alpha)}$, $\alpha>0$, does not contain the highest power of σ_1 and is represented as

$$P_{(\alpha)}(y;\sigma,\zeta)=\sum_{j=1}^{k}\sigma_j P_{\alpha j}(y;\sigma,\zeta)+\sum_{i=1}^{l}\zeta_i P_{\alpha i}(y;\sigma,z)+P_{\alpha 0},$$

where the $(q,1,\dots,1)$-degrees of the symbols $P_{\alpha j}$, $P_{\alpha i}$, and $P_{\alpha 0}$ do not exceed $(m-1)q$, $mq-1$, and $mq-1$, respectively. Relation (3) now follows immediately from (14).

Remark 2. Applying the argument used in the proof of the proposition one can easily show that the symbol in Definition 3 satisfies the conditions of Theorem 3.3.

Remark 3. In case of pluriparabolic operators the method in §2 can be specified to obtain energy estimates in norms that take into account the quasi-homogeneity of the principal part of the operator, and a rather accurate result on the smoothness of the solution to Cauchy's problem for these equations (see Volevich and Gindikin [5]).

3.5. Remarks on Cauchy's problem in spaces of increasing and decreasing functions. It was noted in Section 2:5.7 that for exponentially correct symbols of constant strength we have estimates in the spaces $H^{(s)}_{(\sigma),\gamma}$. Similar estimates also take place under the condition of Theorem B. Moreover, it is possible to generalize the Propositions 1 and 2 in Section 2.1 to the case of the $\|\ \|^{(s)}_{(\sigma),\gamma}$ norm but we shall not dwell on this question.

Theorem. *Under the conditions of Theorem* B *the inequality*

$$c_{s\gamma}\sum_{l=0}^{m}(\gamma_0-\gamma)^l\|P^{(l)}(y,D)u\|^{(s)}_{(\sigma),\gamma}\leqslant\|P(y,D)u\|^{(s)}_{(\sigma),\gamma},$$

$$\gamma\leqslant\gamma(s,\sigma),\quad u\in H^{(\infty)}_{\gamma+}, \tag{22}$$

holds.

An analogous inequality can also be derived for the adjoint operator P^*. A simple reduction of Theorem A makes it possible to prove the solvability of Cauchy's problem in the spaces $H^{(s)}_{(\sigma),\gamma+}$.

Proof. We set $u = (1+|y|^2)^{-\sigma/2}v$. We shall derive two estimates (for sufficiently large $-\gamma$; for the notation see Section 2.1):

$$\sum_{l=0}^{m}(\gamma_0-\gamma)^l\|P^{(l)}(y,D)u\|^{(s)}_{(\sigma),\gamma} \leqslant \text{const}\sum_{l=0}^{m}\sum_{\beta\geqslant 0}(\gamma_0-\gamma)^l\|P^{(l)(\beta)}(y,D)v\|, \quad (23)$$

$$-\operatorname{Im}{}^+[(1+|y|^2)^{\sigma/2}P(y,D)u,(1+|y|^2)^{\sigma/2}P^{(1)}(y,D)u]^{(s)}_{\gamma} \geqslant c\{v\}^2_{(s),\gamma}. \quad (24)$$

We note that under the conditions of Propositions 1 and 2 in Section 2.1 one can prove a somewhat stronger estimate (as compared to (2.15)) following from (2.17):

$$c_s\sum_{l=1}^{m}\sum_{\beta\geqslant 0}(\gamma_0-\gamma)^{2l-1}\|P^{(l)(\beta)}(y,D)v\|^2_{(s),\gamma} \leqslant \{v\|^2_{(s),\gamma}. \quad (25)$$

Comparing (23) and (25) with (24) we obtain (22).

Inequality (23) is a trivial consequence of Leibniz' formula:

$$(1+|y|^2)^{\sigma/2}P^{(l)}(y,D)((1+|y|^2)^{-\sigma/2}v) = \sum_{\beta\geqslant 0}((1+|y|^2)^{\sigma/2}D^{\beta}(1+|y|^2)^{-\sigma/2})/\beta!)$$
$$\times P^{(l)(\beta)}(y,D)v \overset{\text{def}}{=} \sum h_{\sigma\beta}P^{(l)(\beta)}(y,D)v.$$

Similarly, the indicated formula makes it possible to rewrite the left-hand side of (24) as

$$-\operatorname{Im}{}^+[Pv+\sum h_{\sigma\beta}P^{(\beta)}v, P^{(1)}v+\sum h_{\sigma\delta}P^{(1)(\delta)}v]_{(s),\gamma}$$
$$= -\operatorname{Im}{}^+[Pv,P^{(1)}v] - \sum_{\substack{\beta,\delta\geqslant 0\\ |\beta|+|\delta|>0}}{}^+[h_{\sigma\beta}P^{(\beta)}v, h_{\sigma\delta}P^{(\delta)}v]. \quad (26)$$

According to (2.10), the first term on the right-hand side is estimated by means of $\{v\}^2_{(s),\gamma}$.

The argument in Section 3.2 implies that

$$\{P^{(\beta)},P^{(\delta)}\}(y,\eta,\gamma) \leqslant \varepsilon(\gamma)\{P,P\}(y,\eta,\gamma),$$

where $|\beta|+|\delta|>0$ and $\varepsilon(\gamma)\to 0$ as $\gamma\to-\infty$. In view of this, the second term on the right-hand side of (26) can be estimate from above by means of $\varepsilon(\gamma)\{v\}^2$, whence follows inequality (24). The theorem is proved.

Remark. Under the conditions of Theorem B it is possible to derive estimates in norms involving exponentially increasing (decreasing) weights. For more detail see Volevich [1].

§4. Cauchy's problem for dominantly correct differential operators

4.1. In this section we shall prove that a dominantly correct symbol $P(y;\xi,\tau)$ solved with respect to the highest power of τ:

$$P(y;\xi,\tau) = \tau^m + \sum_{\beta<m} P_{\alpha_1\ldots\alpha_n\beta}(y)\xi_1^{\alpha_1}\ldots\xi_n^{\alpha_n}\tau^\beta \tag{1}$$

satisfies all conditions of Theorem B, and hence Cauchy's problem is uniquely solvable for the corresponding differential operator $P(y;D)$.

Recall that, according to the definitions in Section 3.4.2, symbol (1) is said to be dominantly correct if the following conditions hold:

(i) the polygons $\delta(P(y))$ do not depend on y;

(ii) $\forall y^0 \in \mathbb{R}^{n+1}$ the polynomial $P(\xi,\tau) = P(y^0;\xi,\tau)$ is dominantly correct.

For symbols satisfying (i) and (ii) we shall establish a strengthened version of Theorem 3.3, in which the conditions of the theorem are supplemented with a condition of "equivalence" of the variables $\xi_1,\ldots,\xi_n$. We state the necessary definitions.

Consider a symbol

$$H_P(y,\eta,\gamma) = \sum h_{\alpha_1\ldots\alpha_n\beta r}(y)\xi_1^{\alpha_1}\ldots\xi_n^{\alpha_n}\sigma^\beta\gamma^r. \tag{2}$$

We denote by $\Delta(H(P(y)))$ the polyhedron in $\mathbb{R}^3$ spanned on the triples $(|\alpha|,\beta,r)$ for which $h_{\alpha_1\ldots\alpha_n\beta r}(y) \neq 0$ for $\alpha_1+\cdots+\alpha_n = |\alpha|$ and on their projections on the coordinate axes. As usual, let $\Delta(H_P)$ denote the convex hull of the union of all $\Delta(H_P(y))$ over $y \in \mathbb{R}^{n+1}$. We have

Theorem C. *For a dominantly correct symbol* (1) *the following assertions hold:*

(a) $\Delta(H_P(y)) = \Delta(H_P) \quad \forall y \in \mathbb{R}^{n+1}$;

(b) $\exists\gamma_0$, $c>0$ *such that the estimate from below*

$$H_P(y;\eta,\gamma) > c \sum_{(\alpha,\beta,r)\in\Delta(H_P)} |\xi|^\alpha|\sigma|^\beta|\gamma|^r, \qquad \gamma \leqslant \gamma_0, \tag{3}$$

holds;

(c) $\forall y \in \mathbb{R}^{n+1}$ *and for any polynomials* $Q_1(\xi,\tau)$, $Q_2(\xi,\tau)$, $\Delta(Q_1) \subset \Delta(P)$,[1] $\Delta(Q_2) \subset \delta(P)$, *there is a function* $\varepsilon(\gamma)$ *such that* $\varepsilon(\gamma) \to 0$ *as* $\gamma \to -\infty$ *and*

$$|\{Q_1,Q_2\}(\eta,\gamma)| < \varepsilon(\gamma)H_P(y;\eta,\sigma). \tag{4}$$

It is clear that symbol (1) satisfying the conditions (a), (b), and (c) satisfies the conditions of Theorem 3.3 and, consequently, the conditions of Theorem B.

The proof of Theorem C is based on an equivalent description of dominantly correct polynomials in terms of the functions H_P, and it is this question that is treated in the present section.

---

[1] For the notation $\Delta(P)$ see the Introduction to Chapter 2.

4.2. The description of dominantly correct and stable-correct polynomials in terms of H_P. We have

Theorem 1. *For a polynomial $P(\xi,\tau)$ solved with respect to the highest power of τ the following conditions are equivalent.*

(I) *$P(\xi,\tau)$ is a dominantly correct polynomial (i.e. the equivalent conditions of Theorem 3.4.1 hold for it).*

(II) *The following conditions are fulfilled:*

(a) *the polyhedron $\Delta(H_P) \subset \mathbb{R}^3$ is reconstructed uniquely from the polygon $\delta(P)$.*

(b) *$\exists \gamma_0$, $c > 0$ such that the estimate from below (cf. (3))*

$$H_P(\eta,\gamma) > c \sum_{(\alpha,\beta,r)\in\Delta(H_P)} |\xi|^\alpha|\sigma|^\beta|\gamma|^r, \qquad \gamma \leqslant \gamma_0, \tag{5}$$

holds;

(c) *For any polynomials $Q_1(\xi,\tau)$, $Q_2(\xi,\tau)$, $\Delta(Q_1) \subset \Delta(P)$, $\Delta(Q_2) \subset \delta(P)$, there is a function $\varepsilon(\gamma)$, $\varepsilon(\gamma) \to 0$ as $\gamma \to -\infty$, such that*

$$|\{Q_1, Q_2\}(\eta,\gamma)| < \varepsilon(\gamma) H_P(\eta,\sigma). \tag{6}$$

The implication (II)$\Longrightarrow$(I) is an immediate consequence of Theorem 3.4.1. Indeed, (5) implies that the polynomial P is correct in Petrovskiĭ's sense. Further, let $Q(\xi,\tau)$ be a polynomial and let $\Delta(Q) \subset \delta(P)$. In view of the relation

$$H_{P+Q} = \{P+Q, P+Q\} = H_P + \{P,Q\} + \{Q,P\} + \{Q,Q\} \tag{7}$$

and condition (c), there is $\gamma(Q)$ such that

$$H_{P+Q}(\eta,\gamma) > 0 \quad \text{for} \quad \gamma < \gamma(Q).$$

Hence, the polynomial $P+Q$ is also correct in Petrovskiĭ's sense, whence it follows that the original polynomial is dominantly correct.

The proof of the implication (I)$\Longrightarrow$(II) is rather cumbersome and occupies the entire remaining part of this section. As in the case of Theorem 3.4.1, the central point here is the proof of the corresponding assertion for the case $n = 1$, while the proof for the case of $n > 1$ is obtained from the former by passing to polar coordinates ($\xi = \rho\omega$).

In the course of the proof of Theorem 1 we shall obtain an analogous description for the stable-correct polynomials as well. We shall prove

Theorem 2. *For a polynomial $P(\xi,\tau)$ solved with respect to the highest power of τ the conditions below are equivalent.*

(I) *$P(\xi,\tau)$ is a stable-correct polynomial (i.e. it satisfies the equivalent conditions of Theorem 2.4.5; also see Theorem 3.4.3).*

(II) *Conditions* (a) *and*(b) *of Theorem* 1 *are fulfilled as well as the strengthened version of condition* (c):

($c_{\max}$) $\forall Q_i(\xi,\tau), \Delta(Q_i) \subset \Delta(P)$, $i = 1,2$, $\exists\gamma_1$

$$|\{Q_1, Q_2\}(\eta,)| < \text{const}\, H_P(\eta,\gamma), \qquad \gamma \leqslant \gamma_1. \tag{8}$$

Condition ($c_{\max}$) and relation (7) imply that all polynomials $P + \varepsilon Q$, where $\Delta(Q) \subset \Delta(P)$, are correct in Petrovskiĭ's sense for sufficiently small ε. Therefore the polynomial P is stable-correct, i.e. the implication (II)$\Longrightarrow$(I) has been proved.

Relations (c) and ($c_{\max}$) are purely geometric conditions. The second of them is equivalent to the condition

($c'_{\max}$) if $\Delta(Q_i) \subset \Delta(P)$, $i = 1,2$, then

$$\Delta(\{Q_1, Q_2\}) \subset \Delta(\{P, P\}).$$

As to the first of these conditions, to investigate it we need the following

Definition. A point $(\alpha,\beta,r) \in \Delta(H_P)$ is said to be minor if there is a point $(\alpha',\beta',r') \in \Delta(H_P)$ such that $\alpha \leqslant \alpha'$, $\beta \leqslant \beta'$, and $r < r'$. The convex hull of the minor (integral) points of $\Delta(H_P)$ will be denoted $\delta(H_P)$.

A geometrical equivalent of condition (c) is the condition

(c′) If Q_1 and Q_2 are polynomials and $\Delta(Q_1) \subset \Delta(P)$, $\Delta(Q_2) \subset \Delta(P)$, then

$$\Delta(\{Q_1, Q_2\}) \subset \delta(H_P).$$

The equivalence of (c) and (c′) follows from a simple lemma that will be presented below.

Let $Q(z_0, z_1, \ldots, z_k)$ be a polynomial in $k+1$ variables with Newton's polyhedron $N(Q)$. A point $(\alpha_0,\ldots,\alpha_k) \in N(Q)$ is said to be minor if $\exists(\alpha'_0,\ldots,\alpha'_k) \in N(Q)$, $\alpha_0 < \alpha'_0$, $\alpha_j \leqslant \alpha'_j$, $j = 1\ldots,k$. The points of $N(Q)$ that are not minor are called senior. The set of the senior points will be denoted as $\pi N(Q)$. The senior points belong to those faces of $N(Q)$ which do not lie in coordinate planes and do not contain segments of straight lines parallel to the axis $\{\alpha_0\}$.

Lemma. *The condition*

$$|z^\alpha| \leqslant \varepsilon_\alpha(z_0) \sum_{\beta\in N(Q)} |z^\beta|, \quad \varepsilon_\alpha(z_0) \to 0 \quad \text{as} \quad |z_0| \to \infty, \tag{9}$$

is fulfilled if and only if

$$\alpha \in N(Q) \setminus \pi N(Q).$$

Proof. Sufficiency. If $\alpha \notin \pi N(Q)$, then a straight line parallel to the axis $\{\alpha_0\}$ can be drawn through the point α, and let $\overline{\alpha} = (\overline{\alpha_0}, \ldots, \overline{\alpha_k})$, $\overline{\alpha_0} > \alpha_0$, be the point of intersection of this line with the boundary of the polyhedron $N(Q)$. Then we have

$$|z^\alpha|/\Xi(z) \leqslant \text{const}\,|z^\alpha|/|z^{\overline{\alpha}}| = \text{const}\,|z_0|^{\alpha_0 - \overline{\alpha_0}}.$$

Necessity. A point $\alpha \in N(Q) \setminus \pi N(Q)$ is characterized by the property that no supporting plane $\langle q, \beta\rangle = c > 0$, $q = (q_0, \ldots, q_k)$, $q_0 > 0$, $|q_1| + \cdots + |q_k| > 0$, can be drawn through it. It now remains to note that if such plane passes through α, then (9) cannot hold. Indeed, if there is $q = (q_0, \ldots, q_k)$, $q_0 > 0$, such that

$$\langle q, \alpha\rangle \geqslant \langle q, \beta\rangle \quad \forall \beta \in N(Q),$$

then condition (9) is violated along the curve $z_j(t) = t^{q_j}$, $j = 0, \ldots, k$.

4.3. The general scheme of the proof of the Theorems in Section 4.2. As has already been mentioned, the most laborious part of the proof of the assertions stated in Section 4.2 is their verification for $n = 1$.

According to Theorem 3.2.3, a dominantly correct polynomial $P(\xi, \tau)$, $\xi \in \mathbb{R}$, has the form

$$P(\xi,\tau) = \widehat{P}(\xi,\tau) + Q(\xi,\tau), \qquad N(Q) \subset \delta(P), \tag{10}$$

$$\widehat{P}(\xi,\tau) = \tau^b R(\xi,\tau) G(\xi,\tau), \tag{11}$$

where

$$R(\xi,\tau) = \prod_{k=1}^{\mu} (\tau - a_j \xi^{b_k}), \quad \operatorname{Im} a_j > 0, \quad b_k \text{ are even numbers}, \tag{12}$$

$$G(\xi,\tau) = \prod_{k=1}^{h} (\tau - c_j \xi), \quad \operatorname{Im} c_j = 0, \quad c_j \neq c_k \quad \text{for} \quad j \neq k. \tag{13}$$

It is obvious that if all assertions in Theorem 1 are proved for the polynomial $\widehat{P}$, then they will also be true for polynomial (10) with any Q, and the relation

$$N(H_P) = N(H_{\widehat{P}}) \tag{14}$$

will hold.

Theorem 2 in Section 4.2 with $n = 1$ corresponds to the simpler case $G \equiv 1$. Thus, we shall prove the following assertions.

Proposition 1. *Let a polynomial $\widehat{P}$ have the form (11)–(13). Then $\exists \gamma_0$ and $c > 0$ such that*

$$H_{\widehat{P}}(\eta,\gamma) > c \sum_{(\alpha,\beta,r)\in N(H_{\widehat{P}})} |\xi|^\alpha |\sigma|^\beta |\gamma|^r, \qquad \gamma \leqslant \gamma_0, \tag{15}$$

where the constants γ_0 and c in (15) depend on $\max|a_j|$, $\max|c_j|$, $\max(\operatorname{Im} a_j)^{-1}$, $\max\limits_{j\neq k}|c_j - c_k|^{-1}$ *(on b and the numbers $b_1, \ldots, b_k$) solely.*

Proposition 2. (i) *Let a polynomial* $\widehat{P}$ *have the form* (11)–(13) *and let* C_1 *and* C_2 *be polynomials such that* $N(C_1) \subset N(\widehat{P})$ *and* $N(C_2) \subset \delta(\widehat{P})$. *Then*

$$N(\{C_1, C_2\}) \subset \delta(H_{\widehat{P}}).$$

(ii) *If* $G \equiv 1$ *in* (13) *then*

$$N(\{C_1, C_2\}) \subset N(H_{\widehat{P}}), \qquad N(C_i) \subset N(\widehat{P}), \qquad i = 1, 2.$$

Proposition 3. *If a polynomial* $\widehat{P}$ *has the form* (11)–(13), *then the polygon* $N(H_P) \subset \mathbb{R}^3$ *is reconstructed uniquely from the polygon* $\delta(P)$.

Assuming that Propositions 1, 2, and 3 have already been proved we shall complete the proof of Theorems 1 and 2 in the foregoing section.

As in Chapters 2 and 3, we put $\xi = \rho\omega$, $\rho \geqslant 0$, $|\omega| = 1$, and associate with the polynomial $P(\xi, \tau)$, $\xi \in \mathbb{R}^n$, the set of polynomials

$$P_\omega(\rho, \tau) = P(\rho\omega, \tau). \tag{16}$$

According to Theorem 3.4.1, if the polynomial P is dominantly correct, then all polynomials (16) (in the variables ρ and τ) are also dominantly correct, and we have

$$\delta(P_\omega) = \delta(P) \quad \forall \omega \in S^{n-1}. \tag{17}$$

In case the polynomial P is stable-correct, polynomial (16) is N-stable correct, and we have

$$N(P_\omega) = \Delta(P) \quad \forall \omega \in S^{n-1}. \tag{17'}$$

Using (17) and Proposition 3 we conclude that the polyhedra $N(H_{P_\omega})$ do not depend on ω and are uniquely determined by the polygon $\delta(P)$:

$$N(H_{P_\omega}) = \Delta(H_P). \tag{18}$$

By virtue of Proposition 1, for each $\omega \in S^{n-1}$ we can write the estimate from below

$$H_{P_\omega}(\rho, \sigma, \gamma) > c(\omega) \sum_{(\alpha,\beta,r)} |\rho|^\alpha |\sigma|^\beta |\gamma|^r, \qquad \gamma \leqslant \gamma_0(\omega). \tag{19}$$

We can select unified constants $c(\omega)$ and $\gamma_0(\omega)$ for all $\omega \in S^{n-1}$.

Indeed, if P is a dominantly correct polynomial, then the polynomials $P_\omega(\rho, \tau)$ have the roots

$$\tau_j(\rho, \omega) = a_j(\omega)\rho^{b_j(\omega)} + o(\rho^{b_j(\omega)}), \quad \rho \to \infty, \quad j = 1, \dots, k(\omega),$$
$$\tau_j(\rho, \omega) = c_j(\omega)\rho + o(1), \quad j = 1, \dots, h(\omega),$$
$$\tau_j(\rho, \omega) = O(1), \quad j = 1, \dots, b(\omega),$$

where the numbers $b_j(\omega)$, $k(\omega)$, $h(\omega)$, and $b(\omega)$ are reconstructed uniquely from the polygon $\delta(P_\omega)$ and, according to (17), do not depend on ω. The coefficients $a_j(\omega)$ and $c_j(\omega)$ are determined from the polynomials $P_\omega^{[j]}$ in §1.1. Since the coefficients of these polynomials are smooth functions of ω, it can be shown that the numbers $|a_j(\omega)|$ and $|c_j(\omega)|$ are uniformly bounded from above and the numbers $\operatorname{Im} a_j(\omega)$ and $|a_j(\omega)-a_k(\omega)|$, $j \neq k$, are uniformly bounded from below by nonzero constants. Thus, it can be assumed that γ_0 and c in (19) do not depend on ω. Noting that

$$H_{P_\omega}(|\xi|, \sigma,) = H_P(\xi, \sigma, \gamma)$$

we obtain assertion (b) in Theorem 1. Assertion (c) and, the more so, assertion ($c_{\max}$) readily follow from Proposition 3.

4.4. The proof of Proposition 1 in Section 4.3. According to Theorem 5.2.2, the fulfilment of (15) is equivalent to the property that for any nonnegative vector $q = (q_1, q_2, q_3)$, $q_3 > 0$, we have

$$(H_{\widehat{P}}(\xi, \sigma, \gamma))_q \neq 0 \tag{20}$$

provided that

$$\gamma < 0, \quad \xi \neq 0 \quad (\text{if } q_1 > 0); \quad \sigma \neq 0 \quad (\text{if } q_2 > 0). \tag{21}$$

However, it is difficult to determine from Theorem 5.2.2 the character of the dependence of the constants c and γ_0 in (15) on the coefficients of polynomials (11). In this connection we shall prove inequality (15) in two stages. We first estimate $H_{\widehat{P}}$ from above and below via a positive function $T(\xi, \sigma, \gamma)$ not depending on the coefficients a_j and c_j in (12) and (13) and after that show that under conditions (21) we have

$$(T(\xi, \sigma, \gamma))_q \neq 0, \quad q = (q_1, q_2, q_3); \ q_1, q_2 \geqslant 0, \ q_3 > 0. \tag{22}$$

Proposition. *There are constants $\overline{c}$, $\overline{c}'$, and $\overline{\gamma}_0$ depending on the same parameters as c and γ_0 in (15), such that the inequality*

$$\overline{c}T(\xi, \sigma, \gamma) \leqslant H_{\widehat{P}}(\xi, \sigma, \gamma) \leqslant \overline{c}' T(\xi, \sigma, \gamma), \gamma \leqslant \overline{\gamma_0}, \tag{23}$$

holds, where

$$\begin{aligned} T(\xi, \sigma, \gamma) = {} & |\gamma|(\sigma^2 + \gamma^2)^{b-1}(\sigma^2 + \gamma^2 + \xi^2)^h \prod_{j=1}^{\mu}(\sigma^2 + \gamma^2 + \xi^{2b_j}) \\ & + (\sigma^2 + \gamma^2)^{b+h} \sum_{j=1}^{\mu}(|\gamma| + |\xi|^{b_j}) \prod_{k \neq j}(\sigma^2 + \gamma^2 + \xi^{2b_k}) \end{aligned} \tag{24}$$

for $b \geqslant 1$ and

$$T(\xi,\sigma,\gamma) = |\gamma|(\sigma^2+\gamma^2+\xi^2)^{h-1}\prod_{j=1}^{\mu}(\sigma^2+\gamma^2+\xi^{2b_j})$$
$$+(\sigma^2+\gamma^2)^h\sum_{j=1}^{\mu}(|\gamma|+\xi^{b_j})\prod_{k\neq j}(\sigma^2+\gamma^2+\xi^{2b_k}) \qquad (24')$$

for $b = 0$. *The constants* $\overline{c}$, $\overline{c}'$, *and* $\overline{\gamma}_0$ *depend on* $\max|a_j|$, $\max|c_j|$, $\max(\operatorname{Im} a_j)^{-1}$, *and* $\max_{j\neq k}|c_j - c_k|^{-1}$.

Proof. Replacing the derivative $\partial P/\partial\tau$ in the expression $H_{\widehat{P}} = -\operatorname{Im}\widehat{P}\overline{\partial\widehat{P}/\partial\tau}$ by

$$\tau^b R\frac{\partial G}{\partial\tau} + b\tau^{b-1}RG + \tau^b G\frac{\partial R}{\partial\tau}$$

we represent $H_{\widehat{P}}$ as a sum of three nonnegative terms:

$$H_{\widehat{P}} = |\tau|^{2b}|R|^2 H_G + b|\gamma||\tau|^{2b-2}|RG|^2 + |\tau|^{2b}|G|^2 H_R. \qquad (25)$$

Lemma 1. *If a polynomial* R *has the form* (12), *then for* $\gamma \leqslant 0$ *we have*

$$d_1 \leqslant |R(\xi,\tau)|^2 / \prod_{j=1}^{\mu}(\sigma^2+\gamma^2+\xi^{2b_j}) \leqslant d_1', \qquad (26)$$

$$d_2 \leqslant H_R(\xi,\sigma,\gamma) / \sum_{j=1}^{\mu}(|\gamma|+\xi^{b_j})\prod_{k\neq j}(\sigma^2+\gamma^2+\xi^{2b_j}) \leqslant d_2', \qquad (27)$$

where d_1 *and* d_2 *depend on* $\max|a_j|$ *and* $\max(\operatorname{Im} a_j)^{-1}$.

Proof. In view of the $(1, 1/b_j)$-homogeneity, it is easily shown that

$$d \leqslant |\tau - a_j\xi^{b_j}|^2(\sigma^2+\gamma^2+\xi^{2b_j})^{-1} \leqslant d', \qquad (28)$$

where d depends on $(\operatorname{Im} a_j)^{-1}$ and $|a_j|$. Multiplying these inequalities we obtain (26). Estimating each factor in the expression

$$H_R = \sum(-\gamma + \operatorname{Im} a_j\xi^{b_j})\prod_{k\neq j}|\tau - a_k\xi^{b_k}|^2$$

by means of (28) we derive (27).

Lemma 2. *If G is a strictly hyperbolic polynomial, then*

$$d_3 \leqslant H_G(\xi,\sigma,\gamma)/|\gamma|(\sigma^2+\gamma^2+\xi^2)^{h-1} \leqslant d_3',$$

where d_3 depends on $\max_{j\neq k}|c_j - c_k|^{-1}$ and $\max|c_j|$.

This assertion is contained in Proposition 3.4.

With account of (26)–(28), the right-hand side of (25) is estimated from above and below by means of

$$\begin{aligned}
&|\gamma|(\sigma^2+\gamma^2)^b(\sigma^2+\gamma^2+\xi^2)^{h-1}\prod(\sigma^2+\gamma^2+\xi^{2b_j}) \\
&\quad + b|\gamma|(\sigma^2+\gamma^2)^{b-1}|G(\xi,\tau)|^2\prod(\sigma^2+\gamma^2+\xi^{2b_j}) \\
&\quad + (\sigma^2+\gamma^2)^b|G(\xi,\tau)|^2\sum(|\gamma|+\xi^{b_j})\prod_{k\neq j}(\sigma^2+\gamma^2+\xi^{2b_k}) \\
&= J_1 + J_2 + J_3. \qquad (29)
\end{aligned}$$

We derive estimate (29) in the case $b \geqslant 1$ leaving the simpler case $b = 0$ to the reader. Since all the terms in (29) are nonnegative, the right-hand side of this expression can be estimated from below by means of $J_1+\varepsilon_2 J_2+\varepsilon_3 J_3$, where $\varepsilon_2, \varepsilon_3 < 1$. We note that

$$|G(\xi,\tau)|^2 > |c_1 \ldots c_h||\xi|^{2h} - \text{const}(\sigma^2+\gamma^2)(\sigma^2+\gamma^2+\xi^2)^{h-1}.$$

Under the substitution of this inequality into $\varepsilon_2 J_2$ the second term on the right-hand side yields an expression that can be estimated via J_1 for a sufficiently small ε_2. In view of this, the factor $|G|^2$ in the expression for J_2 can be replaced by ξ^{2h}. Therefore the sum of J_1 and J_2 is estimated from above and below by means of the first term on the right-hand side of (24).

Similarly, in J_3 we replace $|G|^2$ by

$$(\sigma^2+\gamma^2)^h - \text{const}\,|\xi|^2(\sigma^2+\gamma^2+\xi)^{h-1}.$$

Since $b_1 \geqslant \ldots \geqslant b_\mu \geqslant 2$, we have

$$\xi^2\sum(|\gamma|+|\xi|^{b_j})\prod_{j\neq k}(\sigma^2+\gamma^2+\xi^{2b_k}) \leqslant \mu\prod_{k=1}^{\mu}(\sigma^2+\gamma^2+\xi^{2b_j}),$$

and therefore for a sufficiently small ε_3 the sum $J_1+\varepsilon_2 J_2+\varepsilon_3 J_3$ can be estimated from above by means of (24). The proposition is proved.

Expressions (24) and (24′) are polynomials in each of the octants of $\mathbb{R}^3$, and therefore the polyhedron $N(T) \subset \mathbb{R}^3$ can naturally be defined. Since $H_{\widehat{P}}$ is estimated from above and below by means of T, we have

$$N(T) = N(H_{\widehat{P}}).$$

It follows that to prove Proposition 1 in Section 4.3 it suffices to verify the fulfilment of (22).

If we open the parenthesis in expression (24) for T, this results in an expression of the form of

$$T(\xi,\sigma,\gamma)=\sum c_{\alpha\beta r}\xi^{2\alpha}\sigma^{2\beta}|\gamma|^{r},$$

which does not vanish outside the planes $\{\xi=0\}$ and $\{\sigma=0\}$ (we remind the reader that, according to (21), we have $\gamma<0$). Thus, (22) has been proved for the case $q_1,q_2>0$.

Now let $q_2=0$. Since σ is involved in (24) only via expressions of the form of $\sigma^2+\gamma^2$, the elimination of the monomials containing σ in (29) does not result in a reduction of the q-degree, that is

$$\deg_q T(\xi,\sigma,\gamma)=\deg_q T(\xi,0,\gamma),$$

whence

$$T_q(\xi,\sigma,\gamma)\geqslant T_q(\xi,0,\gamma),$$

and it suffices to show that the right-hand side does not vanish. In application to $T(\xi,0,\gamma)$ we can repeat the above argument. If $q_1>0$, then $T_q(\xi,0,\gamma)\neq 0$. In case $q_1=0$, it suffices to note that

$$T(0,0,\gamma)=(\mu+1)|\gamma|^{2b+2h+2\mu-1}\neq 0,\qquad \gamma<0.$$

Similarly, in the case $q_1=0$, $q_2>0$ it should be noted that

$$T_q(\xi,\sigma,\gamma)>T_q(0,\sigma,\gamma)\neq 0,\qquad |\sigma|\neq 0,\ \gamma<0.$$

Proposition 1 in Section 4.3 has been proved completely.

4.5. The proof of Proposition 2 in Section 4.3. To begin with, we prove the simpler assertion (ii). By virtue of Lemma 4.2, the inclusion relations in the proposition are equivalent to the corresponding inequalities for the monomials. In view of the lemmas in Section 4.4, it is more convenient to deal with the inequalities.

By the linearity, it suffices to consider the case when Q_1 and Q_2 are monomials:

$$Q_1=\tau^{\mu-j}\xi^{\alpha},\quad Q_2=\tau^{\mu-k}\xi^{\beta},\quad \alpha\leqslant B_j,\quad \beta\leqslant B_k,$$

where

$$B_j=b_1+\cdots+b_j.$$

Since

$$\{Q_1,Q_2\}=-[(\mu-k)\tau-(\mu-j)\bar{\tau}]\tau^{\mu-j-s}\bar{\tau}^{\mu-k-1}\xi^{\alpha+\beta},$$

the verification of (ii) reduces to the proof of the inequalities

$$|\gamma||\tau|^{2\mu-2j-2}\xi^{2B_j}\leqslant \text{const}\, H_R(\xi,\tau,\gamma), \tag{30}$$

$$|\tau|^{2\mu-j-k-1}\xi^{B_j-B_k}\leqslant \text{const}\, H_R(\xi,\tau,\gamma). \tag{31}$$

According to Lemma 1 in Section 4.4, we have

$$\text{const}\,|\gamma|\prod_{l=1}^{\mu-1}(|\tau|^2+\xi^{2b_l})\leqslant H_R(\xi,\tau,\gamma).$$

If the terms $|\tau|^2$ are discarded in the factors corresponding to $l=1,\ldots,j$ and the terms ξ^{2b_l} are discarded in the other factors, we obtain inequality (30).

Before proving (31) we note that from the definition of the numbers B_j it follows that

$$B_j+B_k<B_{j-1}+B_{k+1}.$$

Thus, (31) is a consequence of the inequalities

$$|\tau|^{2\mu-2l}\xi^{B_l-B_{l-1}}\leqslant\text{const}\,H_R(\xi,\sigma,\gamma),\tag{30'}$$

$$|\tau|^{2\mu-2l-1}\xi^{B_{l+1}+B_{l-1}}\leqslant\text{const}\,H_R(\xi,\sigma,\gamma).\tag{30''}$$

According to (27), we have

$$H_R>\text{const}\,\xi^{b_l}\prod_{j\neq l}(|\tau|^2+\xi^{2b_j})>\text{const}\,|\tau|^{2(\mu-l)}\xi^{2B_{l-1}+b_l}=\text{const}\,|\tau|^{2(\mu-l)}\xi^{B_{l-1}+B_l}.$$

Similarly,

$$\begin{aligned}H_R&>\text{const}\,\xi^{b_{l+1}}\prod_{j\neq l+1}(|\tau|+\xi^{b_j})\prod_{j\neq l+1}(|\tau|+\xi^{b_j})\\&>\text{const}\left[|\tau|^{\mu-l-1}\xi^{B_l+b_{l+1}}\right]\left[|\tau|^{\mu-l}\xi^{B_{l-1}}\right]=\text{const}\,|\tau|^{2\mu-2l-1}\sigma^{B_{l-1}+B_{l+1}}.\end{aligned}$$

We now turn to the proof of (i). Let us divide the polygon $N(P)\ni(\alpha,\beta)$ by means of the lines $\beta=b+h$ and $\beta=b$ into the three polygons

$$\begin{aligned}N_1&=\{(\alpha,\beta)\in N(P),b+h\leqslant\beta\leqslant b+h+\mu\},\\N_2&=\{(\alpha,\beta)\in N(P),b\leqslant\beta\leqslant b+h\},\\N_3&=\{(\alpha,\beta)\in N(P),0\leqslant\beta\leqslant b\},\end{aligned}$$

and let δ_1, δ_2, δ_3 be the corresponding partition of $\delta(P)$. Denote by (a_l,b_l), $l=1,2,3$, the points of N_1, N_2, and N_3 and by (c_λ,d_λ), $\lambda=1,2,3$, the points of δ_1, δ_2, and δ_3. As in the case (ii), the proof of (i) reduces to the proof of the inequalities

$$|\gamma||\tau|^{b_l+d_\lambda-2}|\xi|^{a_l+c_\lambda}\leqslant\varepsilon(\gamma)H_{\widetilde{P}}(\xi,\sigma,\gamma),\quad d_\lambda=b_l,\tag{32}$$

$$|\tau|^{b_l+d_\lambda-1}|\xi|^{a_l+c_\lambda}\leqslant\varepsilon(\gamma)H_{\widetilde{P}}(\xi,\tau,\gamma),\quad d_\lambda\neq b_l,\tag{33}$$

The verification of these inequalities is quite simple and is based on the explicit form of the function (24). We shall consider some typical situations leaving the rest to the reader as a simple exercise.

1) Let $l = \lambda = 1$. Then $b_1 = b + h + \mu - j$, $d_1 = b + h + \mu - k$, $(j \neq k)$, $a_1 \leqslant B_j$, and $c_1 \leqslant B_k - 1$, i.e. the left-hand side of (33) is no greater than

$$|\tau|^{2b+2h+2\mu-j-k-1}|\xi|^{B_j+B_k-1}. \tag{34}$$

As was shown in the proof of (ii), the point

$$(B_j + B_k, \beta, r), \qquad \beta + r \leqslant 2\mu - j - k - 1,$$

belongs to $N(H_R)$, and therefore, with account of the second term in (24), expression (34) can be estimated by means of

$$\varepsilon(\gamma)H_R|\tau|^{2b+2h} \leqslant \varepsilon(\gamma)H_{\widetilde{P}}, \qquad \varepsilon(\gamma) \to 0,\ \gamma \to \infty.$$

2) Let $l = \lambda = 2$. Then $b_2 = b + h - j$, $d_2 = b + h - k$, $(k \neq j)$, $a_2 \leqslant B_\mu + j$ and $c_2 \leqslant B_\mu + k - 1$, and the left-hand side of (33) is equal to (or no greater than)

$$\begin{aligned}|\tau|^{2b+2h-j-k-1}|\xi|^{2B_\mu+j+k-1} &= |\tau|^{2b-2}\left[\tau^{2h-j-k+1}|\xi|^{j+k-1}\right]\xi^{2B_\mu}\\ &\leqslant |\tau|^{2b-2}(|\tau|^2+\xi^2)^h|R(\xi,\tau)|^2 \leqslant \text{const}\,|\gamma|^{-1}H_{\widetilde{P}}.\end{aligned}$$

3) Let $l = \lambda = 3$. then $a_3, c_3 \leqslant B_\mu + h$, $b_3 \leqslant b$, and $d_3 \leqslant b - 1$, and as the left-hand side of (33) one should take

$$|\tau|^{2b-2}|\xi|^{2B_\mu+2h} \leqslant |\tau|^{2b-2}|\xi|^{2h}|R|^2 \leqslant c|\gamma|^{-1}H_{\widetilde{P}}.$$

4) Let $l = 1$ and $\lambda = 2$. Then on the left-hand side of (33) we shall have

$$|\tau|^{2b+2h+\mu-j-k-1}|\xi|^{B_j+B_\mu+k-1} = \left[|\xi|^{B_\mu}\right]\left[|\tau|^{\mu-j}|\xi|^{B_j}\right]|\tau|^{2b-2}\left[|\tau|^{2h-k+1}|\xi|^{k-1}\right].$$

The first and second square brackets are estimated by means of $|R|^2$ and the last square bracket is estimated via $(|\tau|^2 + \xi^2)^h$, the whole expression being estimated by means of $\text{const}\,|\gamma|^{-1}H_{\widetilde{P}}$. The other cases are considered in the same simple manner.

4.6. The proof of Proposition 3 in Section 4.3. Let $b_1 > b_2 > \cdots > b_k \geqslant 2$, $b_{k+1} = 1$, and $b_{k+2} = 0$ be the degrees of the roots of the polynomial P and let $\mu_1, \ldots, \mu_{k+1}$ be the multiplicities of the roots. As was proved in Section 3.2.1, if the polygon $\delta(P)$ has a vertical side, then $\mu_{k+2} \geqslant 2$, and all the numbers (b_j, μ_j), $j = 1, \ldots, k+2$, are reconstructed uniquely from $\delta(P)$. Since the behavior of function (24) depends in fact only on the numbers (b_j, μ_j), the polygon $\Delta(H_P)$ is completely determined by the polygon $\delta(P)$. A similar situation takes place when the side of $\delta(P)$ adjoining the axis of abscissas intersects it at an angle no greater than $\pi/4$ (in this case μ_{k+2} is sure to be equal to zero).

We now consider the case when the side of $\delta(P)$ adjoining the axis of abscissas intersects it at an angle of $\pi/4$. In this case two different sets of the degrees of the roots and their multiplicities may correspond to the polygon $\delta(P)$, and the degrees of the roots greater than 1 (and the multiplicities of these roots) coincide, but in one of these cases we have

$$\mu_{k+1} = h, \qquad \mu_{k+2} = 1$$

and in the other case

$$\mu_{k+1} = h + 1, \qquad \mu_{k+2} = 0.$$

In the former case $H_{\tilde{P}}$ is estimated from below by means of function (24) in which one should put $b = 1$. In the latter case $H_{\tilde{P}}$ is estimated from below using function (24′) in which h should be replaced by $h+1$. As is easy to see, the indicated functions coincide, which proves the desired assertion.

4.7. Concluding remarks. Dominantly correct and stable-correct polynomials and symbols are characterized by the invariance of their properties under linear transformations of the variables $\xi_1, \ldots, \xi_n$. We shall briefly discuss the generalization of these notions to the case when no such invariance takes place.

Definition 1. A polynomial $P(\xi, \tau)$ solved with respect to the highest power of τ is said to be N-dominantly correct if it is correct in Petrovskiĭ's sense and if this property is not violated under the addition of the monomials $c\xi^\alpha\tau^\beta$, where $(\alpha_1, \ldots, \alpha_n, \beta) \in \delta^0(P)$ (see Section 3.3).

Definition 1′. A polynomial $P(\xi, \tau)$ solved with respect to the highest power of τ is said to be $\mathcal{H}$-correct if it satisfies conditions (ii) and (iii) of Theorem 3.3.

It is clear that every $\mathcal{H}$-correct polynomial is N-dominantly correct.

The classes of polynomials in Definitions 1 and 1′ were studied in detail by Gindikin [1]. These results imply that the property of $\mathcal{H}$-correctness is stronger as compared to the N-dominant correctness.

We give an example of an $\mathcal{H}$-stable correct polynomial of the first degree with respect to the variable τ. We divide the variables $\xi \in \mathbb{R}^n$ into two groups: $\xi = (\xi', \xi'')$, $\xi' \in \mathbb{R}^k$, $\xi'' \in \mathbb{R}^{n-k}$, and consider the polynomial

$$P(\xi, \tau) = \tau + \langle A, \xi'' \rangle + Q(\xi') + iH(\xi'),$$

where $H(\xi')$ is an N quasi-elliptic polynomial bounded from below; $\langle A, \xi'' \rangle$ is a real linear form on $\mathbb{R}^{n-k}$; and $Q(\xi')$ is a real polynomial, the integral minor points of $N(Q)$ being minor points of $N(H)$. Then the polynomial P is $\mathcal{H}$-correct.

In Section 5.2.4 the definition of N-stable correct polynomials was stated. These are polynomials correct in Petrovskiĭ's sense which are solved with respect to the highest power of τ, satisfy an inequality of the form of

$$c\Xi(\xi, \tau) \leqslant |P(\xi, \tau)| \quad \text{for} \quad \xi \in \mathbb{R}^n, \quad \operatorname{Im} \tau \leqslant \gamma_0,$$

and have a complete and regular polyhedron $N(P)$.

We present some other definitions of these polynomials.

Definition 2. A polynomial $P(\xi, \tau)$ solved with respect to the highest power of τ is said to be N-stable correct if for any polynomial $Q(\xi, \tau)$, $N(Q) \subset N(P)$, there is ε such that the polynomial $P + \varepsilon Q$ is correct in Petrovskiĭ's sense.

Definition 2′. A polynomial $P(\xi, \tau)$ solved with respect to the highest power of τ is said to be N-stable correct if it satisfies conditions (i) and (ii) in Theorem 3.3, and instead of (iii) a stronger condition is fulfilled:

$$N(\{Q_1, Q_2\}) \subset N(H_P); \qquad N(Q_1), N(Q_2) \subset N(P).$$

In the above-mentioned paper by Gindikin [1] it is proved that Definitions 2, 2′, and 5.2.2 are equivalent.

REFERENCES

M. S. Agranovich and M. I. Vishik.

1. *Elliptic Problems with a Parameter and General Parabolic Problems*, Uspechi Mat. Nauk **19** no. 3 (1964), 53–161. (Russian)

V. M. Borok.

1. *On Numerical Characteristics of Systems Correct in Petrovskiĭ's Sense*, Izv. Vyssh. Uchebn. Zaved. Mat. **1** no. 8 (1959), 16–22. (Russian)

A. D. Bryuno.

1. *Lokal'nyĭ metod nelineĭnogo analiza differensial'nykh uravneniĭ* (*A Local Method for Nonlinear Analysis of Differential Equations*), "Nauka", Moscow, 1979. (Russian)

N. G. Chebotarev.

1. *Teoriya algebraicheskikh funktsiĭ* (*Theory of Algebraic Functions*), "Gostekhizdat", Moscow, 1948. (Russian)

Yu. V. Egorov.

1. *Lineĭnye differentsial'nye uravneniya glavnogo tipa* (*Linear Differential Equations of Principal Type*), "Nauka", Moscow, 1984. (Russian)
2. *On the Solvability of Differential Equations with Simple Characteristics*, Uspechi Mat. Nauk **26** no. 2 (1971), 183–198. (Russian)

G. I. Eskin.

1. *Kraevye zadachi dlya ellipticheskikh psevdodifferentsial'nykh uravneniĭ*, "Nauka", Moscow, 1973 (Russian); English transl. in *Boundary-value Problems for Pseudodofferential Equations*, vol. **52**, Amer. Math. Soc., Providence, R. I., 1981.
2. *Cauchy's Problem for Hyperbolic Convolution Equations*, Mat. Sb. **74** (1967), 262–297. (Russian)

J. Friberg.

1. *Multiquasielliptic Polynomials*, Ann. Sckola Norm. Super. Pisa **21** no. 2 (1967), 233–260.

B. A. Fuks and V. I. Levin.

1. *Funksii kompleksnogo peremennogo i nekotorye ikh prilozheniya. Spetsial'nye glavy* (*Functions of a Complex Variables and Some of Their Applications. Special Chapters*), Gostekhizdat, Moscow, 1951. (Russian)

S. G. Gindikin.

1. *Energy Estimates Relating to Newton's Polygon*, Trudy Moskov. Mat. Obshch. **31** (1974), 189–236. (Russian)
2. *On a Generalization of Parabolic Differential Operators to the Case of Multidimensional Time*, Dokl. Akad. Nauk SSSR **173** no. 3 (1967), 499–502. (Russian)

V. V. Grushin and N. A. Shananin.

1. *Some Theorems on Singularities of Solutions to Differential Equations with Weighted Principal Symbols*, Mat. Sb. **103** no. 1 (1977), 37–51. (Russian)

G. H. Hardy, J. E. Littlewood, and G. Pólya.

1. *Inequalities*, Cambridge Univ. Press, Cambridge, 1934.

L. Hörmander.

1. *Analysis of Linear Partial Differential Operators*. II, III, Springer-Verlag, Berlin, Heidebberg, New York, Toronto, Tokyo, 1983, 1985.
2. *On the Theory of General Partial Differential Operators*, Acta. Math. **99** (1958), 255–264.
3. *Linear Partial Differential Operators*, Springer-Verlag, Berlin, Göttingen, Heidelberg, 1963.
4. *Pseudo-differential Operators and Hypoelliptic Equations*, in: Amer. Math. Soc. Symp. on singular integrals (1966), 138–183.

G. G. Kazaryan.

1. *Estimates for Differential Operators, and Hypoelliptic Operators*, Trudy Mat. Inst. Steklov. **140** (1976), 130–161. (Russian)

J. J. Kohn and L. Nirenberg.

1. *On the Algebra of Pseudo-differential Operators*, CPAM **18** (1965), 269–305.

R. Lascar.

1. *Propagation des singularitiés des solution d'équation pseudo-differentialles quasi homogenes*, Ann. Inst. Fourier **27** no. 2 (1977), 79-123.

J. Leray.

1. *Hyperbolic Differential Equations*, The Institute for Advances Study, Princeton, N.J., 1953.

V. P. Mikhaĭlov.

1. *On the Behaviour at Infinity of a Class of Polynomials*, Trudy Mat. Inst. Steklov. **91** (1967), 59–81. (Russian)
2. *The First Boundary-value Problems for Quasi-elliptic and Quasi-parabolic Equations*, Trudy Mat. Inst. Steklov. **91** (1967), 81–99. (Russian)

L. Nirenberg and F. Trevers.

1. *On Local Solvability of Linear Partial Differential Equations* II. *Suficient Conditions*, CPAM **24** (1971), 459–509.

B. P. Paneyakh.

1. *Some Inequalities for Functions of Exponential Type and a Priori Estimates for General Differential Operators*, Uspekhi Mat. Nauk **21** no. 3 (1966), 75–114. (Russian)

I. G. Petrovskiĭ.

1. *On Cauchy's Problem for Systems of Linear Partial Differential Equations for Non-analytic Functions, in: Izbrannye trudy. Systemy uravneniĭ s chastnymi proizvodnymi. Algebraicheskaya geometriya* (*Selected works. Systems of Partial Differential Equations*), "Nauka", Moscow, 1986, pp. 98–168. (Russian)
2. *On Cauchy's Problem for Systems of Partial Differential Equations in: Izbrannye trudy. Systemy uravneniĭ s chastnymi proizvodnymi. Algebraicheskaya geometriya* (*Selected Works. Systems of Partial Differential Equations*), "Nauka", Moscow, 1986, pp. 34–97. (Russian)

N. A. Shananin.

1. *On Local Solvability of Equations of Quasi-principal Type*, Mat. Sb. **97** no. 4 (1975), 503–513. (Russian)
2. *An Example of a Locally Unsolvable Differential Equation of Quasi-principal Type With Real Weighted Principal symbol*, Mat. Zametki **19** no. 5 (1976), 755–761. (Russian)

G. E. Shilov.

1. *Matematicheskiĭ analiz. Vtoroĭ spesial'nyĭ kurs* (*Mathematical Analysis. The Second Special Course*), "Nauka", Moscow, 1965. (Russian)

M. Taylor.

1. *Pseudo-differential Operators*, Princeton Univ. Press, Princeton, 1971.

L. R. Volevich.

1. *Energy Method in Cauchy's Problem for Differential Operators Correct in Petrovskiĭ's Sense*, Trudy Moskov. Mat. Obshch. **31** (1974), 147–187. (Russian)
2. *Local Properties of Solutions to Quasi-elliptic Systems*, Mat. Sb. **59 (101)** (1962), 3–52. (Russian)

L. R. Volevich and S. G. Gindikin.

1. *Oboshennye funksiĭ i uravneniya v svertkakh* (*Distributions and Convolution Equations*), "Nauka", Moscow, 1992, (in print) (Russian) English transl. in *Distributions and Convolution Equations*, Gordon and Breach science Publishers, London, 1992.
2. *On a Class of Hypoelliptic Polynomials*, Mat. Sb. **75 (117)** no. 3 (1968), 400-416. (Russian)
3. *Pseudodifferential Operators and Cauchy's Problem for Differential Equations with Variable Coefficients*, Funktsional Anal. i Prilozhen. **1** no. 4 (1967), 8–25. (Russian)
4. *Cauchy's Problem for Differential Operators with Dominating Principal Part*, Funktsional Anal. i Prilozhen. **3** no. 3 (1968), 22–40. (Russian)
5. *Cauchy's Problem for Pluriparabolic Differential Equations* I, II, Mat. Sb. **75** no. 1 (1968), 64–105 (Russian) **78** no. 2 (1969), 214–236.
6. *Cauchy's Problem*, VINITI. Sovremennye problemy matematiki. Fundamental'nye napravleniya **32** (1988), 4–98. (Russian)
7. *Newton's Polyhedron and Local Solvability of Partial Linear Differential Equations*, Trudy Moskov. Mat. Obshch. **48** (1985), 211–262, Moscow. (Russian)

L. R. Volevich and B. P. Paneyakh.

1. *Some Spaces of Generalized Functions and Embedding Theorems*, Uspekhi Mat. Nauk **20** no. 1 (1965), 3–74. (Russian)

K. Yosida.

1. *Functional Analysis*, Springer-Verlag, Berlin, Göttingen, Heidelberg, 1965.

INDEX

9 780792 320371